Fred Schuh

the master book of MATHEMATICAL RECREATIONS

BY

FRED. SCHUH

Translated by F. Göbel

Translation edited by T. H. O'Beirne

Dover Publications, Inc.

New York

Published in Canada by General Publishing Company, Ltd., 30 Lesmill Road, Don Mills, Toronto, Ontario.
Published in the United Kingdom by Constable and Company, Ltd., 10 Orange Street, London WC2.

The Master Book of Mathematical Recreations is a new English translation of *Wonderlijke Problemen; Leerzaam Tijdverdrijf Door Puzzle en Spel*, as published by W. J. Thieme & Cie, Zutphen, in 1943.

Standard Book Number: 486–22134–2
Library of Congress Catalog Card Number: 68–28064

Manufactured in the United States of America

Dover Publications, Inc.
180 Varick Street
New York, N.Y. 10014

PREFACE

In this book I have endeavoured to show how pure puzzles (that is, puzzles which differ from crosswords, word-play riddles and the like, in that they are not limited to one or more specific languages) can be solved by systematic trial, with the maximum possible assistance from reasoning. This reasoning may often shorten the solving time considerably. Sometimes the reduction is due to the puzzle being put in a simpler form; we may invert it, for example, and solve it backwards. In many cases the solution of the puzzle is found by breaking it up into simpler puzzles; this is a very important strategy, which plays a significant role in mathematics, too. Much attention has been given not only to puzzles, but also to puzzle games (for two players) and their complete analysis. This topic is developed by various worked examples, which include the game of noughts and crosses (also known by the name of ticktacktoe).

Also, those parts of mathematics that are of importance for puzzles have been treated as simply as possible, avoiding algebraic formulae. In the front rank of these is the enumeration of possible cases, the so-called theory of permutations and combinations (§§125–130); I have tried to make this accessible even to non-mathematicians through a treatment based on examples. This theory is indispensable for the determination of the number of solutions for various puzzles—and frequently for the solution itself, too—while it also plays an important role in the preeminently interesting theory of probability. The basic theorems for calculating probabilities are discussed in detail and exhaustively, again without formulae, while numerous applications have been made, especially to games (§§131–164). I have also considered it a good idea to develop in a simple way (as with probability theory, giving historical details) the concept of a number system (§§78–88), because this, too, is applied in various puzzles and games, e.g., in the weight puzzle (§§89–91) and in the particularly enjoyable game of nim (§§113–123), which is played with piles of matches.

In Chapter XV, I give a short discussion of the basic ideas of mechanics. I have done this to be able to mention some simple-sounding, but really not so simple questions on the motion of objects, among which are some questions that even a professional is likely to answer incorrectly. In addition, a few interesting phenomena of motion have been mentioned. Here I hope to have contributed toward dispelling the idea that mechanics has to be a dry subject.

Only here and there has use been made of algebraic formulae, and even then to a very modest extent. The sections in which this occurs have been marked with an asterisk; they can be omitted. Some sections in which the puzzles or the arguments are somewhat difficult have also been marked with an asterisk. So the reader should judge for himself whether he wishes to skip these sections or, perhaps, even give them special attention.

Many of the puzzles in this book and various of the puzzle games are original. However, all sorts of well-known puzzles and games have been treated as well, but even these have been elucidated in as original a way as possible.

The original puzzles that seem to me to be most successful include:

numbers which are written with the same digits as one of their multiples (§§23–28);

the domino puzzles with the smallest and with the largest numbers of corners (§§37–39);

the multiplication sum with two digits 0, two digits 1, two digits 2, and so on (§§235–237);

the puzzle of sixteen numbers in a square with 24 prime sums (§§244–247);

the puzzle of the multiples of 7 with the maximum product (§§250–254);

the multiplication puzzle which involves the first line of the quotation on page vi (§255);

the repeating division puzzle (§258);

the puzzle with the eight dice (§§263 and 264);

the broken lines across sixteen dots in a square (§§274 and 275);

the road puzzle with concentric circles (§§289 and 292);

the counting-out puzzle 1–2–3 (§§301 and 304).

Among the original puzzle games I would like to mention the 5-subtraction game (§222); the modified subtraction game (§§225–228); the Game of the Dwarfs or "Catch the Giant!" (§§170–180), a simplified version of the soldiers' game which, notwithstanding its extreme simplicity, is not easy at all; as likewise another game that is played on the same board and with the same pieces as the soldiers' game (§191). I can also instance various games with piles of matches

(§§101–112) and an extension of the game of nim (§§122–123). I shall confine myself to these examples, although I could mention many other original puzzles.

I take this opportunity to express my thanks to my assistant, Mr. W. T. Bousché, who did most of the drawings, and to my pupil Mr. M. L. van Limborgh, who did some of the others; their work leaves nothing to be desired. To my former pupil Mr. J. Ploeg, M.E., thanks are due for some suggestions in the field of mechanics. I also thank my friends Mr. J. C. N. Graafland, M.E., Mr. J. Spanjersberg, M.A., and Mr. L. A. de Vries for proofreading; their suggestions have led to many an improvement. I am very grateful to the Publisher, who has taken pains to provide an attractive format.

FRED. SCHUH

The Hague, Autumn 1943

Est modus in rebus; sunt certi denique fines,
Quos ultra citraque nequit consistere rectum.

(There is a measure in things; and there are fixed limits beyond or short of which Right is unable to exist.)

HORACE

CONTENTS

[*Asterisks indicate sections that involve algebraic formulae.*]

Chapter I:

HINTS FOR SOLVING PUZZLES

I. VARIOUS KINDS OF PUZZLES

1. Literary puzzles. There is a great difference between various puzzles, not only in their difficulty, but also in their essential nature. According to their character we can divide puzzles into two classes, which we shall call literary puzzles and pure puzzles.

To the literary puzzles belong crosswords, word-play riddles, proverbs that have to be guessed from certain data, and the like. Of course, skill in solving this kind of puzzle does depend on inborn shrewdness, but still more on knowledge of words and often also on geographical and historical knowledge. This knowledge can be supplemented to a large extent by using an atlas and, especially, an encyclopedia. However, if you do not have the necessary simple knowledge (of proverbs, of less common expressions, etc.), you will not succeed in cracking a somewhat difficult literary puzzle, because you will not be able to use the appropriate aids sufficiently well.

The peculiarity of a literary puzzle is that there is no sure way to attain the answer. You must have a flash of inspiration, or have the luck to guess a phrase from a few fragments, or something of the sort. Also, with puzzles of this kind (unless they are very easy) you can never know for certain that the solution found is the intended one, and whether it is the only solution, since trying out all possibilities is so time-consuming that, in many cases, this would take years (sometimes even centuries). However, usually you will have practical certainty that the intended solution has been found. For example, if part of the task is to find a proverb of some definite length, it would be extremely improbable that there was another proverb of precisely the same length which would also satisfy all given conditions, so improbable that the chance of this can be completely neglected (even though it cannot be numerically estimated).

There is a certain kind of literary puzzle where it is impossible to be sure of having obtained the best solution. As an example, suppose that it is required to compose a sentence of as many words as possible which all begin with the same letter. A second example is the problem of forming the longest possible closed chain of words with the property that the last syllable of each word is the same as the first syllable of the next word. Also, with a puzzle of this kind, the poser is under no obligation to have a more or less suitable solution available.

It is hardly or not at all possible to make general remarks (by way of advice) about the solution of literary puzzles, passing beyond what is done automatically by anyone who regularly solves such puzzles. So we shall not occupy ourselves with these puzzles any further.

2. Pure puzzles. While a literary puzzle is linked to a definite language, or sometimes to a few languages, this is not the case with a pure puzzle. With a pure puzzle, the question can be translated into any language, without the nature of the puzzle changing in any way. When solving a literary puzzle, you need some information that you cannot deduce on your own. With a pure puzzle, too, you often have to have a fund of knowledge, but this is mostly of such a nature that you yourself can discover what you need, if you are sufficiently intelligent.

A pure puzzle usually has a numerical, sometimes also a more or less geometrical content. The puzzle can be solved by pure reasoning alone. It is also typical of such a puzzle that it can always be decided with certainty whether the solution is correct. In many cases this can be determined easily and it is then practically out of the question that a wrong solution will be accepted in place of the correct one. Usually either you do not find a proper solution, or else you obtain the correct solution (or, it may be, one of the correct solutions), although it is always possible that persons lacking in self-criticism will consider a wrong solution to be correct.

Yet there is a kind of pure puzzle where an incorrect solution can be mistaken for a correct one, without this implying lack of self-criticism. As an example we mention the case of a puzzle which has more than one solution, where all solutions are required. It may also happen that the solution is required to be one for which a certain number (for example, the number of corners in a figure to be formed from dominoes that fit together in a certain manner) is as large or as small as possible. In such a case, the possibility exists that something has been overlooked, so that some solutions are found, but not all the solutions. The solution for which the number in question is maximum or minimum might happen to be among the solutions that have been overlooked; the supposed solution is then incorrect, without this necessarily implying a lack of intelligence.

Another example of the same nature occurs when the problem requires the smallest number of operations (for example, in moving cubes) by which a certain result (for example, a given final position of the cubes) can be obtained. We call these operations "moves." In a case like this it may happen that a certain move has been overlooked, or discarded too quickly because, considered superficially, it seemed unpromising.

3. Remarks on pure puzzles. No very sharp line can be drawn between pure puzzles and formal mathematics. In mathematics, too, questions occur that have more or less the character of a puzzle. Whether one speaks of a mathematical problem or of a puzzle, in such a case, depends on the significance of the question (and, to some extent, also on the nature of the reasoning that leads to the solution); and it is partially a matter of taste as well. On the other hand, there are problems that impress everyone as puzzles, but which still allow a fruitful application of mathematics (in particular, of arithmetic). The greater your skill in arithmetic, the less time it will take you to find the solution. The puzzles with squares in §§229 and 230 provide a striking example of this.

A puzzle can be considered as uninteresting when the data and the question can without difficulty be put into the form of equations by anyone who has some knowledge of algebra, to let the unknown(s) be found from solution of the equation(s). A mathematician is not in the least interested in such a puzzle, whereas a good puzzle should satisfy the requirement that it can arouse an experienced mathematician's interest, too. As an example of a puzzle which ought rather to be called a dull problem in algebra, let John ask Peter to think of a number, add 15 to it, multiply the sum by 3, subtract 6 from the product, and divide the difference by 3. Peter then has to announce the quotient, after which John is to determine what number Peter had thought of. Obviously it is 13 less than the quotient mentioned by Peter. The puzzle may be modified by requiring Peter to subtract his first number from the quotient obtained. Without Peter needing to say anything, John will then be able to say that the final result is 13; however, now it is impossible for John to know what number Peter had thought of, since every number leads to 13 after the operations in question.

How little interest these puzzles have, is also evident from the fact that they can be varied endlessly. One can make them as complicated as one pleases, and also allow the other person to think of more than one number.

4. Puzzle games. We imagine a game that is played by two persons, John and Peter; we assume that luck plays no part (no more than with chess or draughts). John and Peter make moves in turn, and the winner is the one who succeeds in achieving a certain result. Often such a game is so complicated (again I mention chess and draughts) that it is impossible for a human being to analyze it completely. The game is then different from a puzzle; making a complete analysis could be called a puzzle, but as this puzzle is unsolvable (in the sense that it is too difficult), it has to be left out of consideration.

If the game is so simple that it can have a complete analysis, we speak of a puzzle game. As an example we mention the well-known children's game of noughts and crosses, which will be discussed in Chapter III. This will show that surprisingly many elegant combinations are contained even in such a seemingly very simple game. The laboriousness of the complete treatment of this game shows quite clearly the impossibility of similarly disposing of much more complicated games like chess and draughts. Chapter III can further be considered as an example of a complete analysis of a puzzle game.

The interest of a puzzle game is, of course, lost when both players have seen through it completely. However, a game like chess will never lose interest in this way, although several puzzles and puzzle games can be derived from it. An endgame that is not too complicated can be considered as a puzzle game. A related puzzle is the problem of the smallest number of moves in which mate (or sometimes the promotion of a pawn) can be enforced by White; here it is assumed, of course, that Black puts up the strongest possible defense, playing in such a way that the number of moves becomes as large as possible.

The last-named puzzle offers another example of a case in which one can easily mistake an incorrect answer for a correct one (cf. §2). For, in a not too simple endgame, it may very well happen that even an experienced chessplayer will wrongly consider a certain sequence of moves to be the shortest possible one.

5. Correspondences and differences between puzzles and games. A game like chess shows a resemblance to puzzle solving in many situations, in particular (as we said before) in the endgame, but also when, through a combination, a certain result has to be achieved that is no way near to a final result. A chess problem (mate in so many moves) is nothing else but a puzzle.

In the opening and frequently also in the middle game, chess cannot be compared to puzzles. There the evaluation of the position plays an important role, at a stage when it is not possible to appreciate exactly how an advantageous position can be turned into victory. Besides, in chess quite different abilities of the players manifest themselves, and, as a consequence, a good chess player is not necessarily an able solver or composer of chess problems, and conversely. Naturally, these two abilities are not completely foreign to each other. A good chess player needs a great talent for combination —supplemented by a good ability to visualize situations (the latter especially when playing blindfold chess). He also needs a good memory to be able to retain the combinations—as well as a feeling for position, a quick activity of the mind to escape from pressures of time, and a strong power of concentration, to enable him to put aside all other things at any moment, with the ability not to have his attention diverted by anything or anybody. Added to this, there is the psychological factor, which consists of knowing and understanding the opponent's ways of playing and his weaknesses, so that a good chess player will play differently against one player and against another, in the same situation. As a consequence of all this, chess playing and puzzle solving can hardly be put on one level. In chess, you have a flesh-and-blood opponent, whose strong and weak points you are well-advised to know. In solving puzzles, if you have an opponent at all, he is imaginary; hence here you are your own opponent, in the end.

The requirements for chess and puzzle solving diverge to such an extent that the puzzle solver in a person may hamper the chess player, and conversely. A good puzzle solver, who has acquired the habit of scrupulously examining all possibilities, runs the risk of wanting to do the same thing in practical chess, where he gets entangled in the multitude of possibilities, is unable to make a choice, and then finally, being pressed for time, makes a bad move. Conversely, an experienced chess player who wishes to solve a puzzle runs the risk of relying too much on his intuition, leaving many possibilities unexamined; sometimes he will do this successfully and find the solution of the puzzle quickly, but often he will not.

With regard to games, in what follows we shall occupy ourselves only with those to which the name "puzzle game" applies.

II. SOLVING BY TRIAL

6. Trial and error. Trial is an activity that is important and useful when solving a pure puzzle. This may invite the belief that solving a pure puzzle is a matter of accident, hence of luck; all that matters, it might be argued, is whether one tries the correct thing earlier or later. This opinion is very widespread, but quite incorrect, at least with regard to good puzzles, by which we mean puzzles which cannot in practice be exhausted by trial, to the extent of finding all the solutions. In solving a puzzle in which an excessive role is played by the element of trial, notwithstanding all ingenuity, you may be lucky; but if you are lucky enough to find a solution, you still do not know whether this is the only solution, let alone how many solutions there are. If these questions form no part of the problem, and you are only required to find some solution, then luck can indeed play its part.

In such a case, however, one should speak of a bad puzzle. Such a puzzle can be composed (that is, put to others) by someone who has no idea of puzzles himself. We should like to illustrate this with an example. Consider a figure composed of a number—40, say—of adjacent regular hexagons. Each of these hexagons is divided into 6 equilateral triangles by 3 diagonals connecting opposite vertices. Two triangles which have a side in common and which belong to different hexagons (which occurs a large number of times in the figure) are painted with the same color. Altogether 3 or 4 colors, say, have been used, while the distribution of colors among the pairs of triangles is quite irregular. Now one cuts out the hexagons and interchanges them in a random way, also giving arbitrary rotations to individual hexagons. The figure thus obtained (with the hexagons in new positions, connected to each other again) is now presented as a puzzle. The problem would then be to cut out the hexagons and rearrange them to produce a figure with the same boundary, with like colors making contact everywhere. The person who poses the puzzle knows that success is possible, but knows nothing of other solutions, when there may be many, perhaps millions. Such a thing is wrong. There is no proper idea behind the puzzle.

The opinion that solving a puzzle is a matter of luck is usually found among people who have only a superficial interest in puzzles, and who have rarely or never taken the trouble to solve a good puzzle by reasoning. The same opinion is sometimes held by people who do take pleasure in solving puzzles, but who are not sufficiently intelligent to replace—or, at any rate, supplement—haphazard trial and error by reasoning. Such a puzzle solver often does not remember which cases have already been tried, and which have not; he does the same job several times and overlooks other possibilities altogether. It is clear that this form of puzzle solving will not guarantee success.

Admittedly, of course, luck may play a role even in solving a pure puzzle by reasoning, if one happens to hit upon an efficient procedure. However, the same thing may occur in purely scientific work. Here, in place of speaking of luck, it would be better to speak of an intuition for finding the right path, or a feeling for the right method.

7. Systematic trial. Solving a pure puzzle by trial has to be done with a fair amount of discretion. One might call it reasoned trial or judicious trial, but the intention is perhaps best represented by the expression "systematic trial." This is trial made in such a way that at any moment you know precisely which cases have already been considered, and which have not, and so at a given moment you are able to say, "Now I have examined all possibilities; these are the solutions and there can be no others."

Of course, when the puzzle is somewhat complicated, and you have had to consider many possibilities, it is not always out of the question that you have overlooked some possibility, but such a thing can also occur in solving a mathematical problem. In these cases, the error made is not due to the method, but to incidental carelessness. Without systematic trial, one makes errors systematically, in a sense, and often one goes round in a circle without noticing it.

In a systematic trial you should record accurately which possibilities have been examined, and also which solutions have arisen from one possibility, and which from another. Only in this way will you be able to check the given solution afterwards and, if necessary, improve upon it, with regard either to the correctness of the results or to the simplicity of the method employed.

8. Division into cases. We would now like to explain how to organize a systematic trial. You begin by dividing the various possibilities into groups; in other words, you begin by successively making various assumptions, each of which represents a group of possibilities. Here it is important to arrange these assumptions according to some system, to ensure that no assumption and no group of possibilities is left unexamined. As an example we take a digit puzzle in which there are certain places where unknown digits have to be filled in; these digits have to satisfy certain conditions which are mentioned in the puzzle. Here you consider some fixed place, and successively work out the ten assumptions that a digit 0, 1, 2, 3, 4, 5, 6, 7, 8, or 9 occurs in that position.

In further working out some assumption, you often have to consider further assumptions, and distinguish sub-cases. In the above-mentioned puzzle these may be assumptions concerning the digit in another place, and then it is often desirable to consider the places, too, in some well-chosen system; obviously, the assumptions can equally well be concerned with something else, for example, the sum of the digits to be filled in.

As a second example we take a puzzle which consists of drawing a broken line which has to go through given points, always from one point to another, where you can continue horizontally, vertically, or diagonally, say. On reaching a point, you have to work out three assumptions each time: you may continue horizontally, vertically, or diagonally. Each of these possibilities is divided into three possibilities on reaching the next point, and this division into several possibilities may be repeated a number of times. The essence of systematic trial consists of this repetitive consideration of assumptions.

The indicated procedure for grouping the various possibilities makes one think of a branching tree. It is therefore quite useful to represent your findings in the form of such a tree; we then speak of a puzzle tree, or, more briefly, of a tree.

In a tree some branches may come to a dead end, because the result is in conflict with given conditions of the puzzle; such branches then do not yield solutions. In writing down the tree we indicate this by the word "dead."

9. Example of a puzzle tree. We now consider numbers with the following property: The digits of the number are all different; each digit is three times the previous digit, or one unit more, or two units more; if this gives a number greater than 9, only the units digit is taken; for the first digit, the "previous digit" is defined as the last digit.

Hence, a digit 3 is followed by 9, 0, or 1; a digit 7 is followed by 1, 2, or 3; if 8 is the last digit, the first digit is 4, 5, or 6; etc. Since the digits are evidently thought of as being arranged in a circle, other numbers can be deduced from any number with the property in question, by a so-called cyclic permutation. For instance, if you have found that 4398 satisfies the requirements, then you know at once that 3984, 9843, and 8439 also do. Therefore, we shall also assume that the first digit is the largest digit of the number. We now require to find all numbers of five digits or fewer, which show the characteristics in question.

As the first digit is the largest digit, it cannot be equal to 1 or 2 (since 1 is followed by at least 3, and 2 by at least 6). We now begin by dividing the desired numbers into groups with the same initial digit. The numbers with the same initial digit are divided according to the second digit; the numbers with the same first two digits are again subdivided according to the third digit, and so on. In this way one finds the following trees:

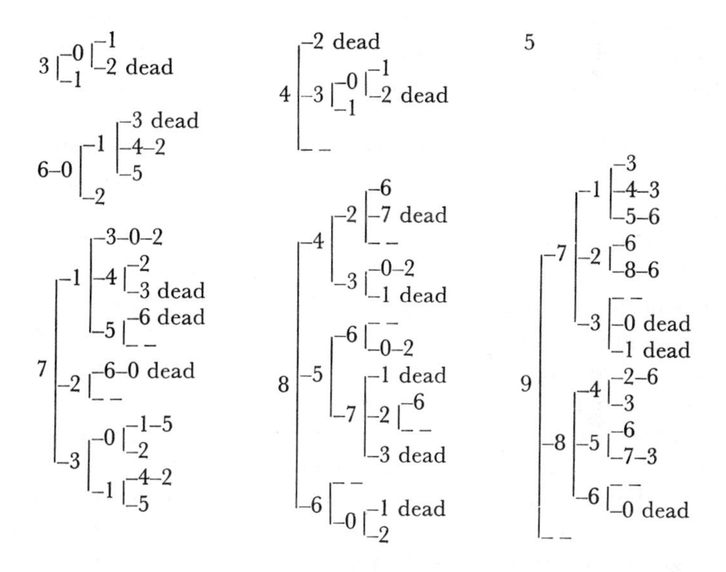

If the first digit is 7, for example, then the last digit can be nothing but 2 or 5; if the first digit is 8, then the last digit must be 2 or 6; and so on. Considerations like this have been taken into account in constructing the trees. Attention has also been paid to the requirement that no digit can occur which is larger than the initial digit, or equal to one of the preceding digits. A final dash indicates a case where a number terminates without requiring a subsequent digit.

It may be noted that the different trees can be considered as branches of one larger tree. The partitioning into different trees has been made only to save space.

The trees show that the following 38 numbers satisfy the requirements:

4, 5, 9; 31, 72, 86; 301, 431, 602, 715, 842, 856, 973, 986;
4301, 6015, 7142, 7302, 7315, 8426, 8572, 8602, 9713, 9726,
9843, 9856; 60142, 71302, 73015, 73142, 84302,
85602, 85726, 97143, 97156, 97286, 98426, 98573.

III. CLASSIFICATION SYSTEM

10. Choosing a classification system. If you divide the various possibilities into groups, with subdivision of these groups and later subgroups, and then make an examination of all possibilities, that is, all cases, sub-cases, etc., you must obtain a solution of the puzzle, but often this would take months or years. For, if it has been necessary to distinguish ten cases four times in succession, say—for example, because in a digit puzzle unknown digits have to be supplied in four places—then this gives as many as $10 \times 10 \times 10 \times 10 = 10{,}000$ possibilities. It is no rare thing to be faced with the task of examining many more cases (sometimes even millions) if you start upon a classification without proper discrimination.

Hence, when solving puzzles, you should not just employ the first classification system that comes to your mind; for instance, in the puzzle of §9 you should not consider all numbers less than 100,000 written with different digits (as this would give 32,490 cases). You should consider carefully what system is the best. The time spent is usually amply repaid by quicker progress thereafter. The difference in usefulness between one system and another may be so large that a better system—hence a different way of choosing the assumptions, because the classification is made according to a different criterion—diminishes the time required for solving by a factor of 1000 or more.

11. Usefulness of a classification system. We shall now endeavor to explain the things which require attention when a classification system is judged for its practical usefulness; hence, how you should proceed in order not to be overwhelmed by the multitude of the various possibilities, which would force you to abandon the puzzle.

If you have a division into five cases, say, it may happen that one of these cases (hence, one of the assumptions that you are thus able to make) can, in a simple way, be seen to lead to an impossibility, due to a conflict with known conditions. This assumption or case has then come to a dead end. It is very advantageous if all cases but one come to a dead end quickly, because then you need continue only with this one remaining case. You then have, so to speak, already solved one part of the puzzle; this partial solution may consist of having found a digit to be inserted in some place, in a digit puzzle.

However, it may also happen that none of the cases into which you have divided the set of all possibilities immediately comes to a dead end, and that you are forced once more to divide each of these cases into groups again—sometimes after having drawn some conclusions—by making assumptions (e.g., concerning left or right, or concerning even or odd). Of course, this can repeat itself once or more often. It happens sometimes that one of the major cases (that is, one of the groups of possibilities corresponding to the division with which you started the solution of the puzzle) eventually comes to a dead end in all its branches. Such a major case has then been removed only after a great deal of reasoning.

If such a thing happens, you cannot help getting the feeling that you have proceeded inefficiently, and often this turns out to be the truth. Hence, if you observe that the cases into which you have made the division do not tend to come to a dead end quickly, you will be well-advised to check first whether it is not better to choose your assumptions (which constantly direct your attention to a certain group of possibilities) on a different basis, related to a different classification system, preferably in such a way that one—or, better still, several—of these assumptions come to a dead end quickly, after not too many branchings.

It is best, of course, if all major cases but one come to a dead end before it has been necessary to proceed to a further division into cases (hence into sub-cases), and if this favorable circumstance repeats itself some more times. The one remaining major case is then treated further by a division into sub-cases, all but one of which come to a dead end before a further subdivision is necessary, and so on. You then head straight for the goal, so to speak, and in this way the solving time is considerably reduced. Admittedly, going straight ahead like this is far from possible for every puzzle, but you can still aim at approaching this state of affairs as much as you can.

We would like to mention another case in which it is possible to suspect that you have chosen an unsuitable division into cases. Suppose that each of the major cases is subdivided into sub-cases, and that in each of the major cases one of the sub-cases comes to a dead end, after a train of reasoning that is roughly the same for each of the major cases. You then get the impression that you are repeating the same work a number of times, and you should wonder whether the results could not have been obtained by a single train of reasoning. In such a case it is very possible, if not probable, that with another classification system (lengthwise instead of breadthwise, so to speak) all those sub-cases would have been joined into one major case, and that the fact that this major case comes to a dead end could have been obtained by a single train of reasoning. Hence, as soon as you can see that some tendency toward the above-mentioned phenomenon (repetition of roughly the same work) is noticeable with some chosen division into major cases, you should look for a better classification. This does not imply, however, that a repetition of a number of similar arguments can always be avoided; this is probably impossible when the arguments show more or less irregular differences. Hence, much depends on the nature of the similarity between the various arguments; some intuition is needed to judge whether the agreement lies in the nature of the puzzle or should be ascribed to an inappropriate division into cases.

12. More about the classification system. The foregoing considerations show that you should not proceed too quickly to a further elaboration of a division into sub-cases. As noted before, you had better think properly about the choice of the classification system first, without starting the actual solution (in the sense of trying out the possibilities). Decisions regarding a classification system are usually best taken (before you have got to the trial stage) without the use of the material aids like dominos, cut-out squares, draughtsmen, and so on, that are involved in the puzzle. It is no rare thing to find that the classification system eventually chosen is such that you do not use these aids at all, except in your mind, but finish the job with paper and pencil. Many a puzzle with dominos, for example, can be composed and solved without possessing a set of dominos and without feeling the need of them when solving the puzzle.

There are undoubtedly puzzles for which it does not help you to think about a classification system and for which, however you attack the puzzle, the number of possibilities to be examined is alarmingly large. As noted already in §6, this may be due to unsystematic features in the puzzle itself which make any simple system hard to find. In such a case you are dealing with a bad puzzle, often a puzzle that has been posed on a purely arbitrary basis.

IV. SOLVING A PUZZLE BY SIMPLIFICATION

13. Simplifying a puzzle. It frequently happens that you can put a puzzle into a simpler form by applying some reasoning before proceeding to a distinction of different cases. The subsequent distinctions then become simpler. Each of the cases is a puzzle in itself, and it is possible that such a puzzle can again be simplified by reasoning, after which, perhaps, a further division into cases follows, and so on.

It is also conceivable that you will manage to simplify the puzzle to such an extent (often through mathematical reasoning, or at least reasoning of a mathematical nature) that a division into cases appears to be no longer necessary. One can then speak of a direct solution of the puzzle. In such a direct solution the puzzle often takes on the character of a mathematical problem.

14. Example of how to simplify a puzzle. As an example of simplification we take the following puzzle, which might be called "the puzzle of the 7 coins." The object is to place a coin at all the vertices but one of the star in *Figure 1* (shown with reduced dimensions).

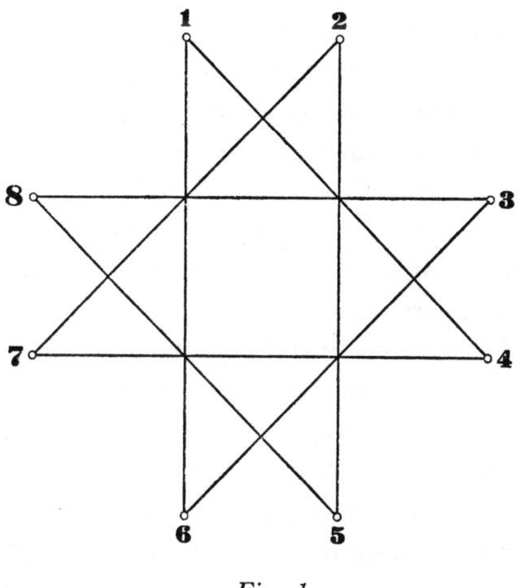

Fig. 1

When placing any coin you should move it along a free line, and put it down at the end of that line. A line is called free if there is no coin at either of its endpoints. The numbers at the vertices are inserted only for the discussion that now follows.

Suppose that we first put a coin on 1 by moving it along the line

4–1, then a coin on 2, moving along 5–2, next a coin on 3 (along 6–3), a coin on 4 (along 7–4), and a coin on 5 (along 8–5). Then we are played out, because there is no free line left; hence, we can place only five of the seven coins in this way.

To solve the puzzle, you might now start a division into cases. You notice, of course, that placing the first coin on 1 is all right, because all vertices play the same role; therefore, with regard to the first coin you do not need to make any division into cases (on 1, on 2, etc.). For the second coin, you have the seven possibilities of placing it on 2, on 3, . . . , on 8; it is irrelevant along which free line you move (when the vertex can still be occupied along two free lines, which is not possible for the vertices 4 and 6). If the second coin has also been placed, you can start to make assumptions about the third coin; it cannot always be placed in six ways, for if the second coin has been placed on 3, then the third coin cannot be placed on 6, because then neither of the lines 1–6 and 3–6 is free any more. Proceeding in this way, you will find all solutions after much sifting. If what you want is to present the puzzle to others, you can stop as soon as a solution has been found; you then note it down and learn it by heart.

However, someone who proceeds thus has not seen through the puzzle, even if he finds a solution. Yet, one might say he has followed the directions about systematic trial and error. Indeed he has, and certainly there are still less efficient ways of solving the puzzle. Many a solver who looks for the solution starts pushing at random, and if, after much pushing, he has been able to place the seven coins, he has the bad luck of not remembering how he did it. This, of course, is puzzle solving of the worst kind.

In order to understand the puzzle completely, you should observe that you ought to try to retain as many free lines as possible, and hence you should make as few lines as possible useless. Placing the first coin on 1 makes two lines useless: 1–4 and 1–6. Placing the second coin makes two more lines useless, except when this coin is placed on 4 or on 6. The correct continuation is, therefore: on 4 or on 6; for example, by occupying 4, only the line 4–7 is made useless, because 4–1 had already dropped out; the third coin should now be placed on 6 or 7, etc. The coins should always be put on adjacent vertices (such as 3, 6, 1, 4); by adjacent vertices we mean vertices that are connected by a line, like 3 and 6.

The puzzle can be simplified considerably (however, only on the face of it, not in essence) by observing that the eight lines form an octagon, and that the shape of the octagon is irrelevant. Hence, instead of *Figure 2* one might also take *Figure 3.* In Figure 2 a different numbering has been chosen from that in Figure 1, in accordance with the concept of "octagon," hence in accordance with Figure 3. If the puzzle is presented using Figure 3 (the numbers at the vertices are then omitted, of course), then anyone will see the solution at once and occupy the vertices (all but one) successively. In essence, however, the puzzle with the star-shaped figure is no different.

15. Remarks on the seven coins puzzle. If you demonstrate the puzzle of §14, in which seven coins have to be placed at the vertices of the star, you should not do this by giving an exact description of what should be done and achieved, but simply show the placing of

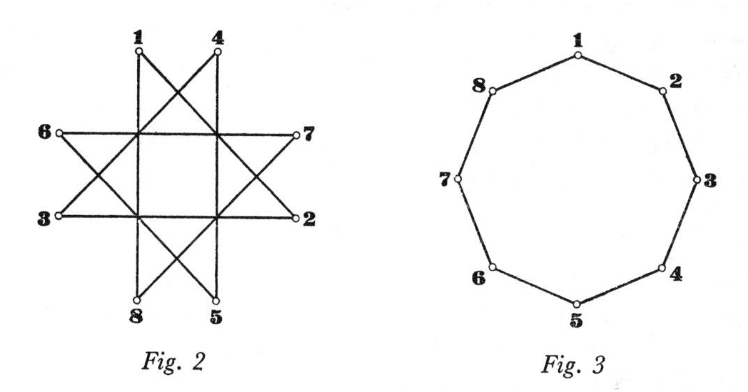

Fig. 2 *Fig. 3*

the seven coins quickly, with the invitation to imitate this. "An exact imitation is not necessary," you might add, "as long as the seven coins are placed by moving along free lines." From the imitation (or rather the attempt at it) it can be seen immediately whether the meaning has been grasped, or whether a further explanation is necessary.

On seeing the demonstration of a solution, everyone will get the impression that the puzzle is plain sailing, so to speak. Most people, however, will get stuck when imitating it, and will manage to place no more than six, or even five, coins. You may safely show the solution once more, so quickly that it is not possible to recall exactly how the moves were made. Even after it has been shown four or five times, there will be many who still cannot do it.

Indeed, the presentation is not of much help, because you naturally will do it differently every time, for example (see the numbers in Figure 1), first in the order 1-4-7-6-3-2-8, then in the order 3-6-8-5-1-2, then in the order 7-2-4-5-1-8-6, etc. Hence, showing solutions in this way confuses rather than helps. As long as someone has not seen through the puzzle, there is little chance of his succeeding in placing the seven coins; in most cases the second coin will already be placed incorrectly, since it can be placed on only two of the seven remaining free vertices. Exactly how great the probability is that someone will succeed in placing all coins by chance, is difficult to tell, but surely it cannot be much greater than 1 in 100 (cf. the end of §135).

If the imitation has not succeeded after the solution has been shown a number of times, you may show the solution in a more amusing way. You can allow each move made to be forbidden once; it then has to be replaced by another move. For example, suppose that you have started on 1 (using the numbers in Figure 1), that you continue with 4, and that this latter move is rejected; then you replace it by 6 (this move cannot be forbidden). If you now continue with 3 and if this move is forbidden, then you replace it by 4.

After the first move has been made, there are two correct moves every time, so that the goal can be reached in

$$2^6 = 2 \times 2 \times 2 \times 2 \times 2 \times 2 = 64$$

ways. Since it is possible to start on any of the eight vertices, the order in which seven out of the eight vertices are occupied can be chosen in $8 \times 64 = 512$ ways. Furthermore, as it may give a different impression when you place the first coin on 5 by running along 2–5 and by running along 8–5, one might even say that placing the seven coins can be achieved in $2 \times 512 = 1024$ ways. So if you make a random choice of the moves in your solutions, you need not be afraid of repeating yourself too soon.

If Peter attempts to place the seven coins, then John, who is watching and who knows the puzzle, can indicate at any moment how far Peter can still go. As soon as Peter, after his first move, makes a move that renders two lines useless, then the total number of coins that can be placed is diminished by one. So if Peter has successively occupied the vertices 1, 4, and 5 (in this case the second move is still correct), then John can say, "Now only three more coins, that is a total of six coins, can be placed." If Peter chooses 3 as his next move, then only a total of five coins can be placed. Peter can always place at least four of the seven coins; he gets no further than four coins, if he occupies the odd-numbered vertices or the even-numbered vertices. The chance that Peter will do this by accident is not great, of course.

It will happen most frequently that Peter manages to place six coins.

16. Reversing a puzzle. It frequently happens that a puzzle consists of obtaining a given final position from a given initial position by making the smallest possible number of moves. One can then start from the final position instead, and try to obtain the initial position by backward moves. If this has succeeded, all one has to do to get the solution of the original puzzle is to perform all moves in the reverse order and in the reverse direction.

It is virtually certain that the new puzzle (the reversed one) will be simpler; otherwise the puzzle would surely have been presented in the reversed form. Hence, in most cases the reversal of the puzzle will produce a simplification.

17. Example of reversing a puzzle. Three dimes (the small circles in *Figure 4*) and two quarters (the larger circles) are placed in a row, with dimes and quarters alternating (Fig. 4, top). Now with every move a dime and an adjacent quarter are moved as one whole, so that they remain adjacent. You are not allowed to reverse the coins; that is, you should not interchange the dime and the quarter. Moving the dime and the quarter is considered as one move. The problem is to get the dimes and the quarters in a row in such a way that the dimes are adjacent, and the quarters, too (Fig. 4, bottom), in as few moves as possible.

As rotation is not allowed, the new row is parallel to the original row. However, it is not necessary that the dimes get to the left of the quarters, as in the lower part of Figure 4; as a matter of fact, if you have managed to get the dimes on the left-hand side, all you have to do to get the dimes to the right of the quarters is to perform the mirror image of each move. If it were allowable to interchange a dime and an adjacent quarter, the final position could be reached without trouble in three moves; the relative order of dimes would not have changed, any more than that of the quarters.

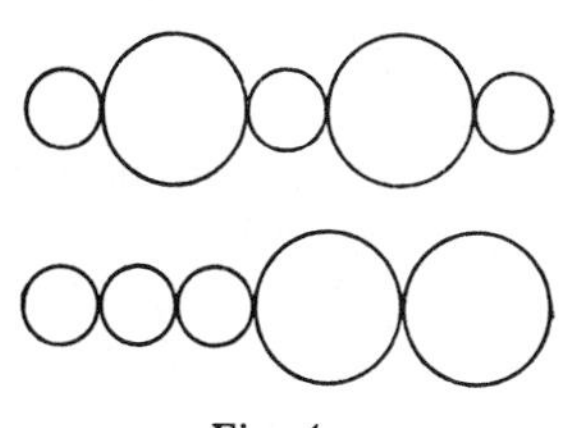

Fig. 4

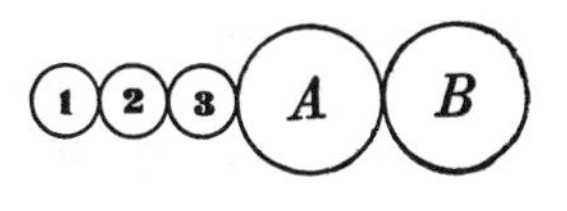

Fig. 5

Since interchanging a dime and a quarter is forbidden, it is slightly more complicated to reach the final position. In the initial position, in which these are four cases where a dime and a quarter are adjacent, a large number of moves can be made. However, from the final position only two moves are possible. Therefore we reverse the puzzle and consider the final position as the initial one; in the new initial position we designate the dimes and quarters as in *Figure 5.* In determining the moves, one should observe that there is no point in performing two consecutive moves with the same pair (because then one can combine both moves into one). Furthermore, one should take care that after every move there will be another pair (consisting of a dime and an adjacent quarter) with which a next move can be made. This leads to the following puzzle tree:

```
                                              ┌─(2 A . . 3 . . . . 1 B)─(2 A 1 B 3)
                                              ├─   (2 A 3 . . . . 1 B)─(1 B 2 A 3)
            ┌─(3 A 1 2 . . B)─(3 . . 2 A 1 B)─┼─(1 B . . 3 . . 2 A)    ─(1 B 2 A 3)
(1 2 3 A B)─┤                                 └─   (1 B 3 . . 2 A)     ─(2 A 1 B 3)
            │                                 ┌─(2 B 1 . . 3 . . A)
            └─(1 2 . . B 3 A)─(1 2 B 3 . . A)─┴─   (1 . . 3 . . A 2 B)
```

The dots indicate free positions. From the position 3 . . 2*A*1*B* infinitely many moves are possible. You can place the pair 2*A* (or 1*B*) wherever you want, because in each case you can make another move with the pair 1*B* (or 2*A*). However, we have only included those third moves (following 3 . . 2*A*1*B*) after which it is possible to make a fourth move which yields the desired final position (the original initial position). Hence, this can be done in four ways; in the first two moves these agree completely and for the rest they come to practically the same thing: the last two moves take the pairs 2*A* and 1*B* to the left of 3, the order being unimportant, while it also makes no difference which of these pairs gets next to 3. The four ways are:

```
        1 2 3 A B                 1 2 3 A B
      3 A 1 2 . . B             3 A 1 2 . . B
      3 . . 2 A 1 B             3 . . 2 A 1 B
2 A . . 3 . . . . 1 B         2 A 3 . . . . 1 B
2 A 1 B 3                 1 B 2 A 3

        1 2 3 A B                 1 2 3 A B
      3 A 1 2 . . B             3 A 1 2 . . B
      3 . . 2 A 1 B             3 . . 2 A 1 B
1 B . . 3 . . 2 A             1 B 3 . . 2 A
1 B 2 A 3                 2 A 1 B 3
```

To obtain the four ways in which you can get from the original initial position to the desired final position with the dimes at the left, you merely have to read this from bottom to top. Of these four ways, the last one seems simplest. If we adapt the order of the numbers and the letters to the original puzzle, then the solution in question is the one shown at the left.

1 *A* 2 *B* 3	1 *A* 2 *B* 3	1 *A* 2 *B* 3
2 *B* 3 . . 1 *A*	2 *B* 3 1 *A*	1 *A* . . 3 2 *B*
3 . . 1 *A* 2 *B*	3 1 *A* 2 *B*	3 2 *B* 1 *A*
3 *A* 2 1 . . *B*	3 . . 2 *B* 1 *A*	3 . . 1 *A* 2 *B*
2 1 3 *A B*	3 *B* 1 2 . . *A*	3 *A* 2 1 . . *B*
	1 2 3 *B A*	2 1 3 *A B*

To the right are two five-move solutions which satisfy the condition that a pair with which a move is made is placed next to a coin, and hence is not placed in a completely detached position. In these solutions, two moves are made with the same pair (1*A* or 2*B*), with one move in between (with 2*B* and 1*A*, respectively). The two moves with the same pair are equivalent to a single move; but splitting up this move into two moves can here be regarded as a means to make the correct preparation for another move.

V. SOLVING A PUZZLE BY BREAKING IT UP

18. Breaking a puzzle up into smaller puzzles. One method that can often help solve a puzzle is to break the puzzle up efficiently into two or more smaller puzzles, which are then successively solved. It is the principle of "divide and conquer," widely applied in mathematics, too, to reach one's goal.

Actually, every efficient distinction of cases involves breaking the puzzle up into smaller puzzles. However, with such a division into cases, the puzzles into which one has made the division are all of the same kind but not all the same, which may obscure what we mean here by efficiently breaking a puzzle up into smaller ones. So we state that what we have in mind here is the division of the puzzle into two or more heterogeneous puzzles. This heterogeneity justifies the conjecture that the division has led us nearer to the goal, because difficulties of different kinds have been distributed to separate puzzles. The division of the puzzle into two or more puzzles that are completely identical is always efficient, too (cf. §§19 and 286). In this case, however, the division rather gives the impression of a simplification.

In the subsequent sections we shall give some simple examples of splitting up a puzzle. In Chapter II we shall give some more complicated examples.

19. Application to the crossing puzzle. A board consists of twenty squares arranged in a rectangle with four rows of five (see *Figure 6*). On squares 1, 6, 11, and 16 there are white pieces, and on squares 5, 10, 15, and 20 there are black pieces placed. With each move one of the 8 pieces is shifted diagonally along an arbitrary number of squares (to a free square); hence, if you choose the piece placed on square 11, then you can move it to any of the squares 17, 7, or 3. By moves of this type you have to reach a final position in which the white pieces are on squares 5, 10, 15, 20 and the black pieces on

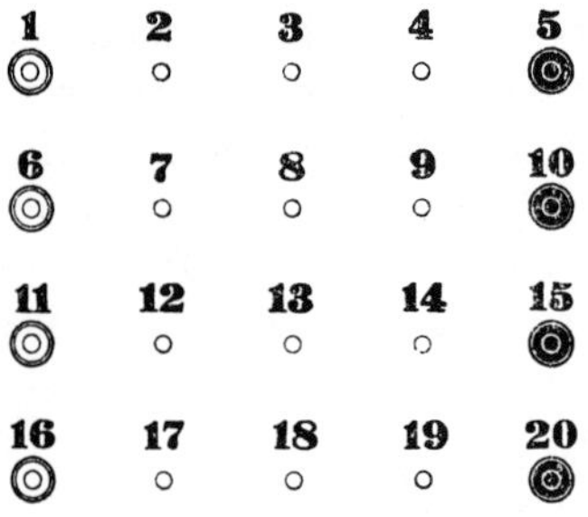

1 2 3 4 5
6 7 8 9 10
11 12 13 14 15
16 17 18 19 20

Fig. 6

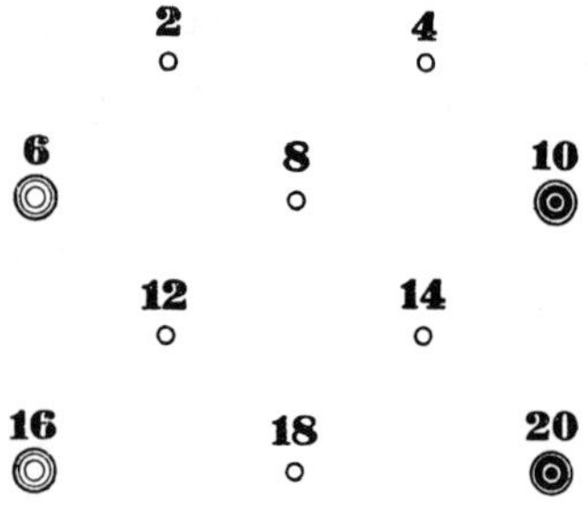

2 4
6 8 10
12 14
16 18 20

Fig. 7

squares 1, 6, 11, 16; thus the pieces must *cross* the board. The problem is to achieve this in the smallest possible number of moves.

We note that a piece from a square having an even number will remain on even-numbered squares; we call it an even piece. The odd pieces always remain on the odd-numbered squares. An even and an odd piece do not interfere with each other. Hence one can partition the puzzle into two parts, that of the even pieces and that of the odd ones. Both puzzles are completely identical, so one could equally well speak of a simplification of the puzzle. However, the circumstance that the two puzzles into which the puzzle has been divided are identical is a more or less accidental one that would not present itself if, for example, one were to take a board with 25 squares, arranged in one large square, with five white and five black pieces.

Figure 7 shows the puzzle with the even pieces; it makes a much simpler impression than the original puzzle, yet it is essentially the same. It is not possible to go from one of the squares 6 or 16 to one of the squares 10 or 20 in one move, hence at least eight moves are required for the crossing. The crossing is certainly not possible in eight moves when a white piece goes from 6 to 10 and a black piece from 10 to 6; for if the white piece tried to do this by the two moves 6–18 and 18–10, and the black piece by the two moves 10–18 and 18–6, then they would be in each other's way. So if you want to try to complete the crossing within eight moves, then one white piece should go (for example) from 6 to 10, the other one from 16 to 20, one black piece from 10 to 16, and the other one from 20 to 6. Then the white pieces go straight across and the black pieces cross obliquely. Alternatively, one may make the white pieces cross obliquely and the black pieces straight, but this comes to the same thing.

We assume that the white pieces go straight across and the black pieces cross obliquely. Within eight moves this is possible only in the way indicated in *Figure 8*. The moves with the white piece that was

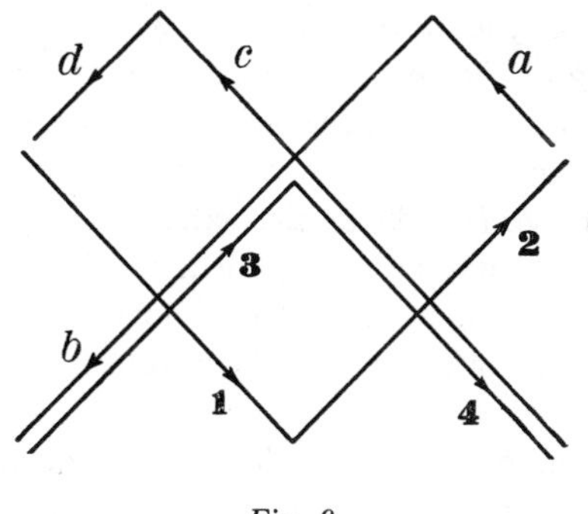

Fig. 8

originally on square 6 have been called 1 and 2, the moves with the other white piece 3 and 4. The moves with the black piece that was originally on square 10 have been termed *a* and *b*, those with the other black piece *c* and *d*. Obviously 2 is preceded by 1, 4 by 3, *b* by *a*, and *d* by *c*, because (for example) move 2 is impossible when it has not been preceded by 1. You should also take care that a white and a black piece are not in each other's way. Move 1 should be made before move *d*, and move *a* before move 2; also, move *c* should be made before more 3, and move 4 before move *b*. For the rest, the order of the moves is arbitrary. The order of the moves can be indicated schematically by *Figure 9*; of 2 moves that are directly connected by a bar, the one at the left should be made first. However, it is unimportant whether (for example) move 2 is made before or after *b* or *c* or *d*. It is true that 2 has been located to the right of *d*, but since 2 is not connected directly to *d*, this is of no importance.

The order of the 8 moves can be chosen in various ways to be in accordance with the diagram: for instance, *ac*132*d*4*b*. Hence the crossing is possible in 8 moves (and not less).

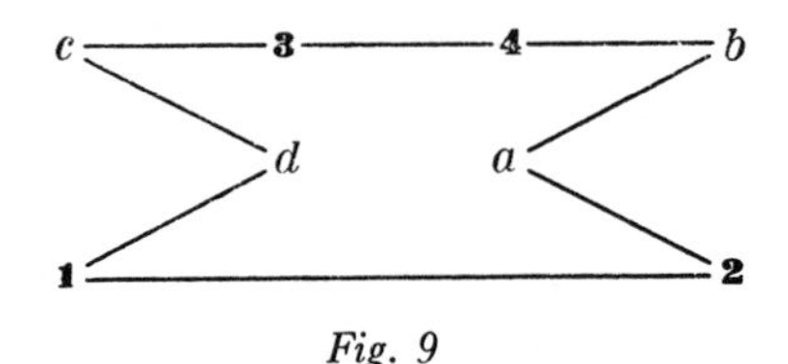

Fig. 9

20. Number of solutions of the crossing puzzle. We consider the simplified crossing puzzle of §19 (with 10 squares, 2 white and 2 black pieces), and assume, as before, that the white pieces move straight across. According to Figure 9, 15 sequences of the moves 1, 2, 3, 4, *b*, *c* are possible. Each of these sequences provides a certain number of places at which move *a* can be performed and a certain number of places at which move *d* can be performed. As the order of moves *a* and *d* is immaterial, you should multiply the 2 numbers to get the number of possibilities pertaining to the order of moves *a* and *d* relative to the assumed order of the other moves. The results have been collected in Table 1.

The first column gives the possible orders of moves 1, 2, 3, 4, *b*, *c*. The second column gives the number of possibilities (as far as order is concerned) of move *a*; this move should be performed before moves 2 and *b*, and hence in the second case (1*c*234*b*) can be performed before move 1, or between moves 1 and *c*, or between moves *c* and 2, which amounts to 3 possibilities. The third column gives the number of possibilities for move *d* (to be performed after moves 1 and *c*). By multiplication one finds, from the second and third columns, the number of order-possibilities of moves *a* and *d* together, to which some cases have to be added corresponding to an exchange of *a* and *d* (fourth column). By addition one then finds the number of all possibilities, which is 297.

The same number of possibilities is obtained when the white pieces cross obliquely, and the black pieces go straight across. Hence, in all, the crossing can be executed in $2 \times 297 = 594$ ways in 8 moves (still

TABLE 1.

order of moves 1, 2, 3, 4, b, c														a	d	a, d
1	2	c			3			4			b			2	4	8
	1	c		2	3			4			b			3	5	15 + 1
	1	c			3		2	4			b			4	5	20 + 2
	1	c			3			4		2	b			5	5	25 + 3
	1	c			3			4			b		2	5	5	25 + 3
		c	1	2	3			4			b			3	5	15 + 1
		c	1		3		2	4			b			4	5	20 + 2
		c	1		3			4		2	b			5	5	25 + 3
		c	1		3			4			b		2	5	5	25 + 3
		c			3	1	2	4			b			4	4	16 + 1
		c			3	1		4		2	b			5	4	20 + 2
		c			3	1		4			b		2	5	4	20 + 2
		c			3			4	1	2	b			5	3	15 + 1
		c			3			4	1		b		2	5	3	15 + 1
		c			3			4			b	1	2	4	2	8
																272 + 25 = 297

for the puzzle of the even pieces only). The same number of possibilities applies in the case of the puzzle with the odd pieces.

We now return to the original crossing puzzle (20 squares and 8 pieces). Obviously there the crossing can occur in $2 \times 8 = 16$ moves. For example, you may first execute the entire crossing of the even pieces, or first the crossing of the odd pieces. Or you can make a move with an even piece and a move with an odd piece alternately. You can have the moves of the even and of the odd pieces succeed each other in many more ways, namely (as we will show in §129) in 12,870 ways in all. The number of ways in which one can cross in 16 moves in the original crossing puzzle is therefore:

$$12{,}870 \times 594 \times 594 = 4{,}540{,}999{,}320,$$

hence, nearly 4,541 million.

21. Restrictive condition in the crossing puzzle. In the simplified crossing puzzle, for example the one with the even pieces, one can impose the condition that the moves with the white pieces and those with the black pieces be made alternatively. We assume that the white pieces go straight across, and denote the moves as in §19; hence we use 1, 2, 3, 4 for the white pieces, and a, b, c, d for the black pieces. We then get a letter and a number in alternation. From Table 1 (first column) we then find the following 10 possibilities for this case:

1a2c3d4b, 1c3a2d4b, 1c3d4a2b, 1c3a4d2b, a1c3d4b2,
1c3a4b2d, c1d3a4b2, c1a3d4b2, c3a1d4b2, c3a4b1d2.

When the moves are denoted by using the numbers of the squares, the first of these solutions reads:

6–18, 10–4, 18–10, 20–2, 16–8, 2–6, 8–10, 4–16.

By rotation one finds from this the following sequence of moves with the odd pieces:

15–3, 11–17, 3–11, 1–19, 5–13, 19–15, 13–1, 17–5.

By first making (for example) 1, 3, 5, or 7 moves from one of the sequences, then all moves of the other sequence, and then the remaining moves of the first sequence, we find a way of crossing in 16 moves for the original puzzle, using white and black pieces alternately.

22. Shunting puzzle. As an example of breaking a puzzle up into two completely identical puzzles, we further take the well-known shunting puzzle illustrated in *Figure 10.* The engine *L* is too heavy to

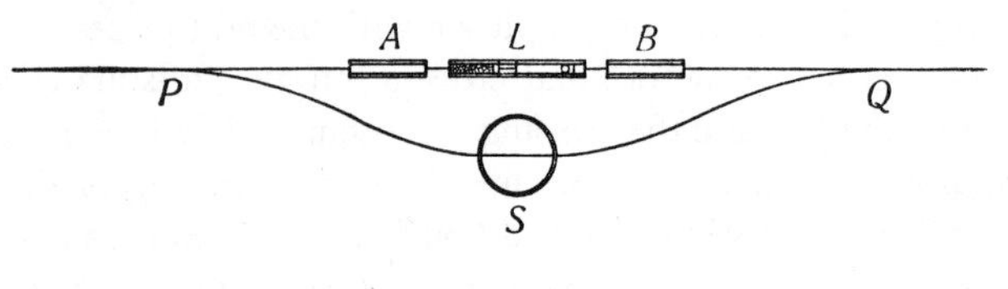

Fig. 10

be allowed on the turntable *S*. The question is to make cars *A* and *B* change places.

If car *B* were not present, and if the engine and car *A* had to change places, then the answer would be easy to find. The engine first pulls *A* to the right of *Q*, and next takes *A* to the turntable. Now *L* moves to the right, past *Q*, then to the left past *P*, and then to the left-hand side of the turntable. The engine *L* then takes *A* to the left, past *P*, and next moves back, with *A*, until it is between *P* and *Q*.

The same thing can be done even with car *B* present. *B* is first taken to the right (past *Q*) sufficiently far that it is not in the way when *A* is being shunted. When *A* has been taken to the right of *L*, the same operation is performed with car *B*. Here car *A* is taken along all the time, so that *L* and *A* together are considered as the engine; here *A* is always to the right of *L*.

The shunting puzzle can be complicated by requiring (for example) that in *Figure 11* cars *A* and *D* be interchanged, and also *B* and *C*.

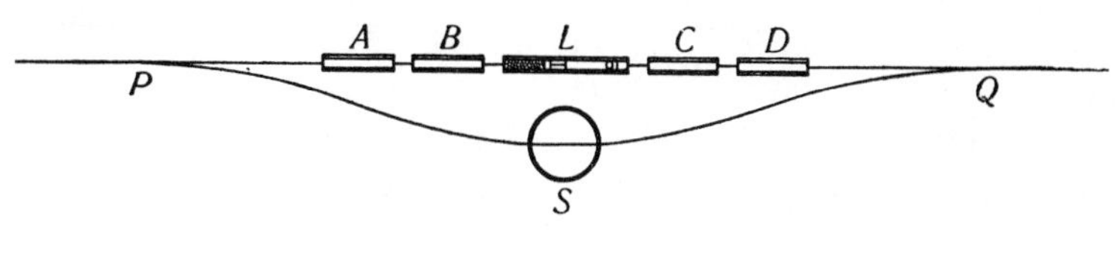

Fig. 11

However, the puzzle has then only seemingly become more difficult because it can now be divided into a (now larger) number of puzzles, each time with one car that has to change places with the engine. In some of these smaller puzzles, the "engine" of the puzzle has to be taken to be the actual engine together with one or more cars connected to it. Cars that would be in the way are first moved aside to the left past *P*, or to the right past *Q*.

How the puzzle is solved by division is indicated in the following manner. The letters in parentheses show what is considered as the engine, and also which cars do not take part in the shunting and are moved aside:

ABLCD—(*A*)(*L*)*B*(*CD*)—(*A*)*C*(*LB*)(*D*)—(*CLB*)*A*(*D*)—*D*(*CLBA*).

For example, in the second puzzle cars *A* and *D* are moved aside, and *C* changes places with the combination *LB*; in the fourth puzzle *D* changes places with the combination *CLBA*. It is clear that the shunting can be varied in a number of ways, and also that every order of the cars and the engine can be achieved.

If in Figure 11 one has to interchange *A* and *C*, and also *B* and *D*, then one can shunt with *A* and *B* combined into one car, and similarly with *C* and *D*. Here, it is not necessary for the turntable to be large enough for *A* and *B* together. When *A* is put on the turntable (with *B* to the right of it), the coupling between *B* and *L* is unfastened; after that the engine goes, via *Q* and *P*, to the left-hand side of the turntable and pulls *A* and *B* together to the left.

VI. SOME PUZZLES WITH MULTIPLES

***23. Trebles puzzle.** We require all five-digit numbers (without initial digit 0) that are written with the same digits as their treble. When these numbers have been found, we shall have obtained at the same time the numbers with fewer digits that show the peculiarity in question, for this property is preserved when one or more zeros are placed to the right of the number.

One could make a division into cases by examining each of the numbers from 10,000 to 33,333 to see whether or not it is composed of the same digits as its treble. Of course, this method is much too laborious, and it would be quite impracticable to solve the puzzle in this way if we had required all 6- or 7-digit numbers, say.

A considerable simplification is obtained at an early stage because (as we shall show) the desired numbers are divisible by 9. To this end we first prove that a number differs by a multiple of 9 from the sum of its digits (its digit-sum). The numbers 10, 100, 1000, 10,000, etc., all differ by a multiple of 9 from 1, because the numbers 9, 99, 999, 9999, etc., are all divisible by 9. Hence 700, say, differs by a multiple of 9 from 7, and 5000 by a multiple of 9 from 5. In this way one sees that the number

$$34{,}752 = 3 \times 10{,}000 + 4 \times 1000 + 7 \times 100 + 5 \times 10 + 2,$$

for example, will differ by a multiple of 9 from $3 + 4 + 7 + 5 + 2$, hence from its digit-sum.

A number that is written with the same digits as its treble has the same digit-sum as its treble. Hence, the number and its treble differ by a multiple of 9 from the same number (the digit-sum in question), and thus they differ by a multiple of 9 from each other. However, the difference between a number and its treble is twice that number, so that this double is a multiple of 9. Hence the number itself is a multiple of 9. As appears from the above, one can see at once whether or not a number is divisible by 9 from its digit-sum. So one may also say that the desired numbers have a digit sum that is divisible by 9.

***24. Breaking up the trebles puzzle.** If the digit 2, say, occurs in a number, this leads (depending on the digit that succeeds 2) to the digit 6, 7, or 8 in the treble of the number; the digit 7 leads to the digit 1, 2, or 3 in its treble, and so on.

Hence, if a number consists of the same digits as its multiple by 3 (in a different order) and if the number contains the digit 2, then it also contains one of the digits 6, 7, or 8, for example 8; then the number also contains one of the digits 4, 5, or 6, for example 5; this digit may lead to a 5 in the treble, but also to a 6 or a 7. Continuing in this way, we obtain a closed sequence of different digits that show the property considered in §9, the difference being that in §9 these digits were understood to be consecutive digits of a number, which is not the case now.

The required five-digit numbers have to be composed of one or more closed sequences of digits of the type considered in §9. Each of the single digits 0, 4, 5, and 9 can be considered to form such a sequence. By this the puzzle has been divided into two puzzles of a completely different nature. The puzzle that has to be solved first requires finding the above-mentioned closed sequences of digits. Next comes the puzzle whose object is to combine these sequences—after an appropriate permutation of the digits—into five-digit numbers that are composed of the same digits as their trebles.

In solving the latter puzzle one can use with advantage the property found in §23, that the digit-sum of the desired numbers is a multiple of 9. This causes most of the combinations of sequences of digits to drop out. It may be further noted that it has to be possible to compose from the sequences of numbers a number less than 33,333, because otherwise its treble would consist of 6 digits (unless for the number 33,333 itself, which does not come into consideration). This causes combinations like 973–4–4, 301–4–9, and 86–0–4–5 to drop out. The remaining combinations (the determination of which is a puzzle in itself) are (cf. §9):

7142–**4**	84**2**–**3**1 dead	**3**1–0–0–5 dead
*8*__*5*__**72**–5	84**2**–0–**4** dead	**3**1–0–5–**9** dead
*30*1–**8**6 *dead*	84**2**–4–**9** dead	**3**1–**4**–5–5 dead
*30*1–0–5	*9*__8__*6*–**3**1 *dead*	**3**1–5–**9**–**9** dead
*30*1–5–**9** *dead*	**3**1–*72*–5	*72*–0–0–0 dead
*4*__3__*1*–5–5 dead	**3**1–**8**6–0 dead	*72*–0–0–**9** dead
602–5–5 dead	**3**1–**8**6–**9** dead	*72*–0–**4**–5
7**15**–**8**6 dead	*72*–*72*–0 dead	*72*–0–**9**–**9** dead
7**15**–0–5 dead	*72*–*72*–**9** dead	*72*–**4**–5–**9**
7**15**–5–**9** dead	*72*–**8**6–**4**	*72*–**9**–**9**–**9** dead

By 7142–4 we mean that in the number a closed sequence of 4 digits occurs (7142 in some order) and a closed sequence of one digit (4). Hence, two 4's occur in the number, a 4 that is followed by a low digit (or by nothing), thus leading to the digit 2 (not increased) in the treble, and a 4 that is followed by a high digit, thus leading to the digit 4 (increased by 2) in the treble; this latter 4 therefore forms a sequence in itself.

A digit of which the treble has been raised by 2, because it is succeeded (in the desired number, not necessarily in the sequence) by a high digit, has been printed in **boldface**; whether a digit has to be bold can be seen from the next digit in the sequence (hence, for the last digit of a sequence, from the first digit). A digit of which the treble has been increased by 1 because, for example, it is followed by a 4 or a 5 in the desired number, has been *italicized.* For the digits in ordinary type the treble is not raised, because, for example, such a digit is the last in the number or is followed by 0, 1, or 2.

A digit that yields a treble that is 20 or more (taking into account a possible raise), is called "high"; hence the high digits are 7, 8, 9 and the bold 6. A digit that (taking into account a possible raise) yields a treble that is 9 or lower is called "low"; hence the low digits are 0, 1, 2 and the ordinary 3 (which, by the way, turns out not to occur). The other digits will be called "moderate" (with a possibly raised treble somewhere between 9 and 20); the moderate digits are the italicized 3, the bold 3, the 4, the 5, the ordinary 6, and the italicized 6.

In the desired number a bold digit is followed by a high digit, an italicized digit by a moderate digit, and an ordinary digit by nothing or by a low digit. Hence, the number of bold digits is equal to that of the high digits, and the number of italicized digits to that of the moderate digits (also, of course, the number of ordinary digits to that of the low digits).[1] The number begins with a low digit at the left and ends with an ordinary digit at the right.

---

[1] For the reader who has some acquaintance with an algebraic notation for numbers, we add the following remarks. In a closed sequence of digits of the kind considered in §9, let h be the number of high digits, m the number of moderate digits, b the number of bold digits, i the number of italicized digits, and s the digit-sum. When all digits of the sequence are multiplied by 3, the same digits reappear (with a shift of one place), provided one diminishes the treble of a high digit by 20 and that of a moderate digit by 10, and increases that of a bold digit by 2, and that of an italicized digit by 1. This yields

$$3s - 20h - 10m + 2b + i = s,$$

hence $2s = 20h + 10m - 2b - i$. This relation also holds for a combination of sequences, where the numbers h, m, b, i, and s refer to the combination, as can be seen by adding the relations for the separate sequences. The relation can be rewritten as follows:

$$2(h - b) + (m - i) = 2s - 9(2h + m).$$

If l is the number of low digits, and o that of the ordinary digits of the combination, then $h + m + l = b + i + o$, because both members are equal to the number of digits of the combination. So one also has

$$2s - 9(2h + m) = 2(h - b) + (m - i) = (h - b) + (o - l)$$
$$= (i - m) + 2(o - l).$$

Hence, if s is divisible by 9, then each of the numbers $2(h - b) + (m - i)$,

For twenty out of the thirty combinations of sequences none of the equations (number of bold digits = number of high digits, and so on) is satisfied. For example, in **60**2–5–5 the numbers of bold and italicized digits are 2 and 0, respectively, and the numbers of high and moderate digits are 1 and 2 respectively. Hence, such a combination cannot lead to a number that satisfies the condition. This has been indicated by the word dead (in roman type).

For the remaining 10 combinations all equations are satisfied. Among these 10 there are 3 that come to a dead end in another way (indicated by *dead* in italic type). In *30*1–**8**6 and *30*1–5–*9* the only high digit, which is also the only bold digit, would have to succeed itself. In *9***8***6*–**3**1 the 1 is the only low digit and the only ordinary digit, so it would have to be both the initial and the final digit of the number.

Each of the remaining 7 combinations of sequences leads to one or more solutions. When looking for the solutions, you can make use of the fact that 2 digits are interchangeable when they are both italicized and both low, or both italicized and both high, etc. When looking for the number, you can then make a choice from these digits; if it leads to a solution, the only thing you have to do to obtain another solution is to permute the digits in question.

In 7*142*–**4**, the digits 1 and 2 (both italicized and both low) are interchangeable. Hence, for the time being, we can assume that the number starts with 1. The second digit is moderate (because 1 is italicized) and is therefore the digit 4 (moderate or bold). The number ends in 4 (printed in ordinary type), preceded by 2 (because 4 is moderate), or ends in 7, preceded by 4. This gives the solutions 14724 and 14247. By interchanging 1 and 2 one further obtains the solutions 24714 and 24147.

In *8***5***7***2**–5 the digits 7 and 8 are interchangeable (both italicized

---

$(h - b) + (o - l)$, and $(i - m) + 2(o - l)$ is divisible by 9. Consequently, if, in addition, $h = b$, then $m - i$ and $l - o$ are both divisible by 9, and hence equal to 0 if the combination contains less than 9 digits. Hence, in the latter event (s always being a multiple of 9, and the combination containing fewer than 9 digits) it follows from $m = i$ that $h = b$ and $l = o$, and from $l = o$ that $h = b$ and $m = c$. Hence, for a combination having a digit-sum divisible by 9, the 3 equations

number of bold digits = number of high digits, etc.

are either all satisfied or all false. However, when determining the desired numbers it is not necessary to know this.

and both high). The number starts with 2 (the only low digit) and ends with 5 (the only ordinary digit). As 2 is bold, the second digit is high, say 7. The last digit but one is italicized (because 5 is moderate), hence it is 8. This gives the solution 27585. By interchanging 7 and 8 one obtains 28575.

In *30*1–0–5, the 1 and the ordinary 0 are interchangeable, but only if we admit 0 as the initial digit. As we do not do this, the number starts with 1, followed by 0 (because 1 is ordinary). The digit 3 is followed by 5 and preceded by the italicized 0. This gives the solutions 10035 and 10350.

Continuing in this way, we find five more five-digit numbers which are written with the same digits as their trebles:

(**3**1–*72*–5)12375, 23751; (*72*–**8**6–**4**)24876;

and

(*72*–0–**4**–5)24750; (*72*–**4**–5–**9**)24975.

The combinations of sequences from which the numbers have been derived are shown in parentheses. In all we find thirteen numbers that satisfy the condition.

If we also admit zeros as initial digits, there are five more similar solutions:

(*30*1–0–5) 00351, 01035, 03501, 03510; (*72*–0–**4**–5) 02475.

From the solutions with the final digit 0 we find the solutions 1035 and 2475 with fewer than five digits, as well as the solution 0351, if we admit 0 as initial digit.

The smallest number without initial digit 0 which is written with the same digits as its treble is 1035. It is quite possible to find this by trial, especially when one first argues that the number has to be a multiple of 9.

***25. Trebles puzzle with larger numbers.** In order to solve the trebles puzzle for six-digit numbers, one must first do this for the problem of §9. In the manner indicated there, one finds (by extending the tree) the following closed sequences of six digits:

714302 dead 715602 dead 730142 dead 843026 dead 8*4*__31__*56* dead *8*__57__*14*__2__
857302 dead 860142 dead 971426 dead 972843 dead 972856 dead 9**7***3***0**2*6*
973156 dead 984273 dead 985173 dead 985726 dead *9*__86__*0*13 dead 986026 dead

Now that we require six-digit numbers (and hence these sequences of digits will not be combined with other sequences), most of them drop out, because the digit-sum is not a multiple of 9; in these sequences no italicized or bold digits have been indicated. Of the remaining sequences two drop out, because the equations discussed in §24 do not hold. Thus, only the sequences *8*57*1*42 and 97*3*02*6* remain. The latter sequence still comes to a dead end, because the two italicized digits coincide with the two moderate digits, and hence the 6 would have to follow the 3, and the 3 the 6. The sequence of digits *8*57*1*42 yields the 6 solutions 142857, 157284, 158427, 271584, 284157, 285714.

The other solutions are obtained by combining 2 or more smaller sequences of digits. Apart from the tenfold multiples of the thirteen numbers found in §24, the other solutions are:

(71*3*0*2*–5) 107235, 123507, 235071, 235107; (97**28***6*–**4**) 276489, 287649 (7*1*4*2*–**31**) 113724, 114237, 123714, 124137, 137124, 137241, 141237, 142371, 237114, 237141, 241137, 241371; (*8***57*2*–**86**) 275886, 285876, 287586, 288576; (*4301*–5–5) 134505, 135045, 143505, 145035, 150345, 150435; (*7*1*42*–0–**4**) 140247, 147024, 240147, 247014; (7*1*4*2*–**4**–**9**) 142497, 149724, 241497; 249714; (*8***57*2*–5–**9**) 275985, 285975, 297585, 298575; (*30*1–84**2**) 103428, 128034, 280341, 281034; (**43***1*–*85*6) 138456, 138546, 143856, 145386, 153846, 154386; (**43***1*–97**3**), 143793, 314379; (**60**2–*85*6) 206856; (**60**2–97**3**) 206793, 207693, 230697, 230796, 306792, 307692, 320679, 320769; (7**15**–*9***8***6*) 189657, 196587; (7**15**–*72*–5) 172575, 175257, 175725, 251757, 257175, 257517; (84**2**–*72*–**4**) 242748, 247428, 248274, 274248, 274824, 282474; (*30*1–0–0–5) 100035; (**3**1–*72*–**86**) 123876, 238761; (**3**1–*72*–0–5) 102375, 237501; (**3**1–*72*–5–**9**) 123975, 239751; (*72*–**86**–**4**–**9**) 248976, 249876; (*72*–**4**–5–**9**–**9**) 249975.

In all this yields 104 six-digit numbers, without initial digit 0, which are written with the same digits as their treble.

If one also admits 0 as an initial digit, one obtains (as well as numbers which arise by placing zeros before numbers already obtained) the following eighteen further solutions:

(71*3*0*2*–5) 071235, 072351; (*4301*–5–5) 034515, 035145, 043515, 045135, 051345, 051435; (*30*1–84**2**) 034128, 034281; (**60**2–*85*6) 068562; (**60**2, 97**3**) 067923, 067932, 076923, 076932; (*30*1–0–0–5) 035001, 035010, 035100

***26. Doubles puzzle with 7-digit numbers.** We now require the 7-digit numbers which are written with the same digits as their doubles.

To this end, we first look for the closed sequences of seven or fewer different digits, such that each digit is the double or the double plus 1 of the preceding digit, possibly diminished by 10. The trees can now only show branchings into two possibilities. If we put the highest digit of the sequence first, the various puzzle trees are:

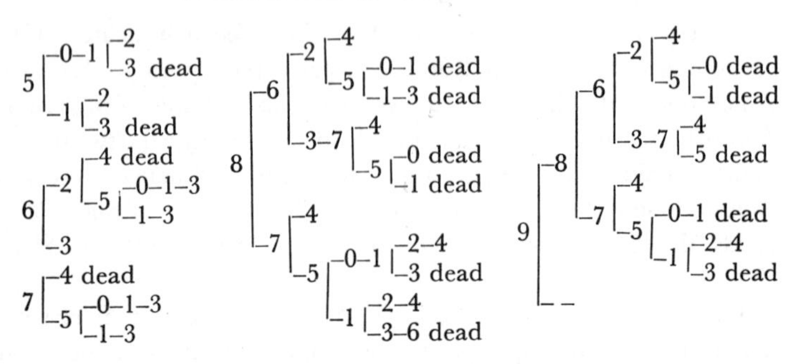

In constructing the trees we can use the fact that the last digit of the sequence is known; it is 2, 3, 3, 4, or 4 when the initial digit is 5, 6, 7, 8, or 9, respectively. Hence, if the sequence starts with 7, it is not necessary to go further when the digit 3 has been reached. We find the following sequences of digits:

0, **9**; **6**3; **512**, **8**74; 5**012**, **7513**, 8624, 9**874**; 6**251**3, **75013**, 8**63**74, 9862**4**; 6**25013**, **875**124, 98**6374**; **8750**124, 9**875**124.

As in §23, it becomes evident that the desired numbers have a digit-sum which is divisible by 9. Hence, only those combinations of sequences for which the digit-sum is divisible by 9 deserve consideration. The seven-digit combinations that satisfy this condition are:

8750124,	9**875**124,	**875**124–0,	**875**124–**9**,
5012–**8**74,	9**874**–**512**,	**512**–**8**74–0,	**512**–**874**–**9**

and the combinations formed from the sequences 0, **9**, and **6**3. A digit of which the double has been increased by 1 has been printed in **bold** type. Hence, the bold digits are the digits which precede an odd digit of the sequence, if the last digit of the sequence is assumed to be followed by the first one again. So as many bold digits as odd digits occur, and as many ordinary digits as even digits, in every sequence.

A digit that has a double greater than 9, will be called high, otherwise it will be called low; a possible increase of the double by one has no influence here. The high digits are 5, 6, 7, 8, 9; the low digits are 0, 1, 2, 3, 4. In a number that satisfies the condition a bold digit is followed by a high digit, and an ordinary digit by nothing or by a low digit.[1] The number begins with a low digit and ends with an ordinary digit.

---

[1] Hence, in a number that satisfies the condition, the number of bold digits is equal to the number of high digits, and the number of ordinary digits to that of

A combination of sequences in which all bold digits are high (and hence all ordinary digits are low) cannot lead to a solution, because a bold digit would always have to be followed by a bold digit again. This causes the combinations that are composed solely of the sequences 0, **9**, and **6**3 to drop out. For the same reason the combinations **875**124–0 and **875**124–**9** drop out.

If in a combination of sequences all high digits but one are bold (so that there is one bold low digit), then, for a number that satisfies the condition, the bold high digits have to be adjacent, immediately preceded by the bold low digit and immediately followed by the ordinary high digit. The remaining low digits can arbitrarily be placed before or after this sequence, or partly before and partly after it, except that the number should not begin with 0. The bold high digits are interchangeable, and so are the ordinary low digits (apart from the restriction mentioned).

In the manner indicated, we find for four of the remaining six combinations of sequences (the first two and the last two) the resulting solutions. They are:

(**875**0124)	4210875(12),	4208751(12),	4087521(12);	
(9**875**12**4**)	2148759(12),	2487591(12),	4875921(12);	
(**5**12–**87**4–0)	4102857(8),	4128570(8),	4285710(8);	2857410(12);
(**5**12–**87**4–**9**)	4129857(12),	4298571(12),	2985741(12).	

Of the numbers that differ by permutations of the digits (bold high and ordinary low digits), only the largest number has been written

---

the low digits. However, this will not help us to conclude that any combination with a digit-sum divisible by 9 must come to a dead end. For, as in the note on p. 28, one has, using the same notation,

$$2s - 10h + b = s,$$

hence $h - b = s - 9h$. So, if the digit-sum s of the combination of sequences is a multiple of 9, then $h - b$ is a multiple of 9, and hence, $h = b$ if the combination consists of less than nine digits; however, it is possible to prove that the latter conclusion can still be drawn when the combination contains 9 or more (but not too many) digits.

Conversely, from $h = b$ one concludes that $s = 9h$, hence that the digit-sum is divisible by 9. For combinations of not too many digits (and, a fortiori, for combinations of fewer than nine digits) the conditions "digit-sum divisible by 9" and "number of bold digits = number of high digits" come to the same thing.

Since in a sequence as many bold digits occur as odd digits, a number that satisfies the condition contains as many low as even digits. This, of course, does not yield a further decrease of the number of possibilities.

down. After this we have indicated in parentheses the number of solutions that the number represents in view of permutations. For example, in 4210875 (which appears in the accompanying sequence) it is possible to interchange the digits 8 and 7, which can be done in two ways (including no change), and also the digits 4, 2, and 1, which can be done in six ways; in all this gives $2 \times 6 = 12$ solutions.

The combination of sequences 5**012**–**8**74 gives the sequences 287 and 05 in the desired number, apart from interchangeability of 2 with 0 and of 7 with 5; for 8 is preceded by one of the digits 0, 2 (since 8 is high), while 8 is followed by one of the digits 5, 7 (since 8 is bold). It is immaterial whether 05 is placed before or after 287. The interchangeable digits 1 and 4 can be placed arbitrarily, even between 05 and 287. This yields:

(5**012**–**874**) 4128705(8), 4287105(8), 4287051(8), 2874105(4),
2874051(4), 2870541(4), 4127085(8), 4271085(8),
4270851(8), 2741085(4), 2740851(4), 2708541(4),

Each of these twelve numbers represents $2 \times 2 \times 2 = 8$ solutions, except for the six numbers that start with a 2; these yield $2 \times 2 = 4$ numbers each, because interchanging 2 and 0 would produce an initial digit 0.

The combination of sequences 9**874**–**512** yields the two series of digits 4859 and 27, or the two series 49 and 2857, or the two series 489 and 257, apart from interchangeability of 8 with 5, or 9 with 7, and of 4 with 2. The digit 1 can be placed before, between, or behind both series. This gives the solutions:

(9**874**–**512**), 1485927(8), 4859127(8), 4859271(8), 1492857(8)
4912857(8), 4928571(8), 1489257(8), 4891257(8), 4892571(8)

Thus one finds, in all, 288 seven-digit numbers without initial digit 0 which are written with the same digits as their doubles; they can be found from the 34 numbers given.

***27. Remarks on the numbers of §26.** From the numbers of §26 that end in 0, one finds the following twelve six-digit numbers that are written with the same digits as their doubles:

125874, 128574, 142587, 142857, 258714, 258741,
285714, 285741, 412587, 412857, 425871, 428571.

They arise from a combination of sequences in which an ordinary 0 occurs, hence from **512**–**8**74–0. Numbers of fewer than six digits with the peculiarity in question do not exist, not even if one admits initial digits 0; the smallest number in which the peculiarity occurs is 125874. Among the twelve six-digit numbers found there are the numbers 142857, 285714, and 428571, which arise by developing 1/7, 2/7, and 3/7, respectively, as recurring decimal fractions. This was to be expected from properties of recurring decimals into which we shall not enter here. The number 142857 shows the peculiarity that it is written with the same digits as its double, treble, quadruple, quintuple, and sextuple. For the three numbers mentioned last, the same succession is obtained when the digits are arranged round a circle; diametrically opposed digits then add up to 9, which implies divisibility of the numbers by 999.

Among the twelve numbers there are three other numbers which are divisible by 999 and which involve a common cycle: 125874, 258741, and 412587. These numbers arise by developing 18/143, 37/143, and 59/143, respectively, as recurring decimals.

The seven-digit numbers with the observed peculiarity evidently all appear to consist of seven different digits. Without finding these numbers, one can see this from the combinations of sequences from which they arise. Lack of coincidence among the seven digits is an accidental circumstance, which is no longer present when, for example, eight-digit numbers with the peculiarity in question are sought. For a solution with seven digits in which a zero occurs leads to a solution with two zeros if a zero is added at the end of the number.

Among the numbers found, there is only one number in which each digit is larger than the preceding digit: 1245789, which arises from the sequence 9**875**12**4**. Even without the consideration of combinations of sequences, it is an easy matter to find all the numbers which have digits in ascending order and which are written with the same digits as their doubles. Below 10,000 there are 29 numbers which have ascending digits and are divisible by 9; 9, 18, 27, . . . , 5679. There are another thirty larger numbers of similar type. These are 123456789 and the numbers that arise from this number by omitting the digits of one of the 29 smaller numbers. It turns out that only the numbers 1245789 and 123456789 satisfy the conditions.

***28. Quintuples puzzle.** We require the seven-digit numbers without initial digit 0 which are written with the same digits as their quintuples. This puzzle can be simplified a great deal by reducing it to the doubles puzzle. For if a number satisfies the condition and its quintiple is multiplied by 2, this gives a tenfold multiple of the original number, formed from the original number by appending a right-hand zero digit. So here the double of the quintuple is written with the same digits as the quintuple, if the quintuple is converted into an eight-digit number by prefixing a left-hand zero digit. Hence, the quintuples of the desired numbers are the eight-digit numbers that are divisible by 5 (hence, end in 0 or 5) and that are written with the same digits as their doubles, provided that they start with a 0, and have a double that does not start with 0 (so that the initial digit 0 is followed by a 5, 6, 7, 8, or 9); the double, after omitting the final digit 0, is a number that satisfies the original question (the one on the quintuples).

If the eight-digit number that is divisible by 5 ends in 0, then it is, after removal of that 0, a number that satisfies the problem of §26, with this difference: the number starts with a 0, but its double does not. So what we have to do is to derive from the numbers in §26, by permuting digits, the numbers that start with a bold 0. These are 0875421 and the numbers that arise from it by permuting the digits 4, 2, 1 and the digits 8, 7 (twelve numbers), as well as the numbers

0874125, 0874251, 0872541 0741285, 0742851, 0728541

and the numbers that arise from these by interchanging 4, 1 and 7, 5. Their doubles are the even numbers that satisfy the original problem (36 in number). These are 1750842 and the numbers that arise from it by permuting the digits 8, 4, 2 and the digits 7, 5, as well as the numbers:

1748250, 1748502, 1745082, 1482570, 1485702, 1457082

and the numbers that arise from these by interchanging 8 and 2, and 4 and 0.

The odd seven-digit numbers that satisfy the original question (the one on the quintuples) are derived from the eight-digit numbers that start with a bold 0 and end in 5, and are written with the same digits as their doubles. Hence, the corresponding combination of sequences should contain a sequence in which a bold 0 occurs; this 0 is preceded by a 5. Hence, we only have to look for the closed eight-digit sequences starting with 5 in which a 0 occurs. As appears from the construction of the tree, there is only one such sequence: 5**0**12**4**9**87**; it has a digit-sum that is divisible by 9. For combinations of sequences individually divisible by 9, the combinations **87**5**0**124–0 and **87**5**0**124–**9** do not lead to any solutions, because they lead to the series of digits 0875 and 09875, respectively (or the series that arise from these by interchanging 8, 7, or 9, 8, 7, respectively) in the solution of the simplified puzzle; this contradicts the requirement that the eight-digit number should start with 0 and end with 5. The remaining combinations of sequences, divisible by 9, lead to solutions as follows:

(9**8750**12**4**)	08792145(4),	08921475(4),	09214875(4);
(5**0**1**2**–9**874**)	08914725(4),	08947125(4),	09147285(4),
	09148725(4),	09471285(4),	09487125(4);
(5**0**1**2**–**8**74–0)	07410285(6),	08741025(6);	
(5**0**1**2**–**8**74–**9**)	07412985(4),	09741285(4),	09874125(4).

The figures in parentheses again show how many solutions are represented by each number. The doubles of the numbers, after removal of the final digit 0, are the 60 odd seven-digit numbers without initial digit 0 which are written with the same digits as their quintuples. These are:

1758429, 1784295, 1842975 (7 and 5 are interchangeable, and so are 4 and 2);
1782945, 1789425, 1829457, 1829745, 1894257, 1897425 (8 and 4 are interchangeable, and so are 9 and 5);
1482057, 1748205 (8, 2, and 0 are interchangeable);
1482597, 1948257, 1974825 (9 and 7 are interchangeable, and so are 8 and 2).

Hence, in all the puzzle has 96 solutions. Of these, there are four that end in 10. Hence, by omitting this 0, one gets the following six-digit numbers that are written with the same digits as their quintuples:

142857, 148257, 174285, 174825;

there are no other solutions, even if one admits zeros as initial digits.

Chapter II:
SOME DOMINO PUZZLES

I. SYMMETRIC DOMINO PUZZLE, WITH EXTENSIONS

29. Symmetric domino puzzle. A striking example of a puzzle that admits a division into smaller puzzles is the following one: to arrange the 28 pieces of a set of dominos within the outline shown in *Figure 12*, so that equal spot numbers occur in groups of four, forming squares. Fourteen squares should thus result, each containing four equal spot numbers.

If a solution has been found, then a second solution ensues from a mirror image of the figure along the horizontal dotted line (or by rotating the figure about the line). However, we consider these solutions to be identical.

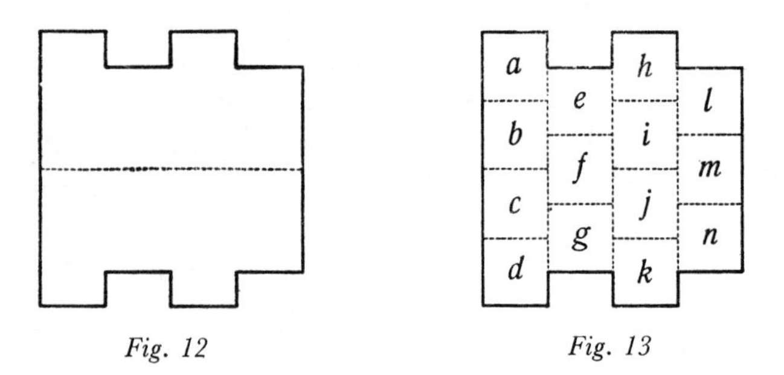

Fig. 12 *Fig. 13*

Since in a set of dominos each spot number occurs once combined with itself and with every other spot number, we find other solutions from a given solution by permuting the numbers 0, 1, 2, 3, 4, 5, 6 among themselves. These solutions are also considered as identical.

A first puzzle is to draw the 14 squares. This yields *Figure 13*; it is so simple that one can hardly call it a puzzle.

In the second puzzle the object is to draw the outlines of the dominos in the figure. This has to be done in such a way that no two equal (adjacent) pieces result. From this it follows that the top half of square *a* is a double, because otherwise we get two equal doubles in this square. Similarly, the top half of square *h* and the lower halves of the squares *d* and *k* are doubles. If the top half of square *l* is a double, then the two other half-dominos of square *l* cannot belong to the same piece, and cannot both belong to vertical pieces (because otherwise these pieces would be equal); so the lower half of square *h* is a domino in that case, which would give two equal doubles in this square. The right half of square *l*, therefore, is a double, and so is, of course, the right half of square *n*. From this it further follows that the right half of square *m* is a double, too. With this the position of the seven doubles has been found.

The position of the remaining pieces now follows simply from the fact that no further doubles can now be allowed to result. If one draws the outlines, working one's way from the outside inwards, it appears that these can only be as indicated in *Figure 14*, in which the

0	0			1	1		
0	0	*e*	*e*	1	1	2	2
b	*b*	*e*	*e*	*i*	*i*	2	2
b	*b*	*f*	*f*	*i*	*i*	3	3
c	*c*	*f*	*f*	*j*	*j*	3	3
c	*c*	*g*	*g*	*j*	*j*	4	4
6	6	*g*	*g*	5	5	4	4
6	6			5	5		

Fig. 14

doubles have been shaded. Because of the interchangeability of the numbers 0, 1, . . . , 6, the spot numbers of the doubles can be chosen arbitrarily; we choose them as in the figure. The position of the pieces 1–2 and 4–5 is now also known. From the spot numbers of the doubles one sees that there is no solution that is also symmetrical when the spot numbers are taken into account. Hence, by rotating the figure about a horizontal line, a different solution always results from a given solution. It is true, however, that the outlines of the dominos produce a symmetrical figure.

Now comes the third and most difficult sub-puzzle, namely filling in the seven still unknown spot numbers in the squares *b*, *c*, *e*, *f*, *g*, *i*, *j*.

While it was simplest with each of the preceding smaller puzzles to work on paper, with the third puzzle there is some advantage in actually starting to use the dominos. However, an experienced puzzle solver does not need the pieces and can work out everything on paper.

The spot number occurring four times in one of the squares b, c, e, f, g, i, j will be indicated by the same letter as the square, so that $g = 2$, for example, will mean that in square g the spot number 2 occurs four times. Since each of the spot numbers is involved in only two squares, the numbers b, c, e, f, g, i, j are all different. So if we write $\neq$ to mean "different," we will have $b \neq c$, and the like.

From the outlines of the pieces and the position of the doubles it follows that:

$$b \neq 0;\ c \neq 6,\ e \neq 0,\ e \neq 1;\ g \neq 5,\ g \neq 6;\ i \neq 2,\ i \neq 3;\ j \neq 3,\ j \neq 4.$$

Since a piece does not occur twice, we further have:

$$b \neq 1,\ b \neq 6;\ c \neq 0,\ c \neq 5;\ e \neq 2,\ e \neq 3;\ f \neq 0,\ f \neq 2,\ f \neq 3,$$
$$f \neq 4,\ f \neq 6;\ g \neq 3,\ g \neq 4;\ i \neq 0,\ i \neq 1;\ j \neq 5,\ j \neq 6.$$

The fact that $e \neq 3$, for example, may be seen from the occurrence of the pieces e–i and i–3, which cannot be the same. The fact that $f \neq 0$ can be seen from the pieces 0–b and b–f, similarly elsewhere.

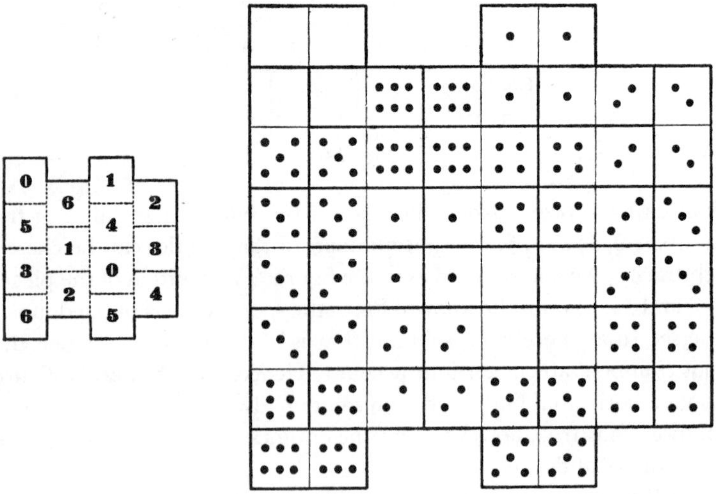

Fig. 15

From the inequalities obtained it follows that $f = 1$ or $f = 5$. The cases $f = 1$ and $f = 5$ come to the same thing because of the symmetry of the figure (disregarding the spot numbers) about the horizontal line, so that we can assume that $f = 1$. Since the piece 1–2 occurs only once, the piece f–j (identified as 1–j) now is evidence that $j \neq 2$. From this it follows that $j = 0$, and from this that $g = 2$. As the pieces 0–3 and 0–4 occur as j–3 and j–4, respectively, it follows from the piece 0–b that $b \neq 3$ and $b \neq 4$. Hence $b = 5$. From the pieces 5–4 and b–e (that is, 5–e) it further follows that $e \neq 4$, so $e = 6$, $i = 4$, and $c = 3$. These results are shown in *Figure 15.*

Besides this solution, there is only the one in *Figure 16,* which can be obtained from the preceding one by interchanging 0 and 6, as well as 1 and 5, and 2 and 4, before rotating the figure about a horizontal line.

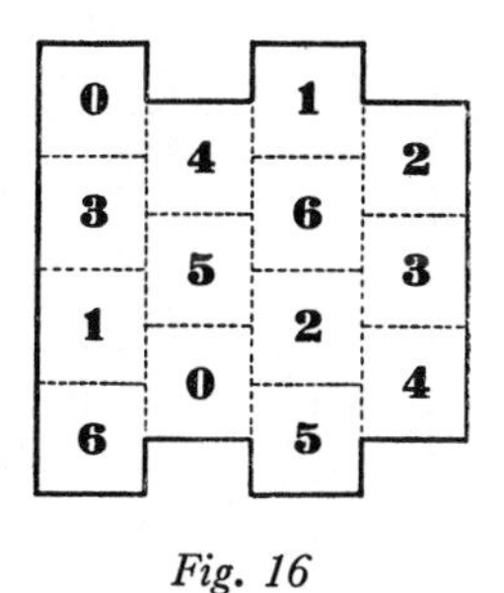

Fig. 16

Other solutions arise from each of the two solutions by filling in the seven doubles differently. In the top left-hand corner we can fill in one of the numbers 0, 1, 2, 3, 4, 5, 6 (seven choices). If we have made a choice from these, and filled in 3, say, then we can still fill in the lower left-hand corner in six ways: from 0, 1, 2, 4, 5, 6 we may, for example, choose 4. For the next double we can make our choice in five ways, etc. To obtain the corresponding solution, we then have to replace 0 by 3, 6 by 4, etc., in the obtained solution. In this way each of the two solutions leads to $7 \times 6 \times 5 \times 4 \times 3 \times 2 = 5040$ solutions. One set of 5040 can be obtained from the other by rotating about a horizontal line.

30. Extended symmetric domino puzzle. We take the same puzzle as in §29, but add the requirement that the total of the spot numbers for the four uppermost squares has to be the same as that for the four lowest squares, as well as for the column of four squares at the left-hand side, the column of three squares at the right-hand side, and the set of three squares in the middle which are completely surrounded by others. If the puzzle is presented in this form from the start, it provides an example of a division into three smaller puzzles which are all somewhat different in nature, and not too simple. Here the determination of the fourteen squares has not been counted as a puzzle (being too simple).

From the results of §29 it appears that the spot numbers are as indicated in *Figure 17*. The letters a, d, h, k, l, m, n represent the numbers 0, 1, 2, 3, 4, 5, 6 in some order. Alternatively, we can place

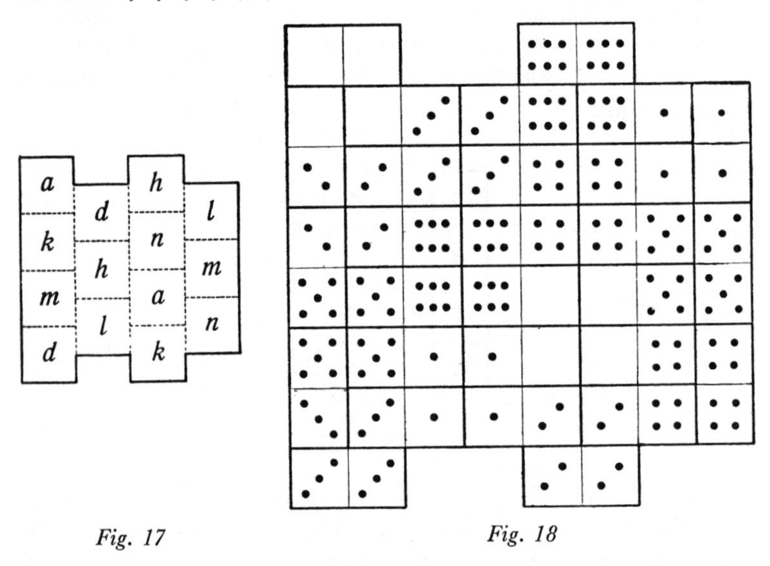

Fig. 17 *Fig. 18*

the letters as for a reflection of the diagram about a horizontal line, but this comes to the same thing. The conditions stated require that

$$a + d + h + l = d + l + k + n = a + k + m + d$$
$$= l + m + n = h + n + a.$$

From this it follows that

$$a + h = k + n = l + m, h + l = k + m, d + l = n, d + k = m.$$

From the last two equations it follows that $d \neq 6$, and that none of d, l, n, k or m can represent 0, so that $a = 0$ or $h = 0$. Furthermore, from

$$a + h + k + n + l + m + d = 0 + 1 + 2 + 3 + 4 + 5 + 6 = 21$$

in conjunction with

$$a + h = k + n = l + m,$$

it follows that

$$3(a + h) = 3(k + n) = 3(l + m) = 21 - d,$$

so that d is a multiple of 3. Since $d \neq 0$ and $d \neq 6$, we have $d = 3$ and hence:

$$a + h = k + n = l + m = 6, m = k + 3, h + l = 2k + 3.$$

From $l + m = 6$ and $m = k + 3$ it follows that $l + k = 3$, hence $h = 3k$, so that $h \neq 0$, hence $a = 0$, $h = 6$, $k = 2$, $l = 1$, $m = 5$, $n = 4$. This leads to the solution in *Figure 18.* Another solution can be obtained by reflecting this one symmetrically along a horizontal line.

***31. Another extension of the symmetric domino puzzle.** We call two of the squares of §29 completely adjoining if they touch each other along a complete edge (that is, with contact between two pairs of half-dominos). Instead of the requirement of §30 we now require that every two completely adjoining squares involve spot numbers that differ by 2, 3, or 4.

We call two numbers complementary, if they add up to 6; thus 0 and 6 are complementary, as well as 1 and 5 and 2 and 4, whereas 3 is self-complementary. The difference of two numbers is equal to the difference of their complements. So another solution is obtained if each number in a solution is replaced by its complement. We call this the complementary rearrangement.

In two completely adjoining squares, if one spot number is 0, then the other spot number can be 2, 3, or 4, and the same is true, if the first spot number is 6. So a solution is changed into another solution by interchanging the numbers 0 and 6. We call this the 0–6 rearrangement. Still another solution is obtained if the 0–6 rearrangement is combined with the complementary rearrangement; this yields the 1–5—2–4 rearrangement, in which each 1 is replaced by a 5, and conversely, and each 2 by a 4, and conversely. In this way each solution gives rise to a group of four solutions.

Figure 19 indicates which pairs of spot numbers belong to completely adjoining squares and hence should have a difference of 2, 3, or 4; these are represented by letters that are directly connected by a line; Figure 17 will show the squares which receive the numbers represented by the letters. As the d and l are connected by a line to the same numbers (h and m), a solution is preserved when the numbers d and l are interchanged; we call this the d–l rearrangement. It may coincide with the 0–6 rearrangement, in which case we may have $d = 0$, $l = 6$, or $d = 6$, $l = 0$; the d–l rearrangement then does not yield new solutions, so that any solution of this type, which we shall term the first type, gives a group of four solutions. If the d–l rearrangement does not coincide with the 0–6 rearrangement (second type), because, for instance, $d \neq 0$ and $d \neq 6$, then each solution belongs to a group of $2 \times 4 = 8$ solutions.

Table 2 shows below the line the numbers that differ by 2, 3, or 4 from each number above the line. Only below 2, 3, and 4 are there four numbers. Since m is connected by a line to four numbers (d, k, l, n), we have $m = 2$ or $m = 3$ (because $m = 4$ comes to the same thing as $m = 2$, as a consequence of the rearrangements). The numbers h and

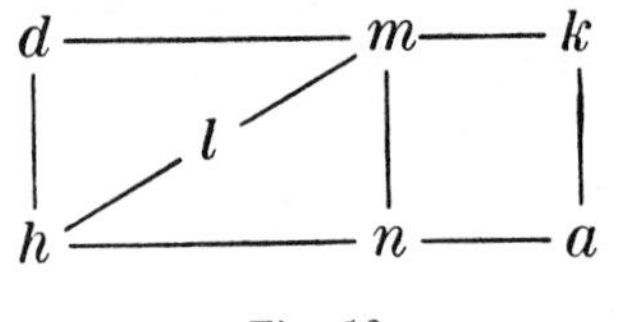

Fig. 19

m are connected by a line to the same three numbers (d, l, and n), so that $m = 2, h = 3$, or $m = 3, h = 2$ (the latter case being equivalent to $m = 3, h = 4$).

TABLE 2

0	1	2	3	4	5	6
2	3	0	0	0	1	2
3	4	4	1	1	2	3
4	5	5	5	2	3	4
		6	6	6		

If $m = 2$, $h = 3$, then the numbers d, k, l, and n, connected by a line to m, are all different from 1, so that $a = 1$. The numbers d, l, and n, connected by a line to h, are not equal to 4 (because $h = 3$), so that $k = 4$. This then leads to $d = 0$, $l = 6$ (first type). The corresponding group of four solutions is shown in *Figure 20*.

If $m = 3$, $h = 2$, the four numbers connected by a line to m are all different from 4, so that $a = 4$. The three numbers connected by a line to h are all different from 1, so that $k = 1$. Because of $a = 4$, we

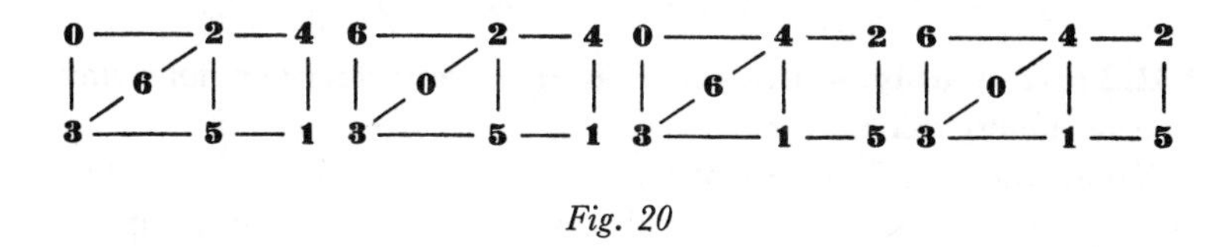

Fig. 20

have $n \neq 4$, hence (for example) $n = 6$, $d = 0$, $l = 5$ (second type). The corresponding group of eight solutions is shown in *Figure 21*.

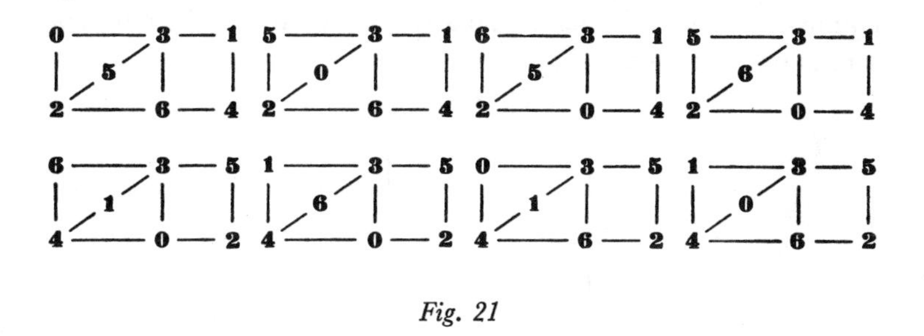

Fig. 21

Two solutions in the same column arise from each other by the complementary rearrangement.

When written into the fourteen squares, the first solutions of the first and second types, together with the solution that arises from the latter through the *d–l* rearrangement, are shown in *Figure 22*.

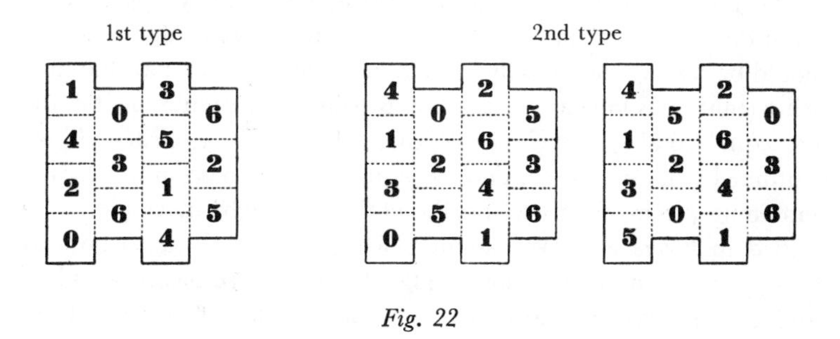

Fig. 22

From each of these solutions, three others arise (through the 0–6 rearrangement, the 1–5—2–4 rearrangement, or both combined). From the twelve solutions thus obtained another twelve solutions are obtained by reflection about a horizontal line.

II. DOUBLY SYMMETRIC DOMINO PUZZLE

***32. First doubly symmetric domino puzzle.** The following puzzle is more difficult:

To arrange the 28 pieces of a set of dominos in an assembly like that of *Figure 23* so that identical spot numbers occur in groups of four, forming squares. In addition, the totals of the spot numbers for the four squares at the top, at the bottom, at the left-hand side and at the right-hand side have to be equal to one another and equal to the total of the spot numbers for the two squares in the middle. We require all solutions. Solutions that arise from each other by reflection or rotation are considered as identical.

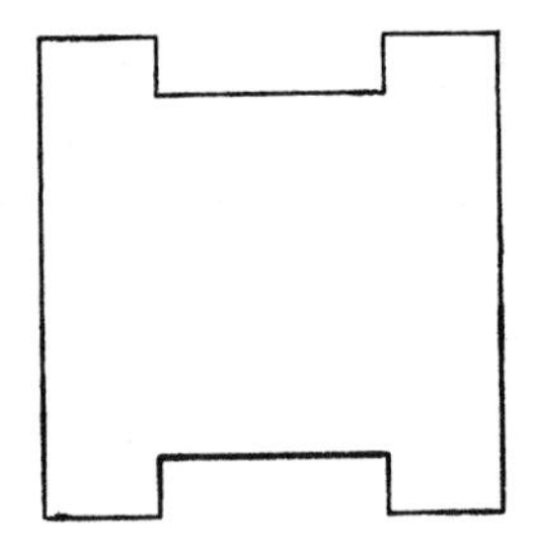

Fig. 23

The position of the fourteen squares is immediately appreciated. From the non-occurrence of equal doubles it is evident that a double should be placed horizontally in each of the four corners; for the time being (that is, as long as no attention is paid to the requirement of the five equal totals), the choice of these doubles is arbitrary. If we continue by noting also that no two equal pieces can occur among the remaining pieces, then we find the outlines of eight more pieces (as shaded in *Figure 24*). We cannot place two vertical pieces both at the upper edge and at the lower edge, because these pieces would be doubles, whereas there are only three doubles left to be placed. So we can assume that there is a horizontal piece in the center of the lower edge. For the same reason we can locate a vertical piece in the middle of the right-hand edge. At this point there are three possible continuations.

Type I. At the left-hand edge we can place two horizontal pieces

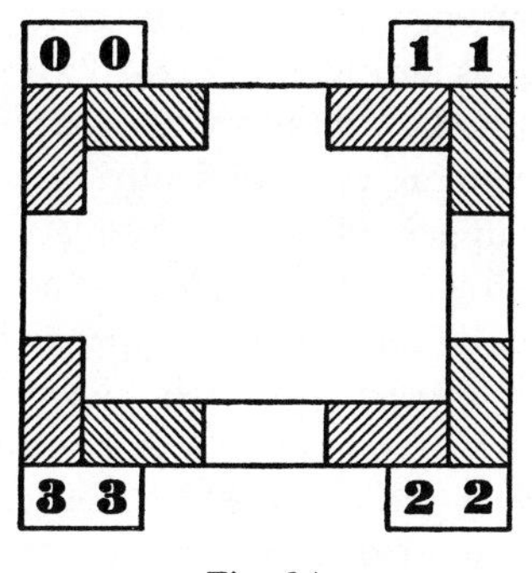

Fig. 24

(doubles, the spot numbers of which can be chosen arbitrarily for the time being); this determines the position of pieces A and B. We can also place piece C, because otherwise we would get another two

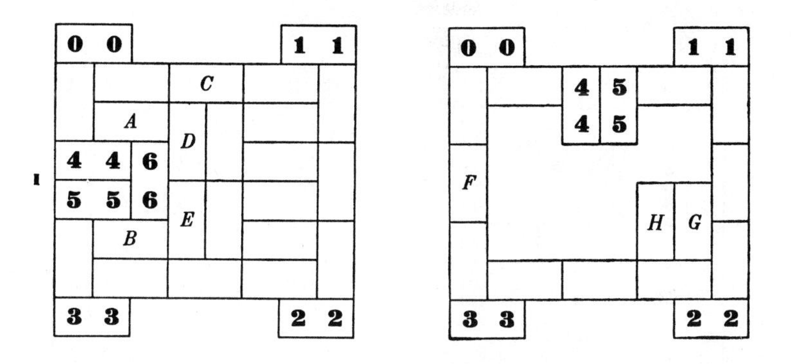

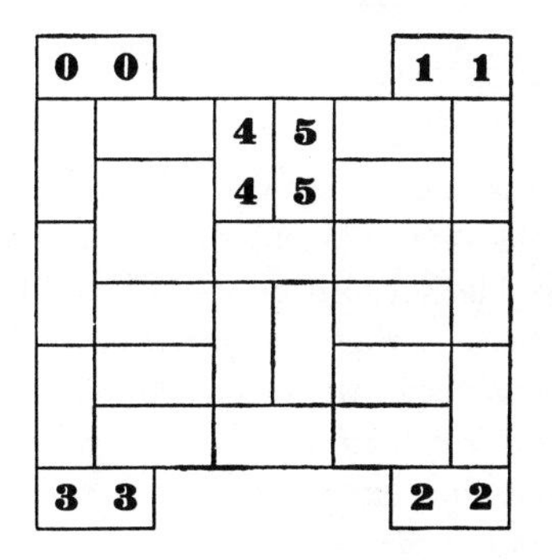

Fig. 25

doubles at the upper edge. As there are no equal pieces, we must place pieces D and E as indicated in *Figure 25* (upper left), so that we find the position of the last double. After that, drawing the outlines of the remaining pieces causes no difficulty.

Type II. At the upper edge we can place two vertical pieces (doubles), after which we can place piece F (because of the doubles). Placing piece G (as a double, Fig. 25, upper right) is wrong, for this entails the placing of piece H (in order to avoid an eighth double), and then next to H we have to place a piece that is the same as H or as the middle piece at the lower edge. Hence, the drawing of the outlines of the pieces must be continued as indicated in Figure 25 if we want to avoid identical pieces. Since only one more double has to be placed, it can be assumed (because of the symmetry) that this does not occur on the right-hand edge. By taking care not to get two adjacent equal pieces, you can fill in the outlines of all but two pieces. This gives two possible dispositions, designated as II and II′ in *Figure 26.*

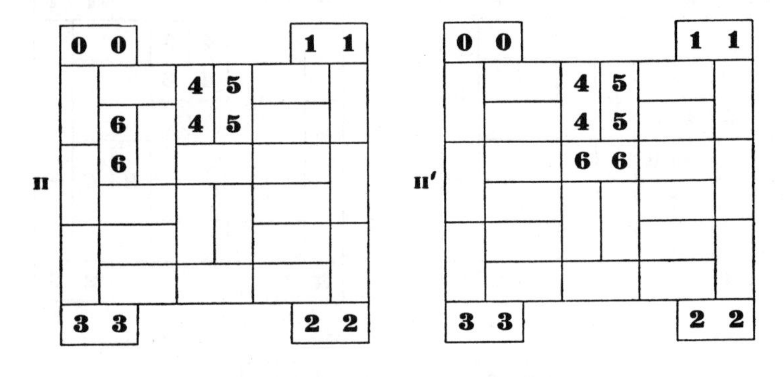

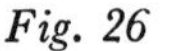
Fig. 26

In type II′, there is one double that is not a double in virtue of its position, but because two adjacent squares are associated with the same spot number. The seventh double cannot be the middle piece on the lower edge, because then we cannot fill in the numbers 2 and 3 elsewhere in such a way that the piece 2–3 occurs. Hence the seventh double has to be the horizontal piece in the middle, because other choices lead to equal adjacent pieces elsewhere.

Type III. We neither place two horizontal pieces at the left-hand side, nor two vertical pieces at the upper edge. Then once more the outlines of the dominos can be chosen in two ways, designated as III and III′ in *Figure 27*. In type III′ three doubles would have to occur that are not doubles in virtue of their positions. However, these can only be pieces *J* and *K*, because otherwise we get two equal adjacent pieces. Since this shows that only two of the three doubles can be placed, type III′ is therefore impossible.

So all that remain are solutions of types I, II, II′, and III.

As in §29, we define the spot number of any of the fourteen squares to mean the spot number common to each of the four half-dominos which it includes. The sum of the spot numbers of the four squares at the upper edge (hence also of the four squares at the lower edge,

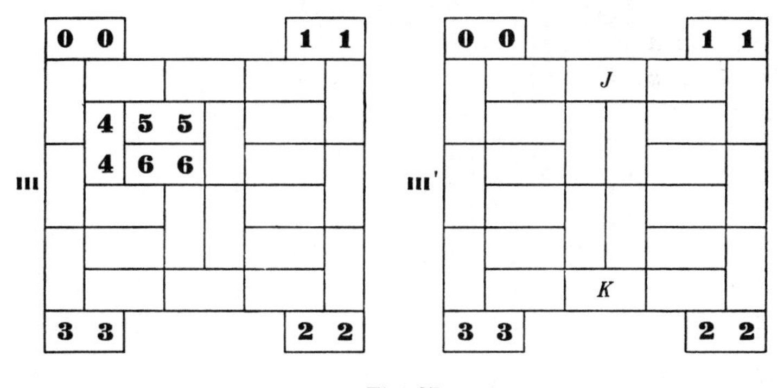

Fig. 27

etc.) will be denoted by x, and the sum of the spot numbers in the four corner-squares will be denoted by y). Since the sum of the spot numbers of the fourteen squares amounts to $2 \times 21 = 42$, we have $5x = 42 + y$. This follows because the spot numbers in the corners are counted twice if we take the sum $5x$ of the spot numbers above, below, left, right, and in the center.

The spot numbers in the four corners are different, and for each of the types I, II, II′, and III, the spot numbers of four squares at the same edge of the figure (above, below, left, or right) can be shown to be different, from the non-occurrence of two equal pieces. In connection with $5x = 42 + y$, this gives the following cases:

1°. $x = 10$ (from $0 + 1 + 3 + 6$ or $0 + 1 + 4 + 5$ or $0 + 2 + 3 + 5$ or $1 + 2 + 3 + 4$ or $4 + 6$ or $5 + 5$), $y = 8$ (from $0 + 1 + 2 + 5$ or $0 + 1 + 3 + 4$);

2°. $x = 11$ (from $0 + 1 + 4 + 6$ or $0 + 2 + 3 + 6$ or $0 + 2 + 4 + 5$ or $1 + 2 + 3 + 5$ or $5 + 6$), $y = 13$ (from $0 + 2 + 5 + 6$ or $0 + 3 + 4 + 6$ or $1 + 2 + 4 + 6$ or $1 + 3 + 4 + 5$);

3°. $x = 12, y = 18$ (from $3 + 4 + 5 + 6$).

Case 3 is not possible, because we would get two squares with six spots in the middle and also at least two squares with six spots at the edge (or in the corners). By systematic trial (which again is a puzzle in itself) we find the possibilities shown in *Figure 28*.

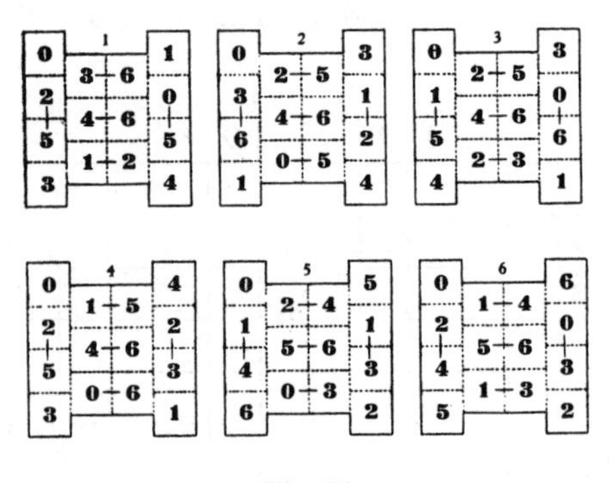

Fig. 28

Here we can still interchange the numbers connected by a small bar (while we can also reflect or rotate the figure, of course).

In cases 1, 2, 3, and 4 (Fig. 28), the doubles other than those at the corners are 2, 5, and 6, and in cases 5 and 6 they are 1, 3, and 4. From the positions of doubles elsewhere than at corners we can see which of the types I, II, II′, or III has to be considered. Type II′, for which two equal numbers occur in the middle, does not lead to solutions, while case 5 does not give any solutions because of the position of the doubles.

In case 1 (which might belong to type I and to type III, in virtue of the doubles 2, 5, and 6), the two squares with the number 6 are in contact either by a side or by a corner, which leads, in either case, to two equal pieces. For the same reason, case 4 does not give any solutions. Case 2 can be made to belong to type II, but also to type III, reflected or rotated as compared with the type III of Figure 27. Type II and the reflected type III yield one solution each, whereas we come to a dead end with the rotated type III. Case 3 can be made to belong to type II (reflected) and in two ways to type III (as in Fig. 27 and reflected). Types II and III give one solution each, whereas we come to a dead end with the reflected type III. Case 6 can be made to belong to type II in two ways, but in both cases we come to a dead end. Thus we find four solutions, two of type II and two of type III, as shown in *Figure 29.*

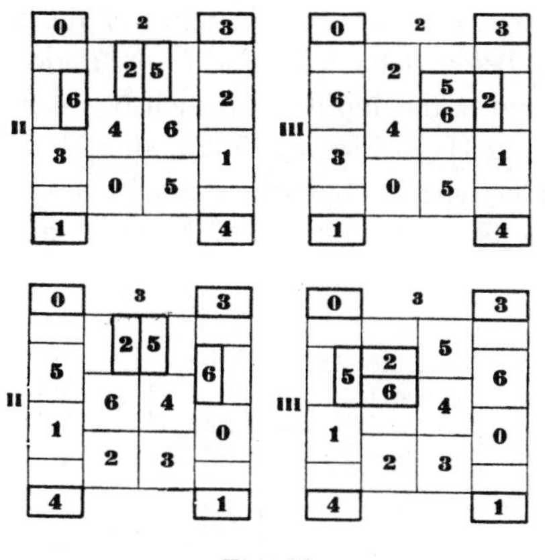

Fig. 29

The positions of the doubles (which indicate the type) have been shown by heavy lines, while the remaining lines identify the fourteen squares.

Solutions II2 and II3 arise from each other by interchanging 0 and

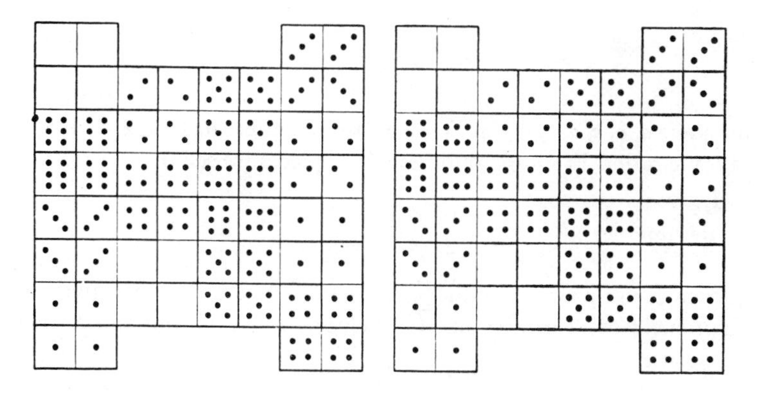

Fig. 30

3, interchanging 2 and 5, and reflecting. In the same way solutions III2 and III3 transform into each other. Completely worked-out solutions II2 and III2 are shown in *Figure 30.*

Each solution arises from the other by changing seven pieces, with no alteration in the positions of the spot numbers.

***33. Doubly symmetric domino puzzle without restrictive condition.** We now require all solutions of the puzzle of §32, dropping the restriction as to the equality of the five sums of spot numbers (above, below, left, right, and in the middle). The location of the dominos and of the doubles can then be taken as indicated by types I, II, II′, and III of Figures 25, 26, and 27.

We first consider type I, where there is symmetry with respect to a horizontal line, so far as the outlines of the pieces are concerned. The

I

0	**0**					**1**	**1**
0	**0**	a	a	c	c	**1**	**1**
4	**4**	a	a	c	c	f	f
4	**4**	**6**	**6**	d	d	f	f
5	**5**	**6**	**6**	d	d	g	g
5	**5**	b	b	e	e	g	g
3	**3**	b	b	e	e	**2**	**2**
3	**3**					**2**	**2**

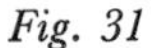

Fig. 31

numbers a, b, c, d, e, f, g in *Figure 31* are the numbers 0, 1, 2, 3, 4, 5, 6 in some order. From the outlines of the pieces we see that

$$a \neq 0, \neq 1, \neq 4, \neq 6;$$
$$b \neq 2, \neq 3, \neq 5, \neq 6;$$
$$c \neq 0, \neq 1, \neq 4, \neq 6;$$
$$d \neq 1, \neq 2;$$
$$e \neq 2, \neq 3, \neq 5, \neq 6;$$
$$f \neq 1, \neq 2;$$
$$g \neq 1, \neq 2.$$

We start by making assumptions about the spot number d which is associated with the only unfilled square that has a symmetric position. Because of the symmetry, $d = 0$ amounts to the same thing as $d = 3$, while $d = 4$ is the same case as $d = 5$. So we only have to consider the cases $d = 0$, $d = 4$, and $d = 6$. If $d = 0$, then $e \neq 4$, hence $e = 1$, $b = 4$, $a = 2$, $f = 6$. If $d = 4$, then $e \neq 0$, hence $e = 1$, $b = 0$, $a = 2, f = 6$. Suppose now $d = 6$; as we cannot simultaneously have $c = 2$ and $e = 1$ (because otherwise the piece 1–2 would occur twice), we can assume, because of the symmetry, that $c \neq 2$; then $a = 2$, $e \neq 0$ and $\neq 4$, and $g \neq 0$ and $\neq 4$, so $e = 1$. Thus we find the following eight solutions of type I:

0 1	0 1	0 1	0 1	0 1	0 1	0 1	0 1
2 3	2 5	2 3	2 5	2 3	2 3	2 5	2 5
4 6	4 6	4 6	4 6	4 4	4 0	4 4	4 0
6 0	6 0	6 4	6 4	6 6	6 6	6 6	6 6
5 5	5 3	5 5	5 3	5 5	5 5	5 3	5 3
4 1	4 1	0 1	0 1	0 1	4 1	0 1	4 1
3 2	3 2	3 2	3 2	3 2	3 2	3 2	3 2

To these we must add the following eight solutions that are obtained from the solutions already stated by interchanging 0 and 3, 1 and 2, and 4 and 5, and rotating the figure about a horizontal line:

0 1	0 1	0 1	0 1	0 1	0 1	0 1	0 1
5 2	5 2	3 2	3 2	3 2	5 2	3 2	5 2
4 4	4 0	4 4	4 0	4 4	4 4	4 0	4 0
6 3	6 3	6 5	6 5	6 6	6 6	6 6	6 6
5 6	5 6	5 6	5 6	5 5	5 3	5 5	5 3
1 0	1 4	1 0	1 4	1 0	1 0	1 4	1 4
3 2	3 2	3 2	3 2	3 2	3 2	2 2	3 2

Next we take type II, where no symmetry occurs (see *Fig. 32*). First of all we have:

$$\begin{aligned} &a \neq 0, \neq 3, \neq 4, \neq 6; \\ &b \neq 0, \neq 3, \neq 4, \neq 6; \\ &c \neq 2, \neq 3, \neq 4, \neq 6; \\ &d \neq 1, \neq 2, \neq 4, \neq 5; \\ &e \neq 2, \neq 3; \\ &f \neq 1, \neq 2, \neq 5; \\ &g \neq 1, \neq 2, \neq 5, \end{aligned}$$

so that $a = 2$ or $b = 2$. Furthermore $c \neq 1$ and $c \neq 5$, since otherwise there would be no place remaining for a 5 or for a 1, respectively.

Hence, $c = 0$, so $e \neq 4$ and $e \neq 6$, whence $f = 4$ or $g = 4$. By further observing that some particular piece still has to be placed (preferably

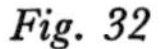

Fig. 32

when this is possible in one way only), we find the following eight solutions of type II:

0 1	0 1	0 1	0 1	0 1	0 1	0 1	0 1
4 5	4 5	4 5	4 5	4 5	4 5	4 5	4 5
6 4	6 4	6 6	6 3	6 6	6 3	6 4	6 4
2 3	2 6	1 3	1 6	5 3	5 6	2 3	2 6
1 6	1 3	2 4	2 4	2 4	2 4	5 6	5 3
0 5	0 5	0 5	0 5	0 1	0 1	0 1	0 1
3 2	3 2	3 2	3 2	3 2	3 2	3 2	3 2

So far as the outline of the pieces are concerned, type II′ (see *Figure 33*) is symmetrical along a vertical line. Since the number 6 has already been located in two places, each of the numbers a, b, c, d, e, and f is different from 6. Further we have:

$$a \neq 0, \neq 3, \neq 4;$$
$$b \neq 0, \neq 3, \neq 4;$$
$$c \neq 2, \neq 3;$$
$$d \neq 2, \neq 3;$$
$$e \neq 1, \neq 2, \neq 5;$$
$$f \neq 1, \neq 2, \neq 5.$$

Hence $a = 2$ or $b = 2$, and $e = 3$ or $f = 3$. Since we cannot simultaneously have $b = 2$ and $f = 3$, we have $a = 2$ or $e = 3$. Because of the symmetry we can assume that $a = 2$. From this it follows that $f \neq 0$ and $f \neq 4$, so $f = 3$. This leads to four solutions, to which we must add the four solutions obtained by interchanging

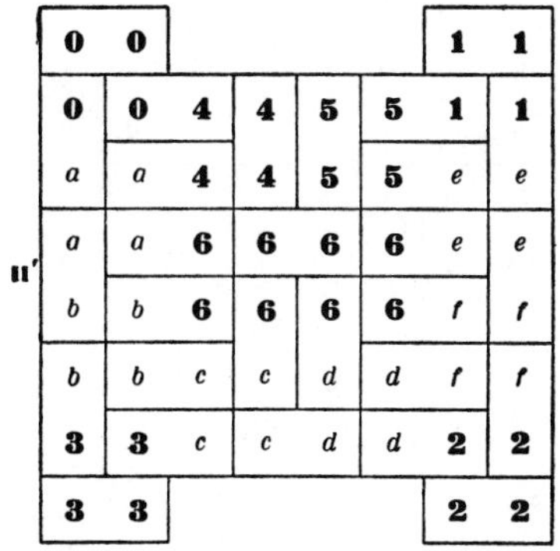

II'

0	0					1	1
0	0	4	4	5	5	1	1
a	a	4	4	5	5	e	e
a	a	6	6	6	6	e	e
b	b	6	6	6	6	f	f
b	b	c	c	d	d	f	f
3	3	c	c	d	d	2	2
3	3					2	2

Fig. 33

0 and 1, 3 and 2, and 4 and 5, followed by rotation about a vertical line. These eight solutions are:

0 1	0 1	0 1	0 1	0 1	0 1	0 1	0 1
4 5	4 5	4 5	4 5	4 5	4 5	4 5	4 5
2 4	2 0	2 4	2 0	5 3	1 3	5 3	1 3
6 6	6 6	6 6	6 6	6 6	6 6	6 6	6 6
1 3	1 3	5 3	5 3	2 0	2 0	2 4	2 4
0 5	4 5	0 1	4 1	4 1	4 5	0 1	0 5
3 2	3 2	3 2	3 2	3 2	3 2	3 2	3 2

For type III (see *Figure 34*), where no symmetry occurs, we have:

$$
\begin{aligned}
&a \neq 0, \neq 3, \neq 4, \neq 6;\\
&b \neq 2, \neq 3, \neq 4, \neq 6;\\
&c \neq 0, \neq 1, \neq 5;\\
&d \neq 1, \neq 2, \neq 5;\\
&e \neq 2, \neq 3, \neq 6;\\
&f \neq 1, \neq 2, \neq 5;\\
&g \neq 1, \neq 2.
\end{aligned}
$$

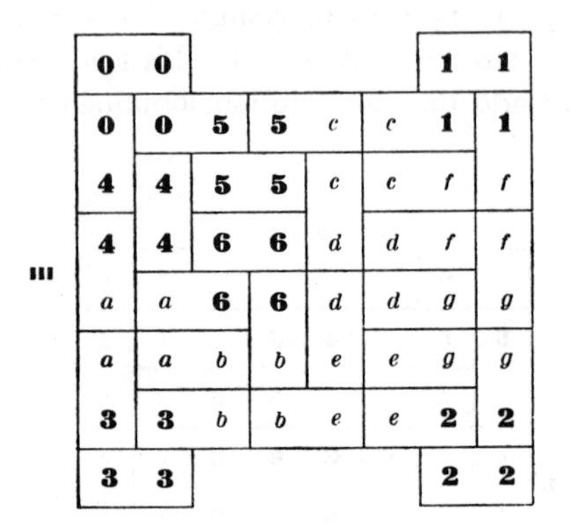

Fig. 34

From this we derive the following fourteen solutions of type III:

0 1	0 1	0 1	0 1	0 1	0 1	0 1
5 2	5 2	5 3	5 3	5 4	5 4	5 6
4 0	4 0	4 6	4 4	4 6	4 3	4 4
6 3	6 6	6 4	6 6	6 3	6 6	6 3
1 6	1 3	2 5	2 5	2 5	2 5	2 5
5 4	5 4	0 1	0 1	0 1	0 1	0 1
3 2	3 2	3 2	3 2	3 2	3 2	3 2

0 1	0 1	0 1	0 1	0 1	0 1	0 1
5 6	5 4	5 4	5 2	5 2	5 2	5 2
4 3	4 6	4 3	4 4	4 0	4 4	4 0
6 4	6 3	6 6	6 3	6 3	6 6	6 6
2 5	2 0	2 0	5 6	5 6	5 3	5 3
0 1	5 1	5 1	1 0	1 4	1 0	1 4
3 2	3 2	3 2	3 2	3 2	3 2	3 2

In all, we obtain $8 + 8 + 4 + 14 = 34$ essentially different solutions. With permutation of the numbers 0, 1, 2, 3, 4, 5, 6, this gives $34 \times 5040 = 171{,}360$ solutions. If we also count the solutions that arise from these by reflection or rotation, we get

$$4 \times 171{,}360 = 685{,}440$$

solutions.

***34. Connection with the puzzle of §32.** We can also solve the puzzle of §32 by temporarily omitting the restrictive condition about the sums of the spot numbers and first solving the puzzle of §33 completely, with a later examination of the 34 solutions found, to sort out those that can satisfy the restrictive condition if we suitably permute the numbers 0, 1, 2, 3, 4, 5, 6. This requirement can only be met for the second solution of type II, and for the ninth solution of type III, and each time in two ways.

We write the second solution of type II with letters, as follows:

$$\text{II2}\qquad \begin{array}{ccc} a & & c \\ & e\,g & \\ b & & e \\ & f\,b & \\ c & & d \\ & a\,g & \\ d & & f \end{array}$$

Here equal numbers have been denoted by the same letter. We have:

$$\begin{aligned} a + b + c + d &= a + e + g + c = d + a + g + f \\ &= c + e + d + f = f + b. \end{aligned}$$

From this we find $a + g = d + f$, $b = d + a + g$, and hence $b = 2d + f$. The five equal sums are at least $1 + 2 + 3 + 4 = 10$, hence f is at least 4. In conjunction with $b = 2d + f$, it follows from this that $d = 1$, $f = 4$, $b = 6$, hence $a + c = 3$. This gives either $a = 0$, $c = 3$, $e = 2$, $g = 5$, or $a = 3$, $c = 0$, $e = 5$, $g = 2$ (two solutions).

For the ninth solution of type III

$$\text{III9}\qquad \begin{array}{ccc} a & & g \\ & e\,b & \\ b & & f \\ & f\,d & \\ c & & a \\ & e\,g & \\ d & & c \end{array}$$

we have

$$\begin{aligned} a + b + c + d &= a + e + b + g = d + e + g + c \\ &= g + f + a + c = f + d, \end{aligned}$$

hence $a + b = d + c$, $f = a + b + c = d + 2c$. From this it follows that $c = 1$, $d = 4$, $f = 6$, hence $a = 0$, $b = 5$, $e = 2$, $g = 3$ or $a = 3$, $b = 2$, $e = 5$, $g = 0$ (two solutions).

For the third solution of type I

$$\begin{array}{ccccc} & a & & g \\ & & e\,d & \\ & b & & f \\ \text{I3} & & f\,b & \\ & c & & c \\ & & a\,g & \\ & d & & e \end{array}$$

we have

$$a + b + c + d = a + e + d + g = g + f + c + e$$
$$= f + b,$$

hence $a + c + d = f$, $a + d = f + c$, hence $c = 0$, hence

$$a + b + d = e + f + g;$$

thus

$$2(a + b + d) = a + b + d + e + f + g = 21,$$

which is impossible.

For the fifth solution of type I

$$\begin{array}{ccccc} & a & & g \\ & & e\,d & \\ & b & & b \\ \text{I5} & & f\,f & \\ & c & & c \\ & & a\,g & \\ & d & & e \end{array}$$

we have

$$a + b + c + d = a + e + d + g = g + b + c + e$$
$$= 2f,$$

hence $b + c = e + g = a + d = f$, so that none of the letters a, b, c, d, e, f, g can represent the number 0.

In a similar way, the second solution of type II′ comes to a dead end.

For the first solution of type II

$$\begin{array}{ccccc} & a & & c \\ & & e\,g & \\ & b & & e \\ \text{II1} & & f\,d & \\ & c & & b \\ & & a\,g & \\ & d & & f \end{array}$$

we have

$$a + b + c + d = a + e + g + c = d + a + g + f$$
$$= c + e + b + f = f + d,$$

hence $a + g = 0$, which is impossible. In a similar way we come to an impasse with the seventh solution of type II.

The remaining solutions found in §33 come to a dead end at once, because two different letters on the edge turn out to represent the same number.

The solution for the puzzle of §32 that we now have given is more laborious than the solution occurring in that section. In contrast to this, for the puzzle of §30 it was more advantageous to seek the solution first, without using the equality of the five sums, and to put in the restrictive conditions afterwards. This is a consequence of the fact that the unrestricted puzzle of §29 has only one solution (if solutions are considered as identical when they arise from one another by interchanges among the numbers 0, 1, 2, 3, 4, 5, 6, combined with reflections or rotations), whereas the puzzle of §33 has no fewer than 34 solutions.

35. Second doubly symmetric domino puzzle. From the 28 dominos we must construct an assembly like that in *Figure 35* such

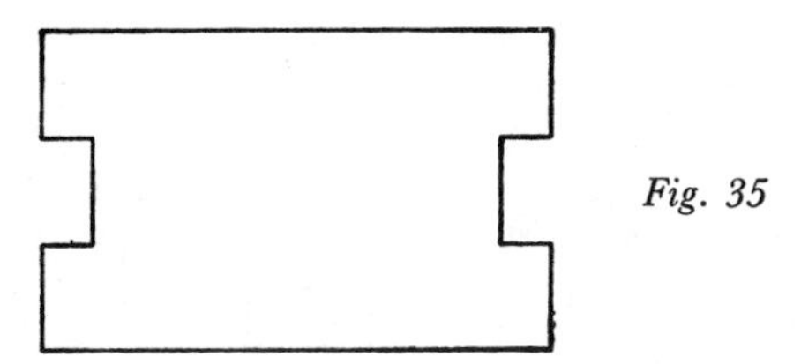

Fig. 35

that equal spot numbers always occur in groups of 4, forming squares. All solutions are required.

There are eleven possible ways of drawing the outlines of the pieces in the figure if we reject cases where equal pieces or more than seven doubles cannot immediately be seen to be present. In one of these ways we get only four doubles (in the corners), and in two of these ways only five doubles; these ways do not lead to solutions. Further, there are four ways by which six doubles can at once be seen to be present; three of these ways lead to solutions. In the remaining four ways we immediately get seven doubles; two of these ways lead to solutions.

The solutions can again be found by filling in the doubles arbitrarily and filling the remaining seven squares with letters among which we then establish inequalities. Even without filling in letters in this manner, the ways to fill in the remaining squares can be found by trial and error. In *Figure 36* the various solutions have been filled in to such a stage that the completion (which can be performed in more than one way) does not cause any difficulty. The capital letters refer to the ways in which the outlines of the dominos have been drawn.

In all, we obtain 29 essentially different solutions.

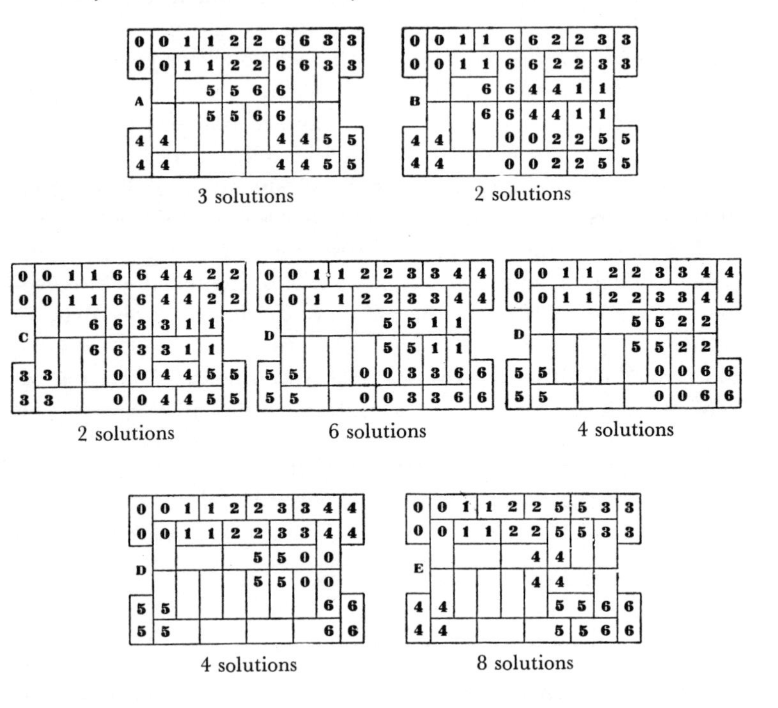

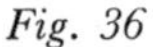

Fig. 36

36. Puzzle with dominos in a rectangle. We arrange the 28 dominos in a 7 by 8 rectangle. The problem is to do this so as to produce rows of four squares (half-dominos) which have the same spot number. All solutions are required.

This is a striking example of a puzzle that can be divided into three puzzles which are not too easy. The first (and easiest) puzzle consists of dividing the rectangle into fourteen smaller rectangles measuring 1 by 4. This can be done in thirteen ways. In one of them, all fourteen rectangles are horizontal, in six cases there are ten horizontal and four vertical rectangles, while in six cases there are six horizontal and eight vertical rectangles (see *Figure 37*).

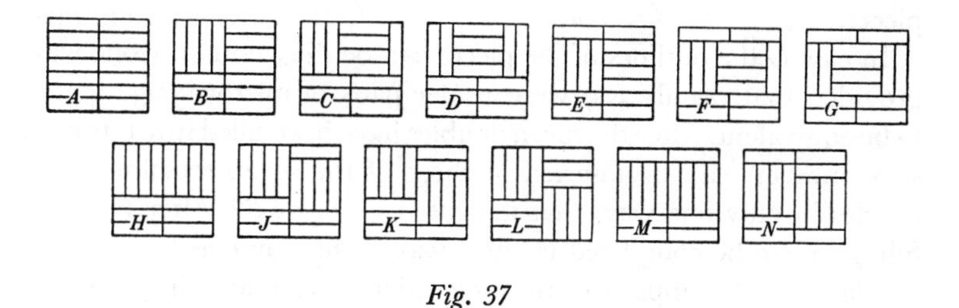

Fig. 37

Next, the outlines of the 28 dominos have to be drawn in the figure, in such a way that not more than seven doubles result directly, and no two equal pieces (hence no two pieces that belong to the same pair of the fourteen rectangles). Case *A* drops out, because in drawing the outlines we soon discover that three doubles arise in each of the four corners. In case *B* we get two doubles in the lower left-hand corner, and three doubles in each of the remaining corners. In case *E* we get three doubles in each of the corners at the right, and one more double in each of the corners at the left. In this way cases *A*, *B*, *E*, *H*, and *L* drop out. For cases *C*, *J*, and *K* we soon observe that we get seven doubles in the four corners; this simplifies drawing the outlines of the pieces, since we must avoid the occurrence of yet another double in the middle; in the course of this procedure the three cases mentioned come to a dead end. This also holds true in case *D*, where we immediately get six doubles in the corners, and have to avoid the formation of two extra doubles. The investigation of cases *F*, *G*, *M*, and *N* is more laborious. In case *F* it proves to be impossible to draw

				0	0	0	0
1							
1				2	3	4	5
1				2	3	4	5
1				2	3	4	5
				2	3	4	5
6	6	6	6				

				0	0	0	0
1			2				
1			2	3			4
1			2	3			4
1			2	3			4
				3			4
5	5	5	5				

Fig. 38

the outlines of the pieces. In case *N* it can only be done in the two ways illustrated in *Figure 38*; in the drawing at the left we reach an impasse when filling in the spot numbers, whereas in the drawing at the right we cannot fill in the seventh double without getting equal pieces.

In case *G* the outlines of the pieces can be drawn in only one way, provided that (as always) we consider pairs of mirror-image figures to be equivalent. After the seven doubles have been filled in arbitrarily, spot numbers can be inserted in three of the seven remaining rectangles in only one way, that indicated in *Figure 39*. After that, the filling in can be completed in three ways. The numbers 3 and 5 can be placed in the upper row in some order; after that, when filling in

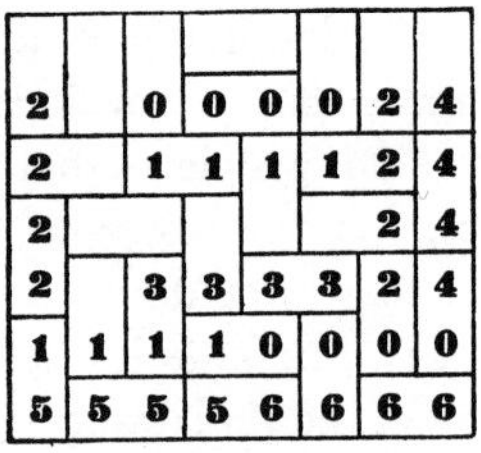

2		0	0	0	0	2	4
2		1	1	1	1	2	4
2						2	4
2		3	3	3	3	2	4
1	1	1	1	0	0	0	0
5	5	5	5	6	6	6	6

3 solutions

Fig. 39

the numbers 6 we must take care that the piece 5–6 does not appear a second time.

In case *M* the outlines of the dominos can be drawn in twelve essentially different ways, as shown in *Figure 40*.

In diagram 1, only four doubles have as yet been indicated. The three remaining doubles have to result from adjacent rectangles with the same spot numbers. However, this does not allow three doubles to be placed without equal pieces resulting. Diagrams 2, 3, and 4 do not yield any solutions either. In diagram 5, only six doubles are indicated by the position of the pieces. The seventh double can only be placed in the second-last row. The numbers 4 and 5 can only be placed in the top row in some order; this leads to four solutions. In diagram 6 the seventh double can only be placed in the top row, after which the numbers 0 and 3 can only be placed in the second-last row; this leads to one solution. In diagram 7 the seventh double can be inserted in two ways, one of which comes to a dead end when the filling in is continued. With the other way, a 2 × 4 rectangle arises which involves equal spot numbers; the numbers 2, 3, 4, and 5 can be inserted in one way only, so that we get two solutions. Diagram 8 also leads to two solutions. Diagrams 9, 10, 11, and 12 come to the same thing as far as the filling in is concerned. The numbers 3 can be placed

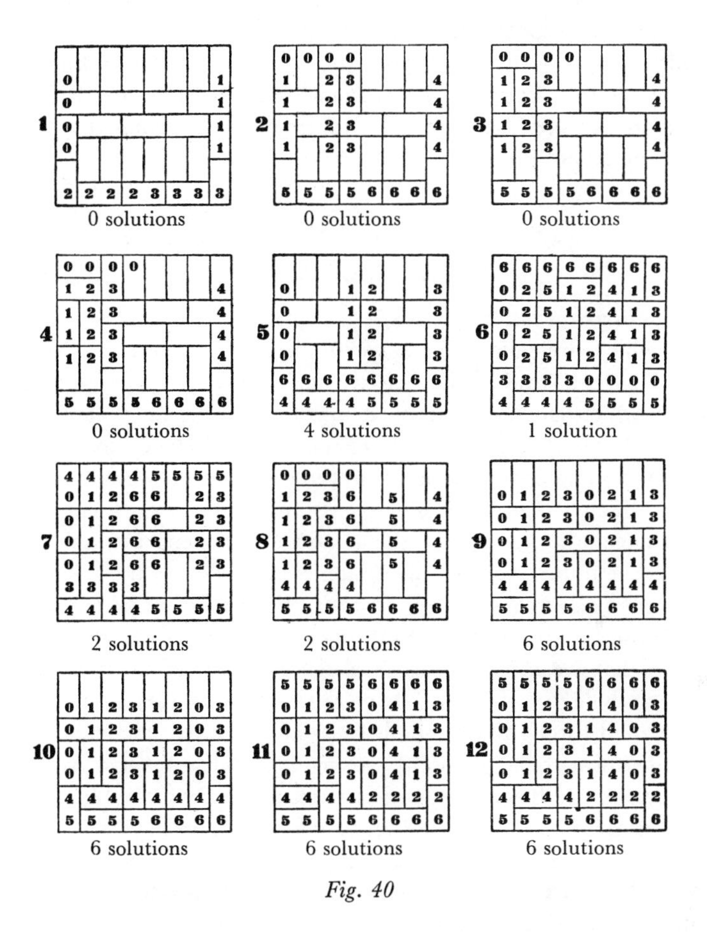

Fig. 40

in one way only. The numbers 2 and 4 can be filled in in two ways only; one of these ways has been chosen in diagrams 9 and 10, the other in diagrams 11 and 12. The numbers 0 and 1, too, can be filled in in two ways; one of these has been chosen in diagrams 9 and 11, the other in diagrams 11 and 12. In diagrams 9 and 10, the numbers 5 and 6 can next be inserted in two ways, in diagrams 11 and 12 in only one way. In all, each of the diagrams 9, 10, 11, and 12 thus leads to six solutions.

So, in all, the puzzle of the fourteen small rectangles in a rectangle has 36 essentially different solutions.

Among the solutions there are only two for which a 2×4 rectangle with eight equal spot numbers results. If the puzzle is presented with an initial requirement for a rectangle of this type, the finding of a solution can then be speeded up somewhat by first arguing that such a solution can occur only if the outlines of the dominos determine fewer than seven doubles.

III. SMALLEST AND LARGEST NUMBER OF CORNERS

37. Salient and re-entrant angles. Figure 12 showed sixteen angles, ten of which are salient (namely 90°) and six re-entrant (namely, 270°). Figures 23 and 35 show twelve angles. Several other figures are possible which can be partitioned into fourteen equal squares, and into which the 28 dominos can be placed in such a way that we obtain the same spot number four times in each of these fourteen squares. Such a figure (which we assume to have no holes in it) cannot have other than 90° (salient) and 270° (re-entrant) angles.

If u is the number of the salient angles and i that of the re-entrant angles, then $u = i + 4$; in other words, the number of salient angles exceeds the number of re-entrant angles by four. To see this, let us imagine that we are running in a clockwise direction around the outline of the figure. When we have returned to the starting point we have rotated through 360° in all. In passing a salient angle, we turn clockwise through 90°, and in passing a re-entrant angle we turn counterclockwise through 90°, so that in all we turn $u \times 90° - i \times 90°$. Since this is equal to 360°, hence to $4 \times 90°$, we have $u - i = 4$. The total number of angles is $2 \times i + 4$, and thus an even number.

38. Puzzle with the smallest number of angles. *Figure 41* exhibits all conceivable figures with ten corners or fewer. No attention has been paid to correct proportions of the figures or to the possibility of partition into fourteen equal squares. We intend only to indicate the various possible successions of salient and re-entrant angles.

We obtain one figure with four angles, one with six angles, four with eight angles, and eight with ten angles. It turns out that only the sixth figure with ten angles leads to a solution (with sets of four equal spot

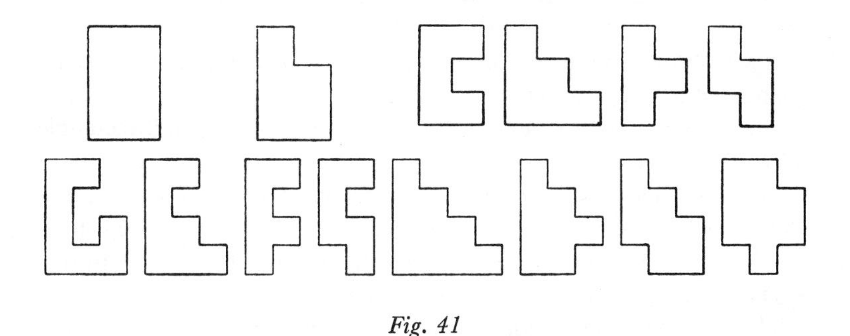

Fig. 41

numbers grouped together in squares). To this end we must give the figure the shape shown in *Figure 42* (figure with the smallest number of angles). When we take note of the position of the seven doubles and of the non-occurrence of two equal pieces, it turns out that the outlines of the dominos cannot be otherwise than indicated at the right of Figure 42. With this, the locations of the seven doubles have been found. The spot numbers of the doubles have been filled in

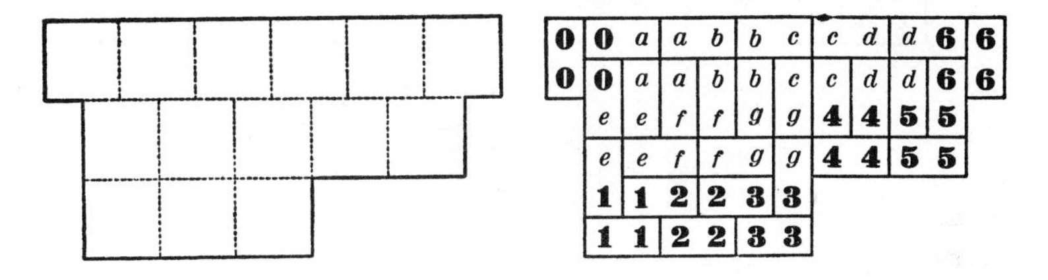

Fig. 42

arbitrarily. The position of the pieces 1–2, 2–3, and 5–6 is then known also. We find further:

$$a \neq 0, \neq 1; \quad b \neq 0, \neq 3, \neq 4; \quad c \neq 3, \neq 4, \neq 5, \neq 6;$$
$$d \neq 4, \neq 5, \neq 6; \quad e \neq 0, \neq 1, \neq 2; \quad f \neq 0, \neq 1, \neq 3;$$
$$g \neq 2, \neq 3, \neq 4.$$

In addition, we have $f \neq 5$ and $\neq 6$, because otherwise we have no place to put the 6 and the 5, respectively. Consequently, $f = 2$ or $f = 4$. The numbers 5 and 6, occurring in squares in which there is a double, occur combined as a piece (namely 5–6) and apart from that only in combination with d. It follows that any solution gives another solution if we interchange the numbers 5 and 6 outside the squares in which there is a double, so when we continue the solution we can make an arbitrary choice between 5 and 6; these numbers 5 and 6 cannot be adjacent, since there is only one piece 5–6.

If $f = 2$, we can place 3 only in square d, and then 1 only in square c, after which only 0 can be in square g. Since 5 and 6 cannot be adjacent, we have $a = 4$, after which the numbers 5 and 6 can be filled in arbitrarily.

If $f = 4$, then $c \neq 2$, because otherwise we could no longer find a place for 1; consequently, $c = 0$ or $c = 1$. The (interchangeable) numbers 5 and 6 have to be placed in the squares e and b, or in the squares e and g, or in the squares a and g.

Thus we find the solutions shown in *Figure 43* and the solutions that arise from these by interchanging the non-bold numerals 5 and 6.

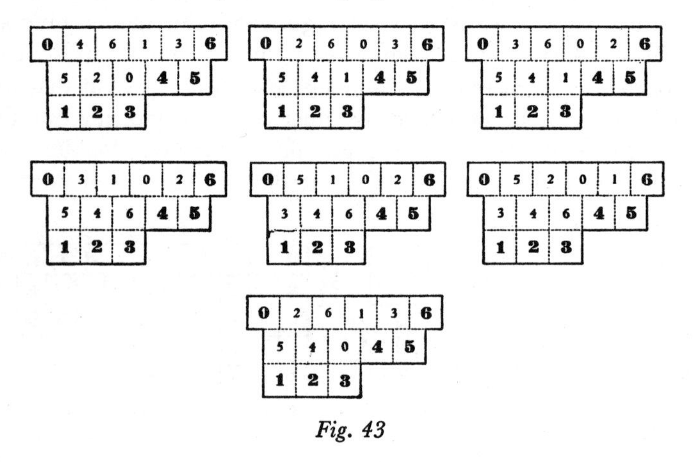

Fig. 43

Hence, in all there are fourteen solutions. In two of these solutions two squares with equal spot numbers are adjacent (see the first of the seven diagrams).

39. Puzzle with the largest number of angles. We modify the puzzle of §38 to the effect that we do not require the figure with the smallest, but that with the largest number of angles. In every salient angle there must be a double. However, it is possible that one double belongs to two salient angles. Since there are seven doubles, the figure can show at most $2 \times 7 = 14$ salient angles. With fourteen salient angles, there are $14 - 4 = 10$ re-entrant angles (see §37), so that the figure can have at most 24 angles. The fourteen salient angles succeed each other in pairs, whereas it is also possible that four (but not six or more) salient angles succeed each other.

In a figure with 24 angles, which also satisfies the condition just mentioned, the ten re-entrant and the fourteen salient angles can still succeed each other in 250 ways. It is hardly feasible to enumerate and to draw all these figures. However, when the puzzle requires dominos to form squares of four equal spot numbers, I have found a 24-corner arrangement (*Fig. 44*) which leads to a solution (and I think this is the only suitable arrangement).

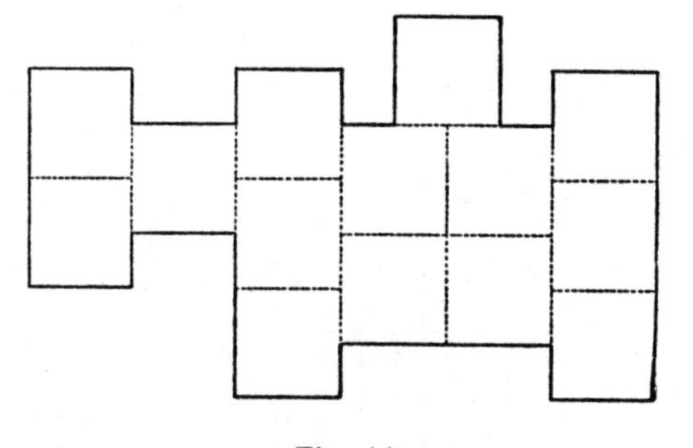

Fig. 44

As is easily seen, the outlines of the dominos can only be as indicated in the *Figure 45*. From the inequality of the pieces it follows that:

$$a \neq 0, \neq 1, \neq 5, \neq 6; \quad b \neq 0, \neq 1, \neq 2, \neq 5, \neq 6;$$
$$c \neq 1, \neq 2, \neq 5; \quad d \neq 2, \neq 3, \neq 4; \quad e \neq 2, \neq 3, \neq 4;$$
$$f \neq 1, \neq 2, \neq 4, \neq 5; \quad g \neq 2, \neq 3, \neq 4, \neq 5.$$

Hence, $a = 2$ (since 2 cannot be filled in anywhere else). If $b = 3$, then 5 cannot be filled in, so that $b = 4$. From $a = 2$ it follows that c and d are not 0, 1, or 6, so that $c = 3$ and $d = 5$. As the piece 0–6

						2	**2**				
0	**0**			**1**	**1**	**2**	**2**			**3**	**3**
0	**0**	*a*	*a*	**1**	**1**	*c*	*c*	*d*	*d*	**3**	**3**
6	**6**	*a*	*a*	*b*	*b*	*c*	*c*	*d*	*d*	*e*	*e*
6	**6**			*b*	*b*	*f*	*f*	*g*	*g*	*e*	*e*
				5	**5**	*f*	*f*	*g*	*g*	**4**	**4**
				5	**5**					**4**	**4**

Fig. 45

occurs only once, it follows that $e = 0, f = 6$, or $e = 6, f = 0$, and $g = 1$. The interchangeability of the numbers 0 and 6 outside the squares of the doubles was to be expected from the disposition of the pieces 0–a and 6–a.

A completely worked-out version of one of the solutions is shown in *Figure 46.* The other one arises from it by interchanging two pieces

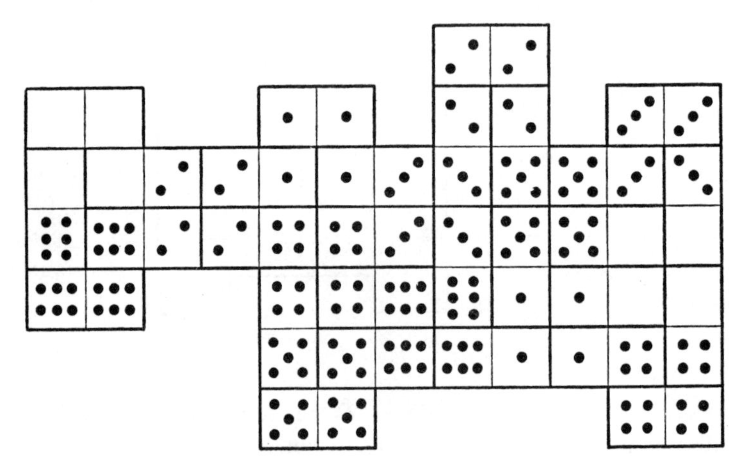

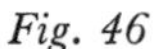

Fig. 46

four times, namely the pieces 1–0 and 1–6, the pieces 3–0 and 3–6, the pieces 4–0 and 4–6, and the pieces 5–0 and 5–6 (an interchange of 0 and 6 outside the squares of the doubles).

Chapter III:
THE GAME OF NOUGHTS AND CROSSES

I. DESCRIPTION OF THE GAME

40. Rules of the game. In the game of noughts and crosses two players take turns to put a mark in a still unoccupied square in a set of nine squares arranged as in our diagram (*Figure 47*). The player who begins will be called John; we assume his mark is a cross. The other player will be called Peter; we assume his mark is a nought. When it is immaterial who begins, we shall call the players by their second names Hook and Crook.

1	**2**	**3**
4	**5**	**6**
7	**8**	**9**

Fig. 47

The numbers in the squares are not required in the game. Their purpose is only to specify a move in the discussion of the game. Thus, for example, move 7 means that John or Peter puts his mark in square 7.

John has to try to get three crosses in a line, either horizontally, vertically or diagonally. For brevity, we shall call this "making a row." Peter too has to try to make a row, but with three noughts. The player who makes a row is the winner; among Dutch children this is accompanied by the exclamation "Boter, melk en kaas, ik ben de baas" ("Butter, milk and cheese, I am the boss"). If neither of the players manages to make a row, which is what usually happens, then the game ends in a draw.

41. Supplement to the game. To get more possibilities, and make the children's game described in §40 somewhat more interesting, we supplement it as follows. If Hook has made a row, the filling in of marks is continued, so that then it is Crook's turn again. If Crook also succeeds in making a row, he has lost singly, otherwise he has lost doubly.

If a player manages to make a second row, or two rows in one move, he has won trebly.

In the following discussion we shall consider the game in this extended version.

42. Consequences of the rules. If Hook was the first to make a row, and if it was a diagonal, then he has won doubly at least. For in such a case there is at least one of Hook's marks in each of the eight rows of the figure, so that a row for Crook is no longer possible.

If Peter was the first to make a row, and if it was horizontal or vertical, then he wins singly. For at the end of the game Peter has only one nought outside the row he has made, so that John will still make a row, too (horizontal or vertical, respectively), no matter how the game is continued.

If John was the first to make a row, and if it is horizontal or vertical, and if Peter can make a row after that, then that row is also horizontal or vertical, respectively. In the first two diagrams in *Figure 48* Peter

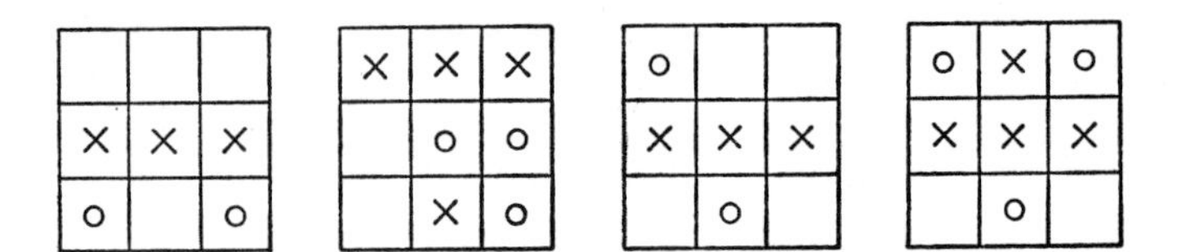

Fig. 48

can make a row immediately after John. However, if Peter cannot do it immediately after John, then John can easily prevent Peter from making a row (as in the third diagram), or else a row for Peter is no longer possible (as in the last diagram).

If one player is to make two rows (winning trebly) five of his marks are necessary. So Peter cannot win trebly, because he fills in a nought only four times. The various cases in which John wins trebly are shown in *Figure 49*.

The heavy cross is on the square common to both rows made. The six diagrams present the only cases of this type, provided figures that arise from each other by reflection or rotation are considered to be identical (as they always will be in what follows). John can only win trebly if Peter fills in his last noughts so badly that it practically never happens.

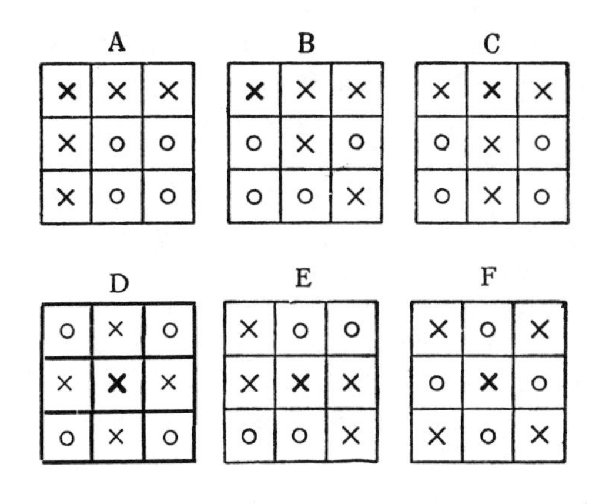

Fig. 49

II. CONSIDERATIONS AFFECTING VALUES OF THE SQUARES

43. Value of a square. In the beginning of the game, especially with his first move, a player is well-advised to make his mark in a square that belongs to as many rows as possible; we shall regard this number of rows as the value of the square or of the move. By making the move with the largest value, Hook makes as many rows as possible useless for Crook, and he makes as many rows as possible available for himself. Hence, at the start of the game, the central square 5 (value 4) looks most profitable. Next come the corner squares 1, 3, 7, and 9 (each having value 3). The border squares 2, 4, 6, and 8 (value 2) are the least advantageous. The safest opening move for John is therefore 5, and the safest reply by Peter is 1, 3, 7, or 9. If John does not begin with 5, then Peter's best move is to reply with 5.

It does not follow that beginning on a square with a larger value necessarily gives an advantage. Indeed, this is not the case. Every opening move by John leads to a draw if good play follows. However, opening on a border square does require more careful play than opening on the central square.

If Hook, on his second move, wants to judge its strength, he should disregard a row containing the square in question, if it is one in which he has already put his mark, since there is then no increase in the number of rows made useless to Crook. Hence, the value that for example John (who is to move) should assign to a still unoccupied square could be put equal to the number of rows to which this square belongs and in which John has not yet filled in a cross. When there are no other factors that decide for or against this move, a player is well-advised to fill in his mark in a square that has the largest value according to this rule.

44. Remarks on the value of a square. In the foregoing we have equated a row in which nothing has been filled in with a row in which your opponent has already filled in his mark. However, these two cases are not equivalent, but it is not possible to express the difference in figures according to a general rule. Sometimes in judging the strength of a move a row in which nothing has been filled in is more important than a row in which your opponent has already put a mark, sometimes the opposite is true. The first case holds when you want to increase your chances of winning, the second when you want to reduce your chances of losing. It depends entirely on the situation to which of these two chances more attention should be given. For example, if you see that you cannot win any more, the best you can do is to make useless to your opponent the rows in which he has already put one mark. This holds to an even greater extent, of course, for a row in which your opponent has already put two marks; then it is generally necessary to occupy the third square of that row.

The rows in which you yourself have already put a mark have not been taken into account, either. In general, such a row increases the strength of a move, because you get two marks in a row and thus your opponent is not free in his reply. However, the advantage of this can vary considerably from case to case.

According to the foregoing, the concept of "value of a square" soon loses its significance. Hence, a player is well-advised to pay attention to combinations that lead to a win or that avert a loss, as soon as two or three marks in all have been filled in. If he does not see such a combination, the values of the free squares at any rate provide some guide, especially when he is not playing for a win, but to avoid a loss.

III. DIRECTIONS FOR GOOD PLAY

45. Semi-row or threat. Making a semi-row must always precede making a row. By a semi-row we mean two like marks on a row, the third square of which is still free; we shall also call this a threat. If one of the players makes a semi-row, we can safely assume that this threat will always be seen by the other player (and also by himself). Hence, if Hook makes a semi-row, then Crook (if he cannot make a row himself) places his mark on the free square of that semi-row (averting the threat). If you stick to this simple rule, you will never lose trebly.

It may happen, however, that it is more profitable (or rather, less disadvantageous) to depart from the above rule, namely when you see that a loss can no longer be avoided, while departure from this rule averts a double loss. This is only possible, if at all, by abandoning the row to your opponent, and replying with a counterthreat. However, this is certain to fail when the (opponent's) threat is along a diagonal (cf. §42). So a diagonal threat should be averted without hesitation.

46. Double threat. The only way for Hook to win is to make two semi-rows with different free squares. Hence, this is what he does, if he gets the opportunity, that is, if there are no threats to be averted, which usually takes priority. We then speak of a double threat; in distinction from this, a single semi-row (as in §45) will also be called a single threat. A player—John, say—achieves a double threat by placing his cross in the square common to two rows, if there is already one cross in each of these rows, but no nought; on his next move John can then make a row.

Peter can only make a double threat on his third move, so that making a row is possible no earlier than on his fourth and last move. John can make a double threat sometimes on his third move and sometimes on his fourth move, so he can make a row either on his fourth or on his fifth (and last) move. However, when a player is concerned with averting a double loss, he may, in exceptional cases, even make a row on his third move.

47. Combined threat. If John makes two semi-rows with a common free square, then this is not a double threat, because Peter can annul both threats simultaneously by occupying this free square. If Peter did not do this, John would win trebly by occupying the square, so that making two such semi-rows is a threat to win trebly.

Such a threat can only be made at the fourth move, and only by John. It is true that with his fourth move Peter can sometimes make the pattern of a threat to win trebly, but this cannot properly be called a threat, because it cannot be followed by any move made by Peter.

It is wrong to make a threat to win trebly, if it is not combined with another threat, since in this event John had an opportunity to make a row and allowed the opportunity to pass. After the threat to win trebly has been annulled, John has only one move left and hence he can no longer make a row.

The cases of a threat to win trebly arise from the six diagrams in Figure 49, by removing from each the heavy cross and one of the noughts. This gives thirteen cases; in seven of these Peter can reply by making a row. The remaining six cases are shown in *Figure 50.*

The letters over the diagrams indicate the particular diagrams of Figure 49 from which they have been derived. The dot indicates the free squares, in which the corresponding diagram of Figure 49 has a bold cross.

In the last three figures the threat to win trebly occurs in combination with another threat. We call this a combined threat. Of course,

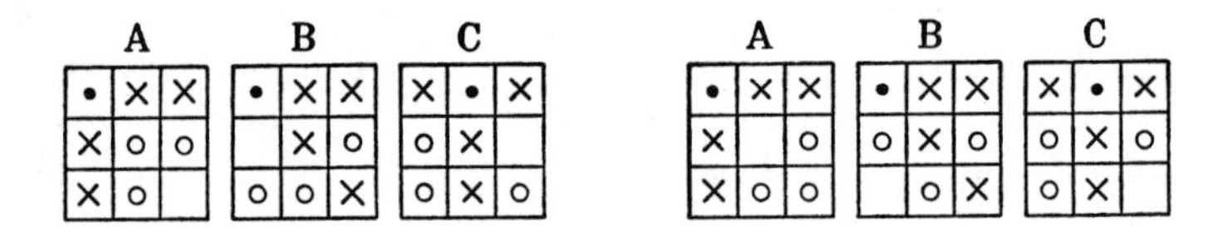

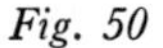
Fig. 50

Peter annuls John's threat to win trebly by putting his nought in the square with a dot; John then wins doubly.

48. Replying to a double threat. In what follows we shall assume that one of the players has to reply to a double threat consisting of two semi-rows, and that he cannot do this by making a row.

If John makes a double threat on his fourth move, then the reply is Peter's last move. Peter then loses doubly. So we can assume from now on that Hook makes the double threat on his third move. Crook then still has to make 2 moves, irrespective of who made the first move, and he should consider how to avoid a double loss, if possible.

This is certainly possible when Crook can reply by a double counterthreat. Hook then makes a row on his fourth move, and Crook makes one on his last move.

If one of Hook's two semi-rows is diagonal (a mixed double threat), then Crook's only chance of success (that is, of avoiding a double loss) is to prevent the diagonal from becoming a row for Hook (see §42). This does not guarantee Crook an escape from a double loss, but he need not consider this when making the move. We would nevertheless like to say something more about the result. If preventing the diagonal row does not impose a counterthreat, it has no effect, because immediately after that Hook can make a row to which Crook cannot reply with a row; this does not imply, however, that the move does have an effect when it is also a counterthreat. If it was Crook who made the first move (so that he thus is John), he avoids a double loss if he averts the diagonal threat; for in making his last move, Peter can do no better than make a horizontal or vertical row, so that he wins singly (see §42); here John's move is automatically a single or double counterthreat.

If it is possible to reply to a mixed threat with a double counterthreat, then this reply is at the same time a cancellation of the diagonal threat; for the double counterthreat leads to a single loss, whereas failure to avert the diagonal threat leads to a double loss. So the reply can be motivated in either of two ways. If Crook notes only that the reply is a double counterthreat, this assures him success without more ado, and whether this can be achieved in another way is immaterial. If Crook notes only that his reply averts a threat along a diagonal, then he sees that no other move can be successful; it can then be left an open question whether his reply avoids a double loss. Both motivations, however, have the same effect, since either argument makes Crook give the reply unhesitatingly, with no need for further consideration.

A diagonal double threat, that is, a threat along both diagonals (which can only result from move 5) always leads to a double win; it is impossible to reply to a diagonal double threat by making a row. Hence, if one of the players can make a double threat in more than one way, and if one of these involves both diagonals, he can then occupy square 5 without hesitation; it is not necessary here to consider any other double threat. The situation in *Figure 51* is the only one in

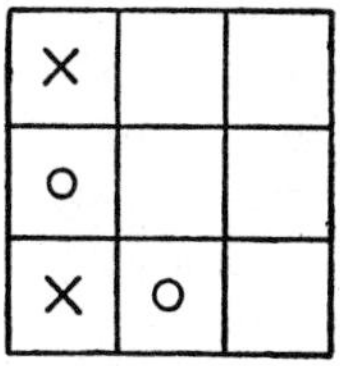

X		
O		
X	O	

Fig. 51

which John has no threat to avert on his third move, but can use this move to make a double threat in more than one way, one of which is along both diagonals; John has another double threat, with move 3, which gives him a single win after it evokes a move 5 in reply, but John should ignore this and select move 5 for his own move.

Suppose that John makes an orthogonal double threat (along a horizontal and a vertical row), and that Peter cannot make a double counterthreat. If Peter does not reply with a counterthreat then John can immediately make a row, after which Peter cannot make a row. If Peter makes a counterthreat that does not cancel either of the two threats, then John can cancel the counterthreat (on his fourth move), after which Peter cannot make a row, though John can. So the only chance for Peter to avoid a double loss is to reply with a counterthreat that cancels one of the two threats. However, there is no guarantee that such a move of Peter will be successful; for if John makes a row or a double threat by cancelling the counterthreat, he wins doubly, but otherwise singly. As Peter has to avoid one of the two threats anyway, Peter has at most two replies (hence none, one or two) by which he avoids a double loss.

If Peter makes an orthogonal double threat, then he cannot achieve more than a single win, however badly John plays (see §42). Hence, John's reply is then completely immaterial; he cannot lose doubly even if he wants to.

The previous arguments also show that the only chance for Peter to win doubly is to make a diagonal double threat.

49. Further directions for good play. When Hook makes a double threat, it does not necessarily follow that Crook could have averted this on his last move without allowing a single threat already present to survive (when he would lose immediately thereafter). Hook's tactics here have been to compel Crook to make a certain move by making a single threat (one semi-row) and after that playing the double threat; Crook's mistake here is farther back. For example, in the position shown in *Figure 52*, John (playing the crosses and having to move) wins singly by playing 4, 7, or 9 (with the respective replies 7, 4, 5) and next making a counterthreat by 5, 9, 7, respectively.

It may happen that Crook sees the danger after the single threat, and also sees that averting the threat leads to a double loss. If Crook can then make a double counterthreat (disregarding the single threat made by Hook), this will then ensure that he avoids a double loss. Hook then sometimes makes a row on his third move. An example is provided by the position of *Figure 53*, in which John (playing the crosses) obviously has the next move. If John plays 4, Peter wins doubly by the diagonal double threat 5; so John makes a double threat by playing 5, and loses singly.

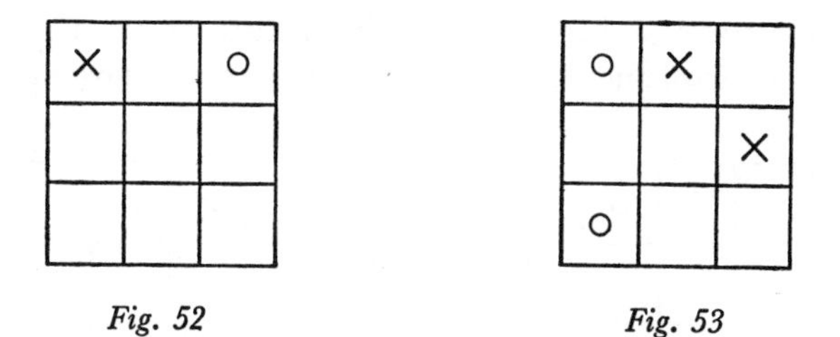

Fig. 52 *Fig. 53*

Also, Crook can sometimes avoid a double loss by ignoring the double threat and replying with a single counterthreat. This applies to the position of *Figure 54*, in which Peter (playing the noughts) obviously has the next move. If Peter plays 4, John wins doubly by the reply 5. Peter plays the threat 5. John then replies with the double threat 8, hoping that Peter will still make an error—9, say. However, if Peter replies 4, then John has to play 9 or risk abandoning his victory; in this way Peter loses singly only.

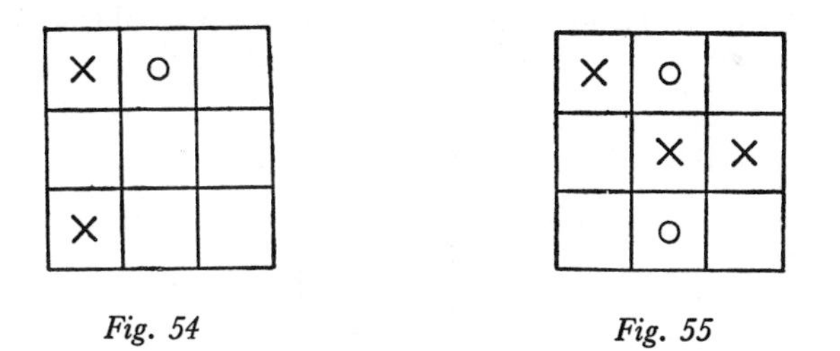

Fig. 54 *Fig. 55*

If John can follow a double threat by making a row on his fourth move, it may be more advantageous to postpone this for one move, and to prevent Peter from making a row first. However, this is only correct if it gives John another double threat. In this way John may achieve a double win instead of a single win. For example, if Peter gives the best possible reply to the double threat in *Figure 55*, namely 9, then John does not play 4 and make a row (by which he would obtain a single win), but instead he makes a combined threat by playing 7, through which he wins doubly.

IV. SOME REMARKS ON GOOD PLAY

50. Remarks on the double threat. If John can make a double threat in two ways, with an orthogonal double threat and a mixed one, then sometimes the one and sometimes the other double threat is more successful, and the result of both threats can also be identical. Thus, in the left-hand diagram of *Figure 56*, the orthogonal double

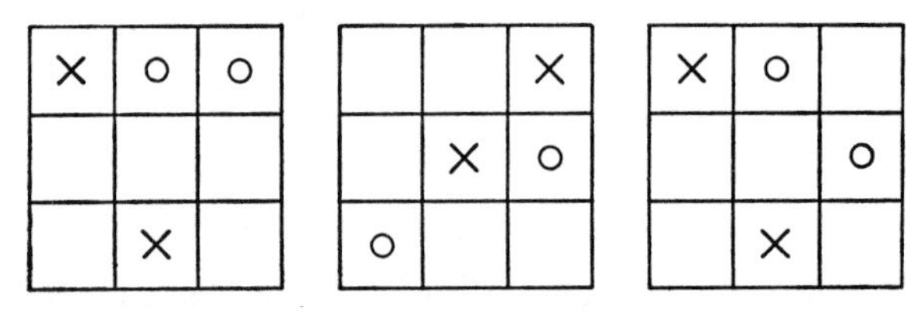

Fig. 56

threat (move 7) leads to a double win, whereas the opposite holds in the diagram at the right. In the middle diagram both double threats (moves 1 and 2) lead to a single win.

John may find that it is not his move, and that he has no semi-row, but that if it were his move he could make a double threat in so many ways that Peter could not simultaneously avert all these threats with a double threat. In *Figure 57* any of John's moves would be a double

X	O	O
		X
	X	

Fig. 57

threat if he were allowed to make a move. When we note that in fact it is Peter's move, John then has one way to make a double threat after each move by Peter (Peter's 4, 5, 7, 9 will be followed by 9, 7, 5, 4, respectively, by John); in every case John obtains a double win.

If John's last move in Figure 57 has been 6, then before this move we had the left-hand diagram of Figure 56; John could then have won doubly in a simpler way by the double threat 9, but the double threat 7 would have been wrong. If 8 was John's last move, then he could have won doubly in a simpler way by the mixed double threat 5, but the orthogonal double threat 4 would have led to a single win only; this provides another case where the orthogonal double threat yields the worse result. If (still in Figure 57) John's last move was 1, then it was a forced move; in this case Peter has made a bad move and driven John into the position of automatically winning doubly with no need to be particularly observant.

51. Connection with the value of a move. If Hook has made seven of the eight rows useless to Crook (so that the values of Hook's moves sum to 7), and if Crook cannot make a row immediately after that, then Hook can no longer lose, for Crook can no longer make a double threat. If Hook has made six rows useless to Crook, and if Crook cannot reply with a row or a double threat, then Hook cannot lose, either; for, after averting a possible single threat, Hook has made at least seven rows useless to Crook.

By examining the various possibilities I discovered also that a player cannot lose when he has made five rows useless to his opponent, provided no row or double threat is possible as a reply; there may be a simple way to show this, but I have not succeeded in finding one.

V. GENERAL REMARKS ON THE ANALYSIS OF THE GAME

52. Preliminary remarks. When children play the game, they often make a semi-row here and there, hoping that their opponent will not notice it. However, we assume that the game is not being played in this naive way, and moves that are too obviously inferior will therefore be omitted in the analysis that now follows.

We assume that Hook sees any threat by Crook, also that Hook always sees any double threat that he can make and any double threat that Crook threatens to make on his next move. In this last situation Hook averts the double threat, if possible, unless he can do something better, such as making a double threat himself; moves that fail to avert a double threat and have no compensating advantages will not be mentioned, except when the double threat is impossible to avert. We also assume that averting a diagonal threat is given the highest priority (in order to avoid a double loss), unless there is also a threat to win trebly (which must be a combined threat); then this latter is averted. More generally, we assume that a player knows how to reply to a double threat.

The number of possibilities, which would otherwise be excessively large, is reduced very considerably by the joint effect of these assumptions.

53. Diagrams. To allow the diagrams to show the order in which the marks have been filled in, we shall record John's marks as 1, 2, 3, 4, 5 (instead of crosses, as in the real game) and Peter's marks as *a*, *b*, *c*, *d* (instead of noughts).

When different continuations all lead to the same result, marks have not usually been filled in following 3 or *c* (sometimes not following *b*, when John's third move is immaterial). It is then easy to examine later possibilities. In some cases in which compulsory moves lead to a draw, the filling in of the moves has been continued to a further stage, occasionally as far as the end.

If the result is a victory by Hook, filling in of the moves has usually been continued until Hook has made a row, to show whether the win is single or double. If there is earlier evidence that the win is single, then we stop earlier when this avoids a further division into cases. Likewise, we make an earlier stop at a diagonal double threat, since this leads to a double win.

Moves that are completely equivalent because they arise from each other by reflection or rotation have been mentioned only once, as the move corresponding to the square with the lowest number. In judging whether moves are equivalent by reflection, the marks 1 and 2, for example, are considered as identical, since both marks represent crosses in actual fact.

A move that does not avert an existing threat has been mentioned only when cancelling the threat would lead to a double loss; the examination of other moves makes no sense when averting the threat leads to a draw or a single loss, because then these other moves cannot possibly produce a better result. So the diagram for a move that does not avert a double threat has been placed after the diagram in which the threat has been averted. Apart from this, the moves that deserve consideration have been dealt with in the order of the numbers of the squares.

Below each diagram the result is given, as follows: d: draw; sw: single win for John; dw: double win for John; sl: single loss for John; dl: double loss for John.

54. Tree derived from the diagrams. To provide a clear survey of the consequence of the various moves contained in the diagrams, we derive a tree from these diagrams. For every move by Hook, the tree presents the replies by Crook that deserve consideration, up to a certain move. From the effects of Crook's replies we can derive the result of Hook's move. This is the result most favorable to Crook among the various replies possible to him, since the result of a move by Hook should be judged in the light of the best defence by Crook.

The merits of a move can be appreciated immediately from the tree by comparing the result with the results of the other moves that deserve consideration. A question mark has been appended to a move that gives a worse result than some other possible move. When a move is still worse than a move that is already bad, it has been provided with two question marks, etc. Occasionally a move has been supplied with an exclamation sign, namely when it is one of the best moves and also when there is the possibility of a bad countermove.

VI. PARTIAL ANALYSIS OF THE GAME

55. John starts with the central square 5, Peter replies with the corner square 1. According to the values of the squares, John cannot do better than start with the central square 5, and Peter cannot do better than reply with the corner square 1. John's next move can be 2, 3, 6, or 9 (because 4 amounts to the same thing as 2, etc.). These possibilities will be indicated by A, B, C, D. Case A, in which Peter's second move 8 is a compulsory move, is subdivided into the cases AI, AII, AIII, and AIV, depending on whether John's third

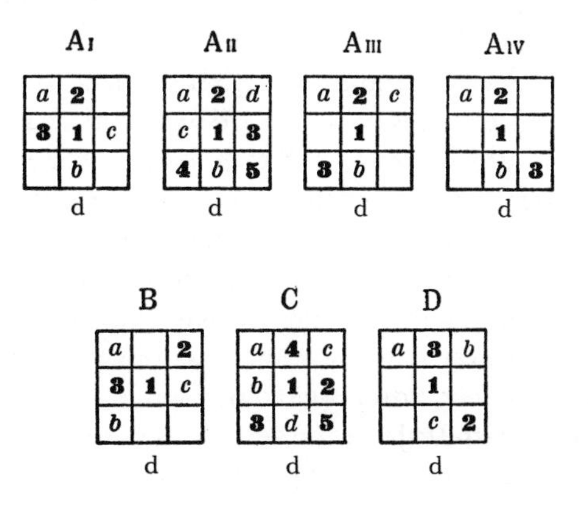

Fig. 58

move is 4, 6, 7, or 9, respectively (move 3 deserves no consideration, since Peter will play the double threat 7). This gives the diagrams of *Figure 58.*

The game results in a draw in every case. If John and Peter do not play too badly after their first moves, no other result is possible.

56. John starts with the corner square 1, Peter replies with the central square 5. If John starts with 1, then (according to the values of the squares) Peter's best reply is 5. After that, there are the possibilities in *Figure 59.*

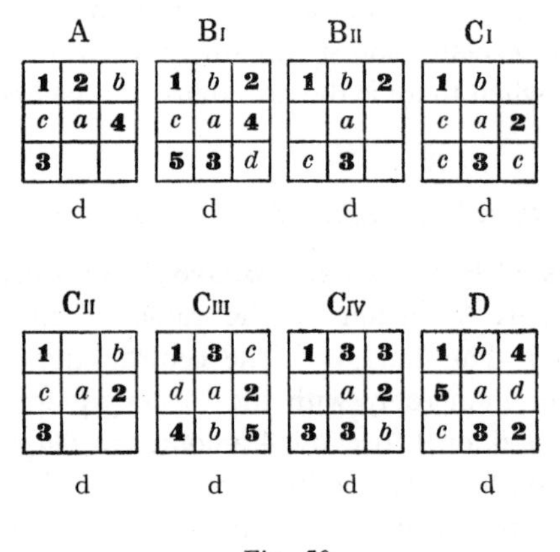

Fig. 59

In case B, Peter's third moves 4 and 6 amount to the same thing, similarly with 7 and 9. In case D, Peter's second moves 2, 4, 6, and 8 are equivalent; Peter's second move 3 (equivalent to 7) deserves no consideration, since it does not avert a double threat. In diagram CI, each of Peter's third moves 4, 7, and 9 deserves consideration (they all lead to a draw), therefore the letter *c* has been inserted in each of the squares 4, 7, and 9. For the same reason, the number 3 has been written in each of the squares 2, 3, 7, and 8 in diagram CIV. This method of economizing in diagrams has often been applied later, also.

In any of these cases, the game results in a draw, and we consequently omit the corresponding tree (as in §55).

57. John starts with the border square 2, Peter replies with the central square 5. The ensuing possibilities are shown in *Figure 60.*

The corresponding tree (in which Peter *a* means "Peter's first move," while John 2 means "John's second move," etc.) is:

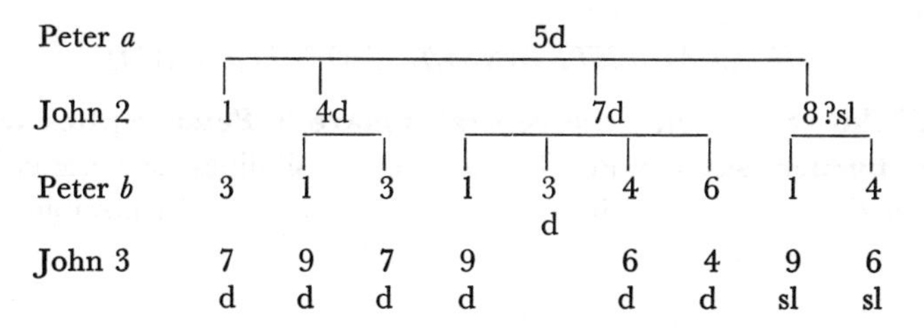

The game results in a draw, unless John plays 8 on his second move, in which case he loses singly. In his first two moves, John has here made the moves with the lowest values.

A		
2	**1**	*b*
c	*a*	**4**
3		

d

BI		
b	**1**	
2	*a*	
		3

d

BII		
c	**1**	*b*
2	*a*	**5**
3	*d*	**4**

d

CI		
b	**1**	
	a	
2	*c*	**3**

d

CII		
3	**1**	*b*
3	*a*	**3**
2		**3**

d

CIII		
	1	*c*
b	*a*	**3**
2		*c*

d

CIV		
c	**1**	**5**
3	*a*	*b*
2	*d*	**4**

d

DI		
b	**1**	**4**
d	*a*	
c	**2**	**3**

sl

DII		
c	**1**	
b	*a*	**3**
d	**2**	**4**

sl

Fig. 60

58. Equitable nature of the game. From the diagrams in §§55–57 it appears that Peter can achieve a draw by replying 1 to John's opening move 5, and by replying 5 to any other opening move. Evidently Peter cannot achieve a better result however John may open. For John is certainly not in a worse position after his first move, whatever it may be, than Peter was at the beginning of the play, so that John, too, can achieve a draw. This shows that with good play on both sides the result is a draw.

Hence, the advantage that John has from making the first move is not sufficiently large to be converted into victory. However, John has a slightly easier game, and a greater chance of winning by a mistake of his opponent. But if Peter plays 5 or 1 on his second move (the latter if John opens with 5), he too then has an easy game.

VII. COMPLETE ANALYSIS OF THE GAME

59. John starts with the central square 5, Peter replies with the border square 2. The ensuing possibilities are shown in *Figure 61.*

AIa

2	*a*	
3	**1**	*c*
4		*b*

sw

AIb

2	*a*	
3	**1**	**4**
c		*b*

sw

AII

2	*a*	*c*
4	**1**	
3		*b*

sw

BI

3	*a*	
2	**1**	*b*
4		*c*

sw

BII

4	*a*	*c*
2	**1**	*b*
3		

sw

C

3	*a*	*b*
4	**1**	
2		*c*

sw

DI

b	*a*	**3**
4	**1**	*d*
c	**2**	**5**

d

DII

	a	
	1	
b	**2**	

d

Fig. 61

The tree derived from this is:

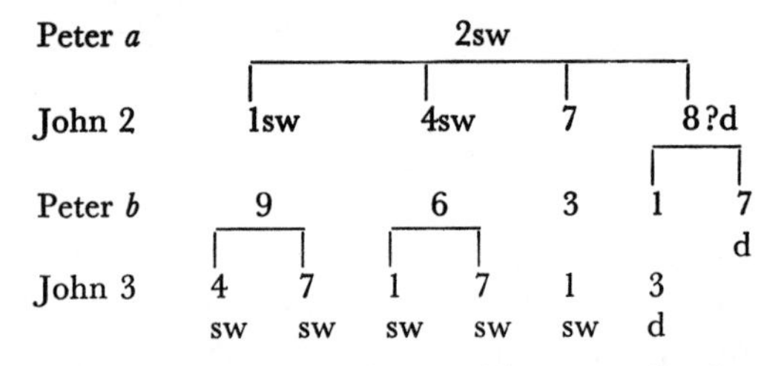

John wins singly, unless he plays 8 (the move having the lowest value) on his second move (which produces a draw).

In connection with §55 one gets the tree:

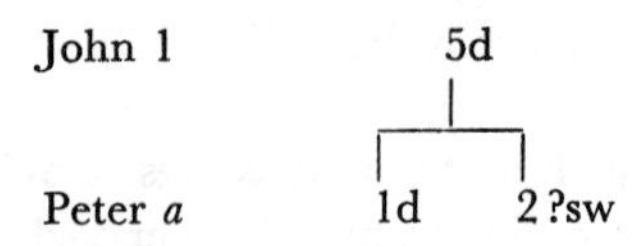

Hence Peter's first move 2 is wrong.

60. John starts with the corner square 1, Peter replies with the border square 2. The diagrams are as in *Figure 62.*

In cases A and D only Peter's second move 5 deserves consideration; after any other move on Peter's part, John makes a double threat. The corresponding tree is:

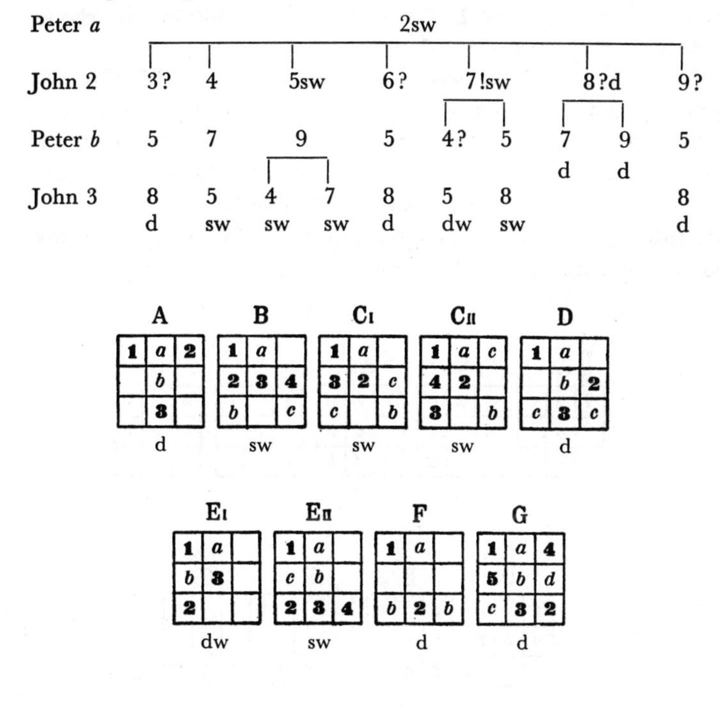

Fig. 62

The result given after Peter's first move 2 is the best result (for John) among the results of John's various second moves. The result given after John's second move 7 is the best result (for Peter) of the second moves 4 and 5 by Peter.

The tree shows the merits of the moves more clearly than the diagrams do. John's second moves 3, 6, 8, and 9 are wrong, since they lead only to a draw, whereas the moves 4, 5, and 7 lead to a single win. Of these moves, 7 (marked with an exclamation point) can be considered as the best. It is true that the result of 7 is the same as that of 4 and 5, assuming that Peter's further play is correct, but by playing 7 John has a chance that the threat will be cancelled by a move 4 by Peter (who after all made an error on his first move), after which John will make a diagonal double threat with 5, and win doubly; if Peter decides to prevent the latter threat by making the counterthreat 5 on his second move, John does not immediately make a row, but instead prefers to have a second chance of a double win by playing 8 (a double threat that cancels Peter's threat); however, this chance is removed if Peter makes the move 4.

Case E (where John plays 7 on his second move) is consequently quite interesting. It is a puzzle in its own right. The first move 1 by John and Peter's reply 2 are given and the problem is: How should John play in order to win, with a chance of a double win, and how should Peter play in order to avoid a double loss?

61. John starts with the corner square 1, Peter replies with the corner square 3. The diagrams are as in *Figure 63.*

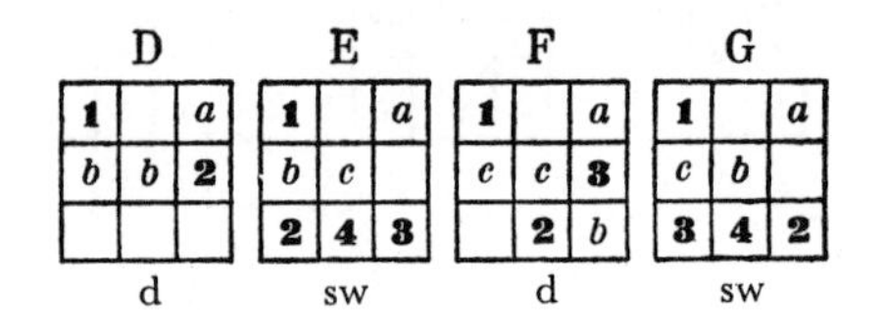

Fig. 63

In diagram F, after his second move, John threatens to make a double threat in 3 ways (on squares 5, 7, and 9). Peter can avert all threats only by the move 9.

The tree derived from the diagrams is:

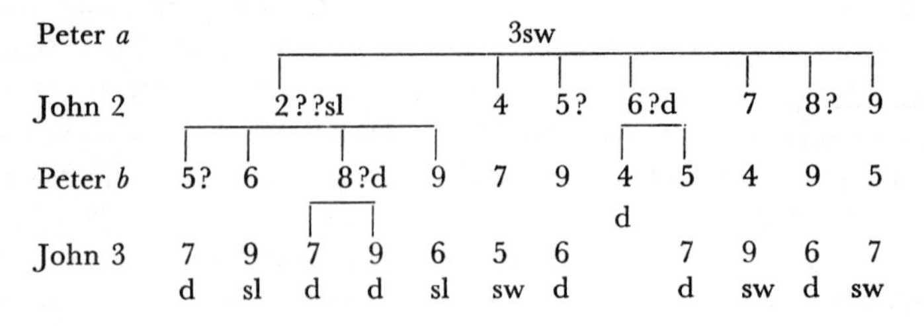

From this tree it is seen that John wins singly by playing 4, 7, or 9 on his second move. John's second moves 5, 6, and 8 lead to a draw only, and hence are wrong. Worse still is the second move 2 by John, which leads to a single loss; however, if Peter commits the error of playing 5 or 8 on his second move, the result is a draw.

62. John starts with the corner square 1, Peter replies with the border square 6. The diagrams are as in *Figure 64*.

In diagram E, if Peter chooses either 3 or 4 for his second move (other moves deserve no consideration at all, except move 9, which comes to the same thing as 3), this leads to a double win for John, as do Peter's third moves 2, 4, and 5 in diagram FII. This has been worked out further in the following tree:

```
Peter a                            6dw

John 2    2???  3?sw     4??d  5?     7dw      8??d   9??

Peter b   3    2?  5       7     9   3    4                5

John 3    9    5   4   3    9    3   9    5                4
          sl   dw  sw  d    d    sw  dw   dw               d

Peter b       2??dw    3?  4??  5       7?sw          9??dw

John 3      7dw   9?   9    5   4     2       5         3

Peter c   4    9   5   5    9   7    3    5   9    2    4    5
John 4    5    3   7   7    3   3    5    3   2    7    5    4
          dw   dw  sw  sw   dw  d    sw   sw  sw   dw   dw   dw
```

The seven possible second moves by John are of six different degrees of merit. The best move is 7; Peter cannot then avert a double loss (see diagram E). Next comes move 3 (diagrams BI and BII), which guarantees a win and gives a chance of a double win; by playing 5 on his second move, instead of averting the threat by playing 2 (which will be answered by a diagonal double threat), Peter avoids a double loss, since he will reply with the threat 2 to John's third move 4; John cannot avert this without abandoning his win. Next in value comes John's second move 5, which yields only a single win, provided Peter does not play too badly (that is, if Peter averts the diagonal threat on his second move). Worse still is John's second move 8 (diagrams FI and FII), which leads to a draw with a chance of a double win, for

Peter could reply with 9 instead of 5. Still a little worse are John's second moves 4 and 9, which lead to a draw without the chance of a win. John's worst second move is 2, which leads to a single loss.

If we assume that Peter plays well after his second move, John's moves are of four possible degrees of merit, because then moves 3 and 5 are equivalent (a single win), as well as moves 4, 8, and 9 (a draw).

If John has chosen 8 as his second move, then no replies by Peter other than 5 and 9 deserve consideration, because only these will prevent a double threat on John's third move. Yet it appears on further consideration that moves 3 and 7 are less bad than 9, because they lead to a single loss only (see the tree). Peter's second move 4

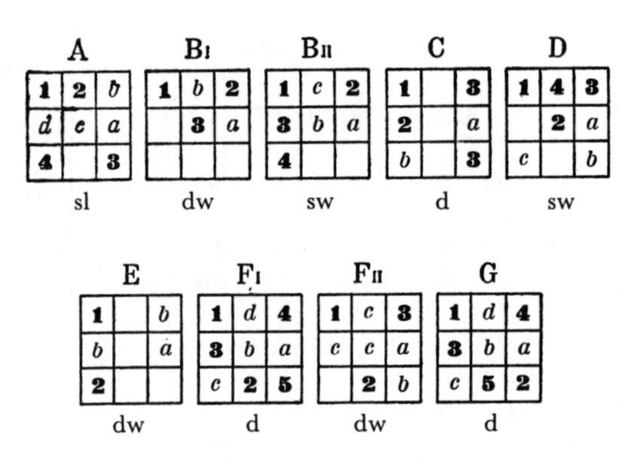

Fig. 64

causes him to lose doubly, and so does his second move 2; by playing the latter move, however, Peter still has a chance that he will lose singly only, namely if John gives the wrong reply 9.

Various curious situations arise in the course of the game. For example, the boldface 9 (corresponding to John's second move 7) yields a treble threat. The remaining three boldface moves in the tree (corresponding to John's second move 8 with 2, 4, or 9 as a reply) produce combined threats. In these situations John forces a double win, and Peter tries, in vain, to escape a double loss.

The first move 1 by John, together with Peter's reply 6, gives rise to the following puzzle, which is possibly even more interesting than the puzzle in §60: How should John continue in order to achieve the best possible result?

63. John starts with the corner square 1, Peter replies with the corner square 9. The possibilities are shown in *Figure 65.*

After John's second move 6 and Peter's reply 8 the situation is precisely the same as after Peter's second move in diagram FII of Figure 64 (assuming a reflection, and an interchange of moves). This type of situation of course occurs repeatedly.

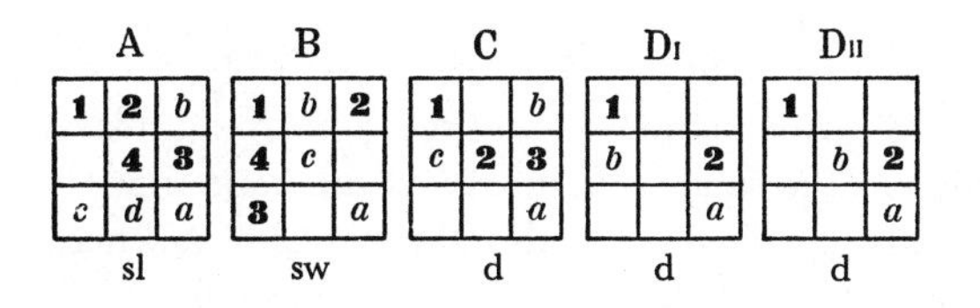

DIII

1	c	
	c	2
b	3	a

d

DIVa

1	5	4
c	d	2
3	b	a

dw

DIVb

1	c	d
5	4	2
3	b	a

dw

DIVc

1	4	
	c	2
3	b	a

dw

Fig. 65

From the diagrams we derive the following tree:

Peter *a*	9sw						
John 2	2??	3	5?	6?d			
Peter *b*	3	2	3	4 d	5 d	7	8?
John 3	6 sl	7 sw	6 d			8 d	7 dw

John's best move is 3 (a single win), and his worst move is 2 (a single loss). Move 6 is slightly better than 5; both moves lead to a draw, but in playing 6 John has a chance to win doubly after a wrong reply 8 by Peter.

64. Results of John's first move 1. The following tree gives a survey of the results of John's first move 1 for different replies by Peter:

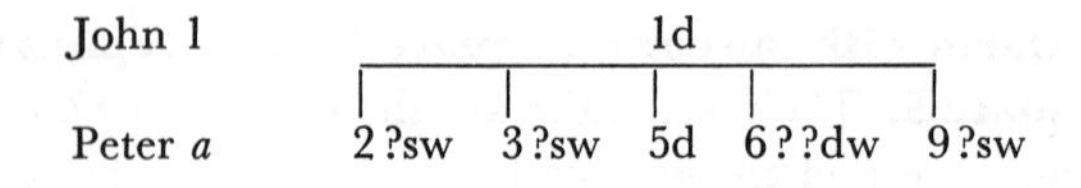

From this tree we can judge the qualities of Peter's various replies. Peter has only one correct reply, namely 5. This leads to a draw, so the result of John's first move also is a draw; in §58 this result has been found in a simpler way. Peter's worst reply is 6; it causes him to lose doubly.

It need not surprise us that 5 is Peter's best reply, because this move has the largest value, namely 4, and the game has just started. Peter's worst reply has the smallest value, 2. At the same time, Peter's reply 2, which leads to a single loss, also has the value 2, whereas moves 3 and 9, which have the value 3, also lead to a single loss for Peter.

65. John starts with border square 2, Peter replies with the corner square 1. We leave the construction of the corresponding diagrams to the reader. The tree derived from the diagrams is:

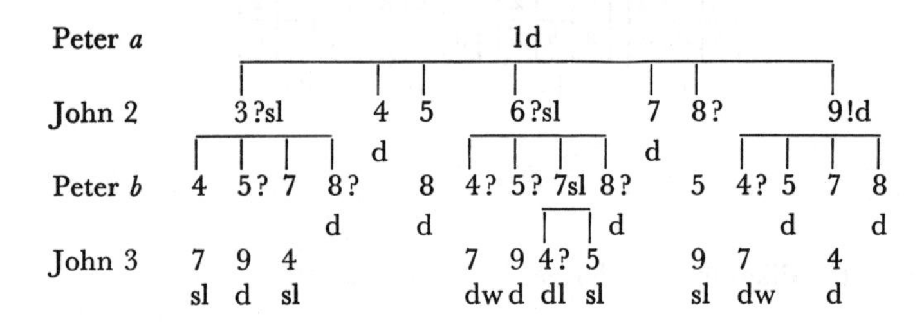

John's best second move is 9; he then achieves a draw with a chance of a double win (if Peter replies with 4). After John's second move 9, the same situation has arisen as in §63 after John's second move 6. The second moves 3, 6, and 8 by John lead to a single loss. After John's second move 3, he still has a chance to draw if Peter plays 5 or 8 on his second move (instead of 4 or 7). To John's second move 6 the best reply is 7; if John then averts the threat by 4, instead of making a double threat by 5, he loses doubly. However, if Peter makes one of the wrong moves 5 or 8 in reply to John's second move 6, then the game results in a draw; should Peter, worse still, play 4, then John wins doubly; in this case, John averts the threat by 7, and then threatens in four ways to make a double threat.

66. John starts with the border square 2, Peter replies with the border square 4. The corresponding tree is:

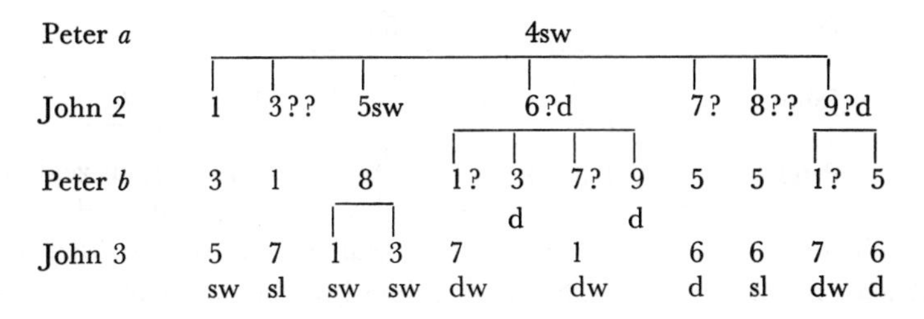

John wins singly by playing 1 or 5, and loses singly after 3 or 8. For the two remaining moves that John can make, the result is a draw; in playing 6 or 9 on his second move, John still has a chance of a double win.

67. John starts with the border square 2, Peter replies with the corner square 7. The corresponding tree is:

```
Peter a                          7sw
John 2    1   3??      4??sl             5?   6??sl        8?? 9?
Peter b   3   1    5?  6?  8??   9sl      8  1sl     4?     5   1
                       d
John 3    5   4    3       9    5   8?   9 4?  5     1      3   4
          sw  sl   d       dw   sl  dl   d dl  sl    dw     sl  d
```

After Peter's first move John wins only by playing 1, in which case he wins singly. If John plays 5 or 9, the result is a draw. After playing 3 or 8 on his second move, John loses singly. This is also the case when John plays 4 or 6 on his second move, but then the result can be more favorable for John if Peter makes a bad move. If 4 is John's second move, then 9 is Peter's best reply; it causes John to lose singly, and even doubly if he is not careful. If Peter replies 5 or 6 to John's move 4, the result is a draw; if Peter plays 8 he loses doubly, for after John's third move 9 we have the previously mentioned situation in which Peter may receive any of four threats none of which he can avert with a double threat (see §§50 and 65). If 6 is John's second move, and if Peter replies with 1, then John loses singly (and even doubly if he is not careful); however, if Peter gives the wrong reply 4, then John still wins, in fact doubly. So on either side there are ample opportunities to make errors, and the advantage can alternate.

68. John starts with the border square 2, Peter replies with the border square 8. The tree that corresponds to this case is:

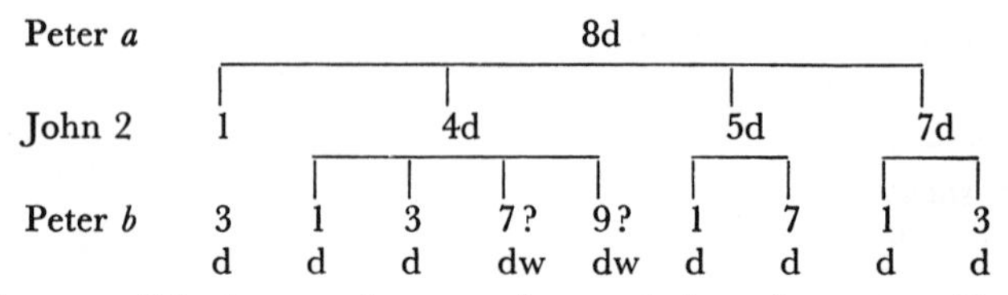

Whatever John's second move, the result is a draw, provided the defender makes no errors. Still, 4 is John's best move, because it gives a chance of a double win.

69. Results of John's first move 2. The following tree represents a survey of the results of John's first move 2 for the various replies by Peter:

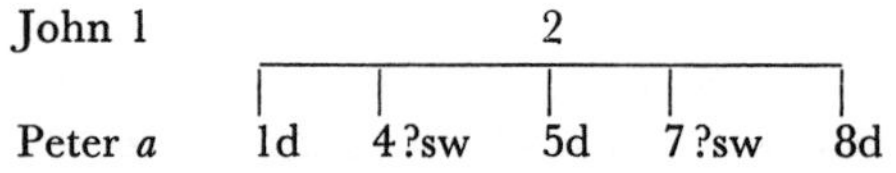

Peter has three correct replies: 1, 5, and 8 (leading to a draw). Of these, 1 can be considered as the best, and 8 as the worst move, because after Peter's first move 5 it is possible that John will make the wrong move 8, so that Peter wins singly (see the tree in §57), whereas after Peter's first move 1, three out of the seven replies that John can make are wrong.

VIII. MODIFICATION OF THE GAME OF NOUGHTS AND CROSSES

70. First modification of the game. In the game of noughts and crosses, a player will soon observe the advantage of occupying the central square 5, if he gets the chance, and a corner square otherwise. When both players see this, the game always results in a draw. In consequence, the fun will soon go out of the game.

The game can be made more attractive by the stipulation that neither player may occupy the central square 5 earlier than on his second move. This does not affect the analysis of the game except for the deletion of the cases in §§55, 56, 57, and 59 (the least interesting cases). From the trees in §§64 and 69 the middle branches (hence 5d) should be deleted. This causes the result of John's first move 1 to change into a single win, whereas that of John's first move 2 remains unchanged, namely a draw. So John can still achieve a single win, However, the chances become equal again if an even number of games is required, with the players taking turns to start.

The rules will make John aware of the advantage of the move 5, if he did not see this of his own accord. This may suggest to him that he should play 5 on his second move. The trees in §§60–63 and 65–68 show that this move is in no case better than all other moves, while the second move 5 will change a single win into a draw in cases 1–3 (John starts with 1, Peter replies with 3), 1–9, and 2–7, and will change a double win into a single win into the case 1–6. So when Peter replies with 3 or 9 to John's first move 1, and with 7 to 2, he has a good chance that the game will result in a draw through the wrong move 5 by John. So the game has become more difficult, with alternations of the chances of winning and losing, and thus has a greater appeal. It would be less attractive to permit the occupation of the central square only on Peter's second move, because then the possibilities for John to make errors would be reduced.

Since a border square is the opening move least favorable to John, the stipulation may also be made that John is required to open on a border square, whereas Peter is not allowed to reply with the central square, these being the only restrictions. This stipulation has the advantage that it leads to a draw in the case of good play. The complete analysis is now given by the four trees in §§65–68 only.

71. Second modification of the game. Another modification of the game, again to prevent an early occupation of the central square 5 (but a modification that is more natural and also more interesting), introduces a rule that a player is allowed to occupy the central square only when his opponent would make a row by occupying the central square. John, who begins, is not allowed a fifth move, and the game is over after Peter's fourth move.

The analysis of the game thus modified does not consist in deleting some of the trees (or branches of trees) corresponding to the original game, but has to be made anew. The construction of the diagrams is left to the reader. The trees derived from these, corresponding to John's first move 1 are:

```
Peter a                                               2sw
John 2              3sw         4?d                         6sw                 7      8?d   9?
Peter b       4    7     8?   7           3?dw             4?   7      8?   9   4    7    9 5
                                                                                     d    d
John 3        9    9     5    8       4?    7      8       9    9     5     4   9           8
                      dw                           dw
Peter c       5    5         6   9   5    4    8           5    5     9     8   5           7
              sw   sw        d   d   sw   dw   dw          dw   sw    dw    sw  sw          d
```

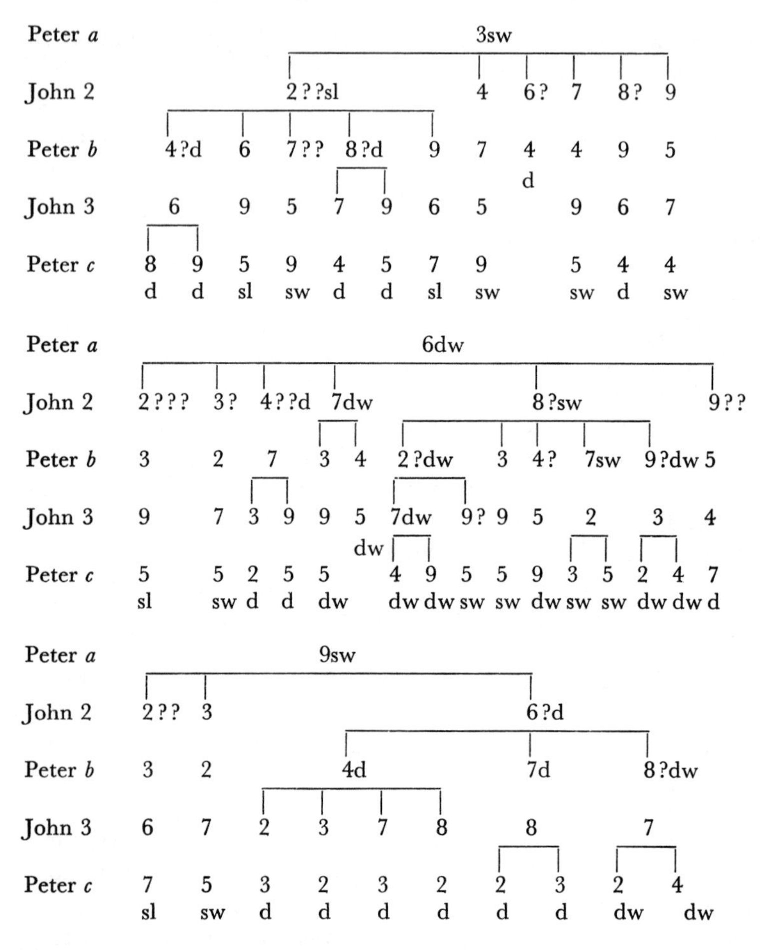

The trees for John's first move 2 are:

Peter *a* 1d

John 2 3?sl 4d 6?sl 7 8? 9d

Peter *b* 4 6? 7 8? 3 6 3? 4??dw 7sl 8? 8 5 4 7 8

d d d d d

John 3 7 4 4 9 7 7 3 4? 9 9 7 4

d dw

Peter *c* 5 9 8 8 3 8 9 9 5 3 7 6

sl sl d d dw dw dw sl dl sl sl d

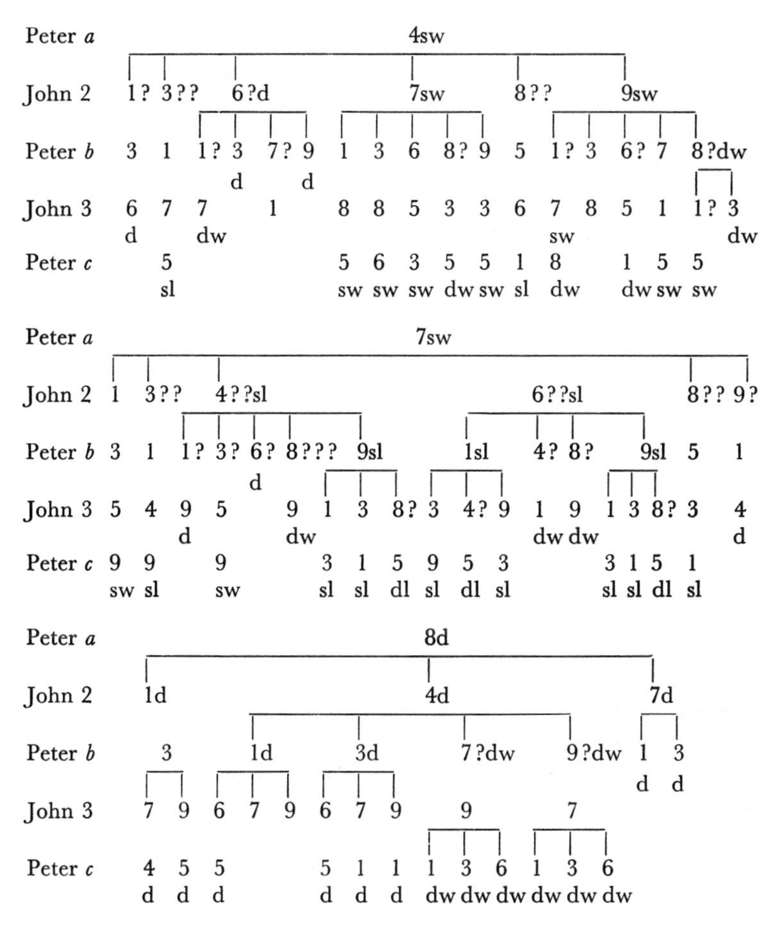

72. Conclusions from the trees of §71. From the trees of §71 we derive the following tree:

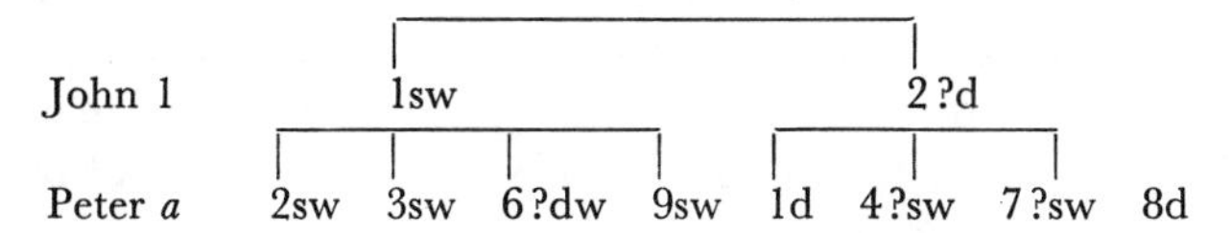

From this tree we determine the merits of the first moves of John and Peter. With correct play, John wins singly.

On the whole, the merits of the various moves have been changed only slightly by the modification defined in §71. Yet here and there we find important differences, which make the modified game more difficult and more interesting. For instance, if John has opened with 1 and Peter has replied with 2, then in the ordinary noughts and crosses John's second moves 3 and 4 lead to a draw and a single win for John, respectively, whereas in the modified game of §71 the converse is true. If John plays 3 on his second move, then, because of the modification, Peter is prevented from making the only correct reply, namely 5; on the other hand, if John plays 4 on his second move, Peter should reply by 7, which prevents John from making the correct move 5. So the restriction is advantageous sometimes to John, sometimes to Peter.

From the first tree of §71 we derive the following puzzle: How should John play in the position of *Figure 66* in order to obtain the

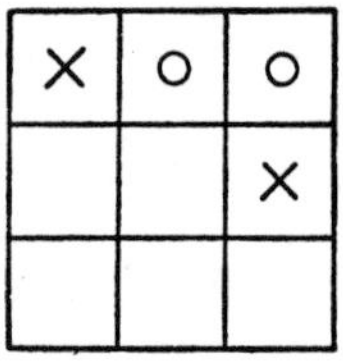

X	O	O
		X

Fig. 66

best result? John is inclined to make a double threat by playing 4; however, after Peter's answer 5 he obtains a single win only. On the other hand, the single threat 7, and also move 8, which does not make a semi-row at all, both lead to a double win. Without the restrictive condition of §71 the puzzle would also have had the obvious solution 5 (a double threat).

IX. PUZZLES DERIVED FROM THE GAME

***73. Possible double threats by John.** We now return to the game as described in §§40 and 41. We suppose that on his second move John makes a double threat (consisting of two semi-rows), and that Peter cannot reply by making a row. As a puzzle, we require all possible cases in which this can occur.

These cases arise from the six diagrams of Figure 49 if in each diagram we delete two crosses from different rows (hence not the heavy cross) and two noughts, in such a way, however, that no semi-row for Peter results. This gives the 36 possibilities indicated in *Figure 67.*

The capital letters refer to the diagrams of Figure 49, from which the cases have been derived, and correspond to the rows along which the double threat occurs. The dots (which represent free squares) indicate the moves by which Peter averts a double loss; hence, they

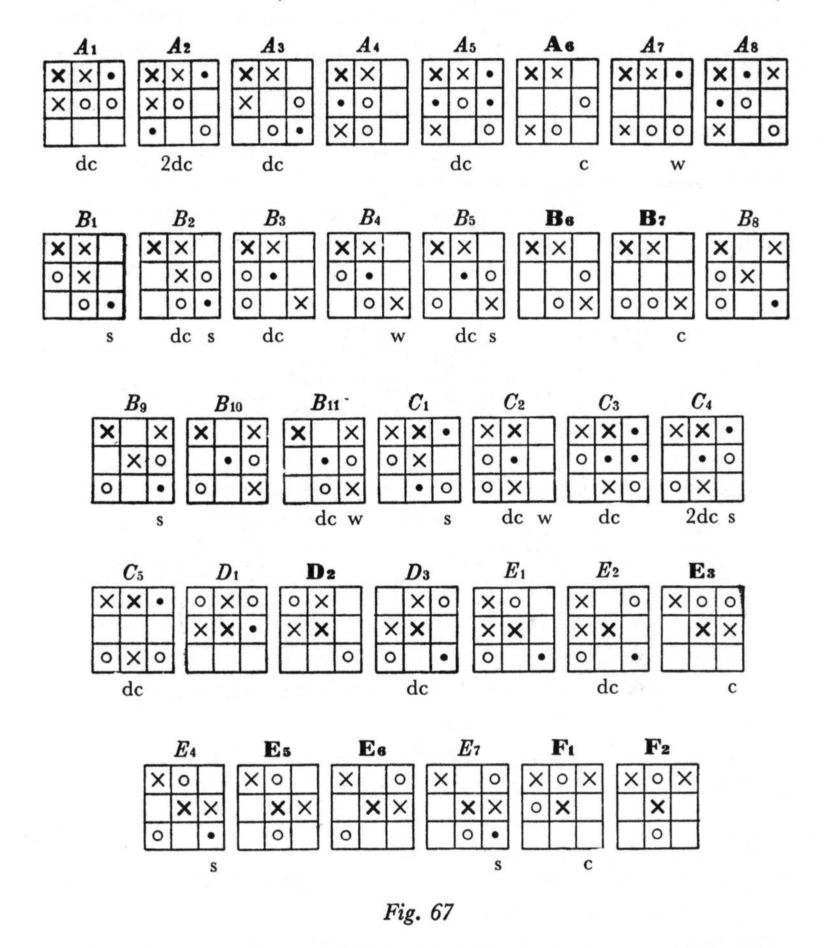

Fig. 67

also indicate how many of Peter's moves are successful. In a diagram in which no dot occurs Peter loses doubly; these cases have been further indicated by printing the capital letter and the number in **boldface**. If a double counterthreat is possible, this has been indicated by "dc" below the diagram. If a double counterthreat is possible in two ways, this has been indicated by "2dc." Hence, we also see immediately how many of Peter's moves that yield only a single counterthreat are successful.

If John's last move has been to draw the heavy cross (which will usually be the case), and if John could have made another double threat, then the letter "w" (wrong) has been placed below the diagram if that other double threat yields a better result for John, the letter "c" (correct) if the other double threat gives a worse result, and "s" (the same) if both double threats lead to the same result (which then turns out to be a single win for John in all cases).

***74. Possible double threats by Peter.** There are fourteen cases in which Peter's third move makes a double threat to which John cannot reply by making a row; these are shown in *Figure 68.*

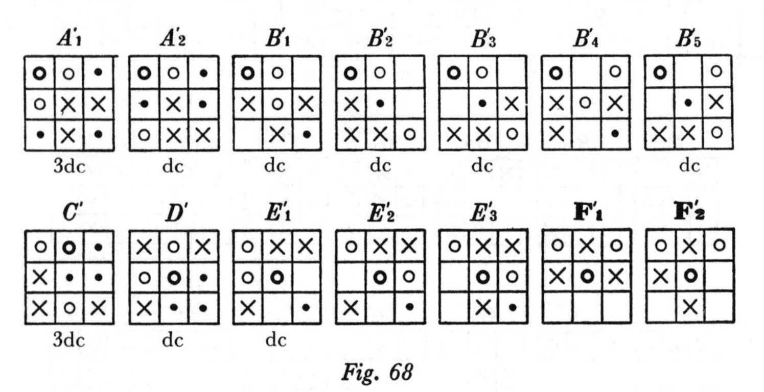

Fig. 68

The notation is the same as in §73. The dots indicate the moves by which John averts a double loss.

***75. Some more special puzzles.** The results in §§73 and 74 produce the following puzzles and their solutions:

CONDITION 1: John makes a double threat on his third move, and Peter cannot reply by making a row.

I. Peter replies in such a way that John can win doubly only by making a combined threat.

II. Peter can make a counterthreat, but, for every reply by Peter, John can still make a row on his fourth move and win doubly.

III. Peter cannot reply with a counterthreat.

IV. Peter can reply with a double counterthreat, in two essentially different ways.

In each case the possible positions are required. Condition 1 has to be satisfied for all 4 puzzles.

CONDITION 2: Peter makes a double threat on his third move, and John cannot reply by making a row.

V. John can reply with a counterthreat, but he still loses doubly.

VI. John cannot reply with a counterthreat.

VII. John can reply with a double counterthreat in 3 essentially different ways.

In each case the possible positions are required. Condition 2 has to hold for each of the puzzles V, VI, and VII.

The solutions are shown in *Figure 69.*

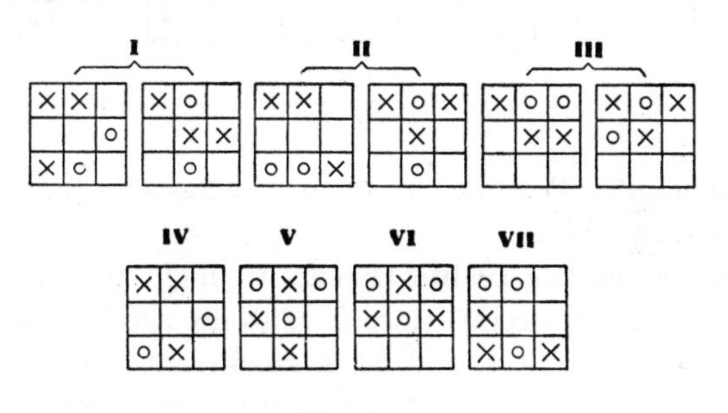

Fig. 69

***76. Possible cases of a treble threat.** It may occur, but only on the third move, that a treble threat is formed, consisting of three semi-rows with three different free squares (three equal marks forming a triangle). If the treble threat is made by Peter, then there are three cases, indicated in *Figure 70.* The free squares indicated by

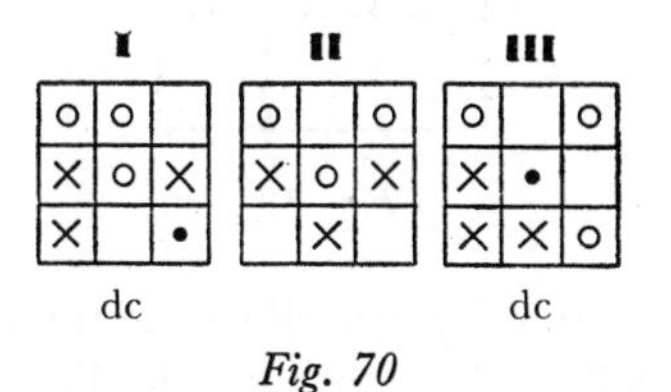

Fig. 70

dots represent the replies (double counterthreat) by which John averts a double loss. In case II John always loses doubly.

If the treble threat is made by John, we must interchange the noughts and the crosses and after that omit one nought. This gives the seven cases in *Figure 71.*

In cases I_3 and III_2, Peter averts a double loss by a double counterthreat (on the square with a dot). In the remaining cases Peter loses doubly, and should take care not to lose trebly; in these remaining cases the dots indicate Peter's moves that avert the treble loss. Peter has to be especially careful in cases I_1, II_1, and III_2, because he has only one move by which to avoid a treble loss.

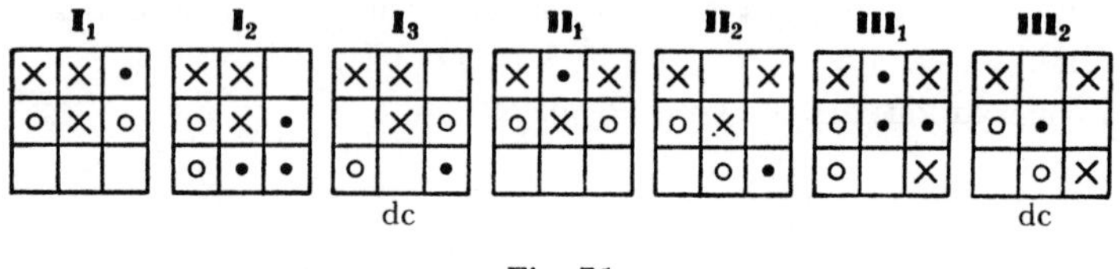

Fig. 71

This shows that Peter can avoid a treble loss after his second move whatever the position.

77. Remark on the treble threat. It might be thought that a treble threat never occurs, since it can only arise when Crook has failed to annul a threat, and when Hook continues by missing an opportunity to make a row. Nevertheless a treble threat is possible, in the case where a player realizes that he is lost and tries to avoid a double loss.

As an example we take the situation in *Figure 72* derived from the

X		
		O
X		

Fig. 72

tree in §62. Peter sees that by cancelling the threat by 4, he will lose doubly, since John will reply with 5 to this. Therefore Peter plays 3, hoping that John will make a row with 4. John, however, makes a treble threat with 9, so that Peter's attempt to avoid a double loss fails. If Peter, discouraged, now haphazardly replies 2 (a move that does not avert a threat and that does not make a threat, and hence is very bad indeed), then he indeed loses trebly, irrespective of the continuation of the game.

Chapter IV:
NUMBER SYSTEMS

I. COUNTING

78. Verbal counting. Anyone who has received any education at all is able to count, which implies pronouncing numbers ("one, two, three, . . . ,") in ascending order. This counting can be performed easily up to a million, although this is such a time-consuming task that in all probability no one has ever done it; before you finish this completely useless labor, it will perhaps have driven you insane, or perhaps you need to be insane even to start it. Just imagine how long this counting to one million would take if you kept on counting day and night. In calculating this, I assume that it takes five seconds to pronounce a number, which is certainly on the low side for numbers above 100,000. On this basis you would have to do nothing else but count for 58 days, without sleeping or eating, otherwise it would take longer still.[1]

79. Numbers in written form. Although everyone is familiar with our system of writing numbers, only relatively few people realize what an extraordinary discovery this simple and perspicuous system represents. This is often the case with simple things, but here it applies preeminently.

The essential feature of our system for writing numbers is the fact that any one digit can make different contributions to the number, depending on the position which this digit has in the sequence. For example, in 270377 the middle 7 (of the three) gives a contribution of seventy to the number. The first 7 (from the left) contributes seventy thousand (7×10^4) to the number, and the last 7 only seven. Therefore, we speak of the positional value of the digits, and the system adopted is called a positional system.

The positional values of the second, third, fourth, fifth digits, and

[1] [The five paragraphs following this one in the Dutch original have not been translated, since they are based on counting practices peculiar to Europe or to the Dutch language. The substance of the chapter is completely unaffected by this omission—TRANSLATOR.]

so on, counted from the right, are found by multiplying the digits by ten and the successive powers of ten (a hundred, a thousand, ten thousand, etc.), respectively. So the notation must make clear which position is occupied by any given digit. To ensure this the digit 0 is used to indicate the unfilled positions. If there were no digit 0, the above-mentioned number 270377 would be written as 27377; this represents a number, it is true, but a considerably smaller one. So an appropriate use of the digit 0 is an essential requirement of a positional system, so essential that we can certainly regard the introduction of the 0 as an important discovery.

With the positional system, it becomes very simple to perform arithmetic operations (addition, subtraction, multiplication, and division), provided you take care to use the same (vertical) column for writing digits corresponding to the same power of ten. This is why we speak of adding up a column of figures and the like. One might also use the term positional calculations.

80. Concept of a digital system. It will be clear to everybody that it was an arbitrary choice to think of the digits as being multiplied by the various powers of 10. Here we can replace the number 10 by any other whole number greater than 1. But we must be consistent and multiply the various digits by the powers of the same number. This number is called the base, while the method of writing numbers with the aid of such a base (hence the positional system) is called a digital system. If 10 is its base, we speak of the decimal system.

It is obvious that 10 was chosen for the base because the fingers are a preeminently suitable aid in counting (and also adding). When someone has used all fingers of both hands, then he has to start the counting anew, remembering that he has reached one "decade." Children who have just begun to learn to count are not the only ones who may be observed counting on their fingers; this offers undeniable advantages to adults, too; hence such expressions as "You can count them on the fingers of one hand."

Counting on the fingers might alternatively have led to the base 5; this happens not to be the case, probably because the base 5 is somewhat small. In counting on the fingers, the base 5 would have the advantage that we could count the units on one hand, indicating every multiple of 5 by a finger of the other hand. In this way, we could count up to 25 on both hands together. In the digital system with the base 5 we have only the digits 0, 1, 2, 3, 4. In this system, the numerals are written down according to altogether the same method, and with as much ease, as in the decimal system, namely in the following way:

1–2–3–4–10–11–12–13–14–20–21–22–23–24–30–31–32–33–34
40–41–42–43–44–100–101–102–103–104–110–111–112–113–
114–120–121–122–123–124–130–131–132, etc.

As you see, numbers having many digits arise earlier than in the decimal system, and this has to be considered as a drawback of the system with the base 5.

From the foregoing it appears that the base of the system is the smallest number that is written with two digits, that is, as 10. The second (third) power of the base is the smallest number that is written with 3 (4) digits, that is, as 100 (1000).

The writing of the numerals occurs according to the same method in any digital system. If the base is larger than 10, then we must devise new symbols, as for ten and for eleven in the duodecimal system. If we take *t* and *e* as these symbols, the initial letters of the corresponding words, then the numerals will be written as follows:

1–2–3–4–5–6–7–8–9–*t*–*e*–10–11–12–13–14–15–16–17–
–18–19–1*t*–1*e*–20–21–22–· · ·–29–2*t*–2*e*–30–31–· · ·–3*t*–
–3*e*–40–41–· · ·–4*e*–50–· · ·–5*e*–60–· · ·–6*e*–70–· · ·–7*e*–
–80–· · ·–8*e*, 90–· · ·–9*e*, *t*0–*t*1–· · ·–*t*9–*tt*–*te*–*e*0–*e*1–· · ·–
–*et*–*ee*–100–101–· · ·–10*t*–10*e*–110–111–· · ·–11*e*–120–121–
–· · ·–12*e*–130–· · ·–140–· · ·–150–· · ·–160–· · ·–170–· · ·
–180–· · ·–190–· · ·–1*t*0–· · ·–1*te*–1*e*0–1*e*1–· · ·–1*ee*–200–
–201–· · ·–20*e*–210–· · ·–21*e*–220–· · ·–230–· · ·–2*ee*–300, etc.

Of course, we forego the pronunciation of the numbers in the duodecimal system, because this is quite unnecessary. Or else we can consider the naming the digits as a way of pronunciation.

II. ARITHMETIC

81. Computing in a digital system. Addition and multiplication in the decimal system (of which subtraction and division are immediate results) are based on a knowledge by heart of the addition table and the multiplication table. So if you want to learn to compute in another digital system, for instance the duodecimal, you must start by constructing the addition and multiplication tables and then learn these tables so well that you can not only say them in order, but can also produce immediate random answers like $7 \times 9 = 53$. When you know both tables in this manner, all difficulty has disappeared, and the calculations are performed as quickly as they are now in the decimal system. If you had never learned to calculate in any but the duodecimal system, you would have the same troubles in calculating in the decimal system that we have now in the duodecimal.

The construction of the duodecimal addition table is based in a simple way on the method of counting in this system. Thus one finds:

$1+1=2$			$e+e=1t$	$t+t=18$	$9+9=16$	$8+8=14$
$2+1=3$	$2+2=4$			$e+t=19$	$t+9=17$	$9+8=15$
$3+1=4$	$3+2=5$	$3+3=6$			$e+9=18$	$t+8=16$
$4+1=5$	$4+2=6$	$4+3=7$	$4+4=8$			$e+8=17$
$5+1=6$	$5+2=7$	$5+3=8$	$5+4=9$	$5+5=t$		
$6+1=7$	$6+2=8$	$6+3=9$	$6+4=t$	$6+5=e$	$6+6=10$	
$7+1=8$	$7+2=9$	$7+3=t$	$7+4=e$	$7+5=10$	$7+6=11$	$7+7=12$
$8+1=9$	$8+2=t$	$8+3=e$	$8+4=10$	$8+5=11$	$8+6=12$	$8+7=13$
$9+1=t$	$9+2=e$	$9+3=10$	$9+4=11$	$9+5=12$	$9+6=13$	$9+7=14$
$t+1=e$	$t+2=10$	$t+3=11$	$t+4=12$	$t+5=13$	$t+6=14$	$t+7=15$
$e+1=10$	$e+2=11$	$e+3=12$	$e+4=13$	$e+5=14$	$e+6=15$	$e+7=16$

The multiplication table in the duodecimal system is derived from the addition table, and is as follows:

$2\times 2=4$				$9\times 9=69$		
$3\times 2=6$	$3\times 3=9$			$t\times 9=76$	$t\times t=84$	
$4\times 2=8$	$4\times 3=10$	$4\times 4=14$		$e\times 9=83$	$e\times t=92$	$e\times e=t1$
$5\times 2=t$	$5\times 3=13$	$5\times 4=18$	$5\times 5=21$			
$6\times 2=10$	$6\times 3=16$	$6\times 4=20$	$6\times 5=26$	$6\times 6=30$		
$7\times 2=12$	$7\times 3=19$	$7\times 4=24$	$7\times 5=2e$	$7\times 6=36$	$7\times 7=41$	
$8\times 2=14$	$8\times 3=20$	$8\times 4=28$	$8\times 5=34$	$8\times 6=40$	$8\times 7=48$	$8\times 8=54$
$9\times 2=16$	$9\times 3=23$	$9\times 4=30$	$9\times 5=39$	$9\times 6=46$	$9\times 7=53$	$9\times 8=60$
$t\times 2=18$	$t\times 3=26$	$t\times 4=34$	$t\times 5=42$	$t\times 6=50$	$t\times 7=5t$	$t\times 8=68$
$e\times 2=1t$	$e\times 3=29$	$e\times 4=38$	$e\times 5=47$	$e\times 6=56$	$e\times 7=65$	$e\times 8=74$

The table of 2 is found by starting off at $2 \times 2 = 4$, and then adding 2 to the result obtained every time, which is done with the aid of the addition table. From $2 \times 3 = 6$ you find 3×3 as $6 + 3 = 9$; to this result you have to keep adding 3 in order to obtain the table of 3. From $3 \times 4 = 10$ you find $4 \times 4 = 10 + 4 = 14$, from which the table of 4 then follows by adding 4 every time, using the addition table. From $4 \times 5 = 18$ you find $5 \times 5 = 18 + 5 = 21$ (because of $8 + 5 = 11$), and so on.

It is clear that you should not think of the multiplication tables in the duodecimal system as being constructed in terms of the decimal system, because in calculating in the duodecimal system you should entirely abandon the decimal system; that is, you should think duodecimally.

Whereas the tables in the duodecimal system are more extensive than in the decimal system, the tables in the system with the base five are much shorter, namely:

$1 + 1 = 2$			
$2 + 1 = 3$	$2 + 2 = 4$		
$3 + 1 = 4$	$3 + 2 = 10$	$3 + 3 = 11$	
$4 + 1 = 10$	$4 + 2 = 11$	$4 + 3 = 12$	$4 + 4 = 13$

$2 \times 2 = 4$		
$3 \times 2 = 11$	$3 \times 3 = 14$	
$4 \times 2 = 13$	$4 \times 3 = 22$	$4 \times 4 = 31$

***82. Changing to another number system.** In order to make a direct conversion of a written number from one number system into another number system you must be able to calculate in at least one of these number systems. If you can calculate in the original number system, then the conversion is effected by using the original system to represent the base of the new system, with repeated division of the given number (also written in the original system) by the new base. As an example we take the (decimal system) number of 955486. In order to convert this into the duodecimal system, we carry out the following divisions:

12⌋955486⌊79623				
84				
115	12⌋79623⌊6635			
108	72			
74	76	12⌋6635⌊552		
72	72	60		
28	42	63	12⌋552⌊46	
24	36	60	48	
46	63	35	72	12⌋46⌊**3**
36	60	24	72	36
10	**3**	**11**	**0**	**10**

From the first division we find that the given number is equal to $12 \times 79623 + 10$. From the second division it follows that $79623 = 12 \times 6635 + 3$, from which we find the original number in the form:

$$12^2 \times 6635 + 12 \times 3 + 10.$$

From the third division we have $6635 = 12 \times 552 + 11$, so that the original number becomes:

$$12^3 \times 552 + 12^2 \times 11 + 12 \times 3 + 10.$$

Continuing in this manner, we find:

$$12^5 \times 3 + 12^4 \times 10 + 12^2 \times 11 + 12 \times 3 + 10.$$

Hence, in the duodecimal system the number is written as $3t0e3t$.

Conversely, this duodecimal notation for the number can be converted into the decimal system by multiplying by 12 (in the decimal system), thus:

3	552	6635	79623
12 ×	0 +	12 ×	12 ×
36	552	13270	159246
10 +	12 ×	6635	79623
46	1104	79620	955476
12 ×	552	3 +	10 +
92	6624	79623	955486
46	11 +		
552	6635		

Both computations, the division and the multiplication, have been written down in some detail, but this has been done only to give a clear indication of the operations which have been performed. The divisions by 12 can be done mentally, and only the digits of the quotient and those of the remainder need be written down. The multiplications by 12, too, can be performed mentally, the multiplication and the subsequent simple addition being combined each time into one operation. The computations then take much less space, and become as follows (see the first two columns):

955486	10	3	3	9	9	$3t0e3t$	6
79623	3	46	10	$7e$	5	47364	8
6635	11	552	0	677	5	5642	4
552	0	6635	11	5642	4	677	5
46	10	79623	3	47364	8	$7e$	5
3	3	955486	10	$3t0e3t$	6	9	9

In the first column (to be read downwards) the divisions by 12 have been performed. Each time the quotient has been written below the dividend and the remainder has been written to the right of the dividend. The digits of the number in the duodecimal system are provided by the remainders, read from the bottom upwards (with 10 replaced by t and 11 by e). In the second column (to be read downwards again), the multiplications by 12 have been performed, and each number to the right of a product has been added to this product; the numbers to the right are decimal representations of the duodecimal numbers. The lowest number gives the original number converted to decimal notation. It will be seen that the second column read upwards is identical to the first column read downwards. Hence, the two computations only differ in that they are performed in opposite directions.

In the third column the decimally written number has been converted into the duodecimal system by multiplications by t (ten), while in the fourth column the opposite has been done by divisions by t. These computations have been performed in the duodecimal system. By repeatedly consulting the tables of §81, this can be done in a manner well known to everyone; however, things go a little less smoothly here, because we do not know the tables by heart and thus have to look up results every time, nevertheless we can easily convince ourselves that the procedure is feasible.

Using the tables of §81, we can directly convert the duodecimally written number $3t0e3t$ into the system with the base 7 (by divisions by 7 performed in the duodecimal system). We then obtain 11056450. If we want to perform the conversion by computing exclusively in the decimal system, then we first convert the duodecimally written number into the decimal system by multiplications (by 12), after which we use divisions (by 7) to convert the decimally written number into the system with the base 7.

III. REMARKS ON NUMBER SYSTEMS

83. The only conceivable base of a number system is 10. If you submit the following assertion to someone who is properly acquainted with the concept of a number system: "The only conceivable base of a digital system is 10," the chances are that he will say: "I don't see why you say that. I can take whatever base I want, can't I?" If you then ask him to mention a number that he would choose as a base, and he replies "seven," say, then you can ask: "But how would you write this number in the digital system that you have in mind?" and he cannot make any other reply than 10. In any digital system the base is written as 10.

Your original assertion, in order to be successful, must be submitted in writing. If you submit it verbally and say the word "ten," then the assertion is false; on the other hand, if you speak of the "system with the base one-zero," you reveal your meaning. If you submit the assertion in writing, you take advantage of the fact that the other person may read "ten" and that his further thoughts will be based on this impression.

For the rest, the assertion is, of course, practically meaningless, and the significance of it therefore lies exclusively in the paradoxical effect that it exercises at first sight. Yet it clearly shows that the base of a digital system cannot be described using the digital system itself. Indeed, this would be circular reasoning, since the digital system only exists when the base has been fixed; one cannot give the base in a digital system not yet existing.

84. Comparison of the various digital systems. The digital system with the smallest base, in certain respects the simplest digital system, is the binary system (the system with the base 2); here there are no other digits than 0 and 1. The tables in this system are restricted to $1 + 1 = 10$, and this does not require much learning. The numerals are written down as follows: 1–10–11–100–101–110–111–1000–1001–1010, etc. Evidently the binary system has the great disadvantage that even small numbers are represented by long sequences of digits, while in addition the sequences present such a monotonous picture that it is easy to make mistakes. The decimal system number 973 becomes 1111001101 in the binary notation; the decimal system number 98927645 reads

$$101111001011000010000011101.$$

The ternary system (with base 3) has the digits 0, 1, and 2. Numbers are written: 1–2–10–11–12–20–21–22–100–101, etc. The tables are restricted to:

$$1 + 1 = 2,\ 2 + 1 = 10,\ 2 + 2 = 11,\ 2 \times 2 = 11.$$

In the system with base 4 (digits 0, 1, 2, 3) the tables are:

$$1 + 1 = 2,\ 2 + 1 = 3,\ 3 + 1 = 10,\ 2 + 2 = 10,$$
$$3 + 2 = 11,\ 3 + 3 = 12;$$
$$2 \times 2 = 10,\ 3 \times 2 = 12,\ 3 \times 3 = 21.$$

In these number systems with small bases, calculation requires little practice, but takes much time because of the many digits.

On the other hand, large numbers are represented by numbers of few digits in a number system with a large base, such as 60. Thus the decimal system number 98527643 becomes 7(36)8(47)(23) in the system which has the base 60, where (36) means the digit that has a value equal to the number 36 of the decimal system, and similarly for

(47) and (23). Calculations in the system with the base 60 proceed quickly even with rather large numbers, provided you first know the tables for this number system.

This, however, is no easy achievement. First of all, you must devise and remember 60 single symbols for the various digits, while the addition table contains no less than 1770 sums and the multiplication table 1711 products (compared with 45 sums and 36 products in the decimal system). It takes a very good memory to remember tables like these.

Hence, when choosing a base, you have to strike the happy mean, and then you will doubtless end up in the neighborhood of ten. So we cannot say that the base ten has been a bad choice, especially when we envisage counting on the fingers. However, in many respects the base twelve would have been a better choice. The tables would have been slightly more extensive, with 66 sums and 55 products. In primary school it would take a little more time for children to master the tables, but they could easily pass this stage, and then computations could be done slightly more quickly. In this respect, there is little difference between the decimal system and the duodecimal system. However, the great advantage of the duodecimal system lies in the fact that 12 has more divisors than 10. It contains (apart from 2) the small divisors 3 and 4 and, besides, the divisor 6, against which 10 has only the divisor 5. Just as divisions by 2 and by 5 proceed with ease in the decimal system, the like is true of divisions by 2, 3, 4, and 6 in the duodecimal system. Moreover, since divisions by small numbers like 3 and 4 occur much more frequently than divisions by larger numbers, calculations in the duodecimal system would in many cases be much easier than they are now in the decimal system.

However, the duodecimal system has still other and considerably greater advantages. The circumference of the circle can be divided into twelve equal parts in a much simpler way than into ten equal parts; a regular hexagon is a much more common and simple figure than a regular pentagon. Similar arguments caused the angle of a square, the so-called right angle, to be divided into 90 rather than 100 degrees; otherwise the frequently occurring triangle with three equal sides and three equal angles (the so-called equilateral triangle) would have an angle of $66\frac{2}{3}$ degrees, instead of the present 60 degrees, which is a rounder number. So 10-part or 100-part divisions have been rejected for angle measurement, and indeed for good reasons. It is true that recently some land surveyors have changed over, to divide the right angle into 100 equal parts, but this has the above-mentioned drawback for equilateral triangles.

All objections disappear when the duodecimal system is used. If we then divide the right angle into 100 degrees (where, of course, 100 is to be understood as a duodecimal number), then the equilateral triangle has angles of 80 degrees.

These arguments show that, if we had a fresh choice, the duodecimal system would quite certainly be selected. However, it is not likely that we will soon adopt so radical a change. Just imagine if everyone, including the less educated, had to abandon their primary school arithmetic and make a fresh start! For the next generation—meaning children who have not yet started primary school—this difficulty would not arise; but for grown-ups it would be a disaster. All the same, if the reform could be carried through, posterity could not fail to be grateful. However, we have a greater debt of gratitude to those who have taught us how to write numbers in a positional system, since the introduction of a system of this type must be considered to make a greater advance than occurs in the change-over to a more suitable base.

85. Arithmetical prodigies. In §84 we remarked that computations are more rapid in a number system with a larger base, once you know the tables for that base. This advantage can also be obtained, however, without leaving the decimal system, by the process of considering the system to have 100 for a base. This is equivalent to adding and multiplying with two digits simultaneously instead of one. However, this requires the learning of the sums and products of all pairs of numbers less than 100, something which certainly not everybody will manage. If you lack a natural talent for this kind of thing, you had better not begin it.

Someone who has an extraordinary memory for numbers (which is in no way connected with mathematical talents) can stagger us in this way with the long multiplications, divisions, etc., that he manages to do mentally in a very short time. Arithmetic prodigies, as they appear now and then in variety theaters, will no doubt make use of the expedient outlined, although they will certainly keep much more in their head so they can find starting points quickly for the various questions they are asked.

IV. MORE ABOUT DIGITAL SYSTEMS

86. Origin of our digital system. Our digital system originated in India in the first centuries of the Christian era. Whereas in Greece geometry rose to unprecedented heights, in India it was arithmetic which flourished. Although the names, and in some cases the works (or fragments of works), of various Indian mathematicians and astronomers of the fifth and sixth centuries have been preserved, it is not known to whom the discovery of our notation for numbers, using digits, is to be ascribed. Even the century in which this notation arose cannot be stated with certainty; presumably it dates from the third or fourth century.

As has been noted before, the essence of the system consists in the positional values of the digits and in using a symbol for zero to indicate the blanks. This use of the zero, which can be considered as the crowning point of the system, is more recent than the positional value of a digit and occurs around the year 400, perhaps under Babylonian influence.

The Indian system was adopted by the Arabs, who translated the Indian writings, and in this way it came to Europe. Its introduction into Europe was first advocated zealously by Gerbert, a meritorious mathematician who later, as Pope, took the name Sylvester II (999–1003).

Above all, however, it was the great Leonardo of Pisa, better known under the nickname of Fibonacci (son of the kind-hearted one), who about 1200 gradually made decimal computing common property by his spectacular works, which gave him access to the palace of Emperor Frederick II. However, it was only in the middle of the sixteenth century that the system became really current in everyday life.

87. Forerunners of a digital system. One should not equate the writing of a number in a digital system with the introduction of names or symbols for groups of units, not even when powers of the same number are used more or less consistently in forming these groups. Thus the introduction of the term dozen for 12 and gross for $12 \times 12 = 144$ is not quite the same as creating a duodecimal system.

The formation of groups of units will naturally be found among all peoples of some degree of civilization. Civilization is inconceivable without the knowledge of rather large numbers, and in order to indicate such numbers the formation of groups—hence the combination of a certain number of units into successively larger units—is a prerequisite, because otherwise one would have to introduce as many separate words as there are numbers that play a role. Often such a formation of groups points in the direction of the decimal system, but sometimes again in the direction of the base 4, 20, or 60, whereas a mixture of various digital systems may also be at the root of it.

Among the ancient Romans, the groups were formed mainly according to the decimal system used at present. However, the very fact that signs for 10 and the various powers of 10 were used, such as X for 10, C for 100, M for 1000, is contrary to the essence of a digital system, which rests on the positional values of the digits. In other respects, too, the method of the Romans is far removed from a digital system.

The ancient Greeks also made use of groups in naming the numbers, and although powers of 10 did play a role, the system followed did not offer the advantages of our present decimal system. It is doubtless due to this inadequate system that arithmetic developed so much less than geometry, in spite of the preeminently scientific and specifically mathematical talent of the Greeks, as exemplified by Pythagoras of Samos (c. 580–500 B.C.), Plato (429–348 B.C.), Aristotle (384–322 B.C.), Euclid (c. 300 B.C.), Archimedes of Syracuse (287–212 B.C.) and Apollonius of Perga (d. 170 B.C.).

Perhaps the arithmetic of the Sumerians, the inhabitants of Babylonia 5000 years ago, came closest to using a positional system with a base. The Sumerian system was mainly duodecimal, although there is an admixture of a decimal system which makes one think of a system with the base 60. The Babylonian system was applied to subdivisions of the circle circumference, which in their turn are closely connected with subdivisions of time. These subdivisions are still in use today, and have not been superseded by any 10-part or 100-part division.

We further observe that something like a binary system seems to have been in existence in ancient China in the twenty-ninth century before Christ. It is not known, however, to what extent this can be considered as a complete or consistent number system.

88. Grouping objects according to a number system. Suppose we take a pile of matches, and we want to determine the number of these matches in the number system with the base 3, say, without first counting the matches in the decimal system and afterwards converting this number to the ternary system. To do this we first form some piles of 3 from the given pile. As soon as we have formed three such piles, we join these to make one pile of 9. Now we start anew, forming piles of 3 from the remaining matches, and joining these to form a pile of 9, as soon as we have three of them. If, continuing in this way, we can form another pile of 9, then we join the three piles of 9 to form one pile of 27. Then we continue to form piles of 3, which again, as soon as there are three of them, are joined to form one pile of 9. Suppose that we can form two more such piles of 9, and that then one match is left. Our matches then give us a total of one pile of 27, two piles of 9, no piles of 3, and one pile of 1, as shown in *Figure 73.*

Fig. 73

In the ternary system the number of matches is thus 1201. To know this, it is not necessary to realize that this amounts to 46 matches (in decimal notation). This would be equivalent to grouping the matches as shown in *Figure 74* (where, of course, the pile of six should be considered as six piles of 1).

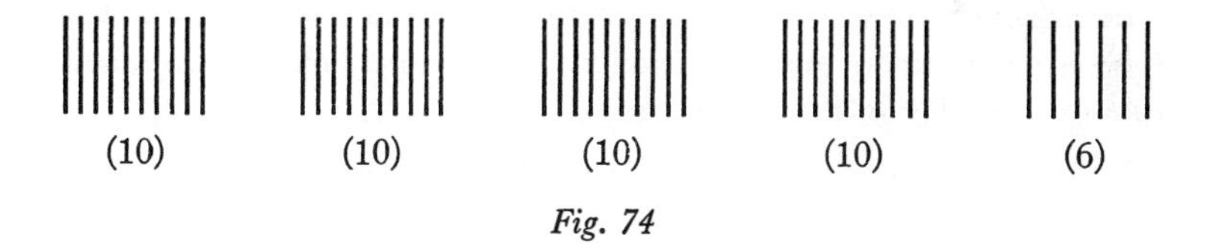

Fig. 74

Grouping the piles according to the binary system is a still simpler procedure. We form a pile of 2 and another pile of 2, and combine both piles into a pile of 4. In this way we form another pile of 4 from the remaining matches and join it to the first pile of 4, obtaining a pile of 8. From the remaining matches we form another pile of 8, if possible and join it to the first pile of 8, obtaining a pile of 16. If we can form another pile of 16, then we combine both piles, obtaining a pile of 32. Suppose now that we can do this and then form a pile of 8, a pile of 4, and a pile of 2, and that then the pile is exhausted. This has been indicated in *Figure 75.*

The number of matches in the binary system is then 101110. This follows from the way they have been grouped; again it is not necessary to know that there are 46 matches (in decimal notation).

After some practice, this method of grouping the matches and thus reading off the number of matches in some digital system is almost as quick as counting the number of matches in the decimal system. At any rate, it is much quicker than first counting the matches in the usual way (that is, in the decimal system) and then converting the number into the desired digital system. With the method outlined

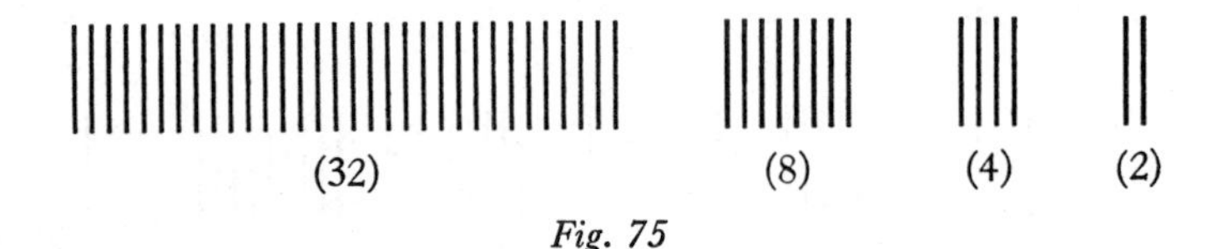

Fig. 75

above, you need only shift matches, and it is not necessary to use paper and pencil.

The foregoing can be successfully applied in various games in which the correct way of playing is based on the writing of a number in a certain digital system; there are cases that cannot, or at any rate can scarcely, be treated without the concept of a digital system. In some puzzles, too, powers of 2 or 3 occur; this is of course closely related to writing a number in a digital system; in other cases there is a looser connection with a digital system, so that, in a way, it gives the impression of having been dragged in. In the next 2 chapters we shall discuss examples of the cases mentioned.

Chapter V: SOME PUZZLES RELATED TO NUMBER SYSTEMS

I. WEIGHT PUZZLES

89. Bachet's weights puzzle. To weigh an object with a pair of scales or a balance, we need a set of weights. For weighings to an accuracy of one gram, the set usually consists of the following weights (in grams): 1–1–2–5–10–10–20–50–100–100–200–500–1000; hence, two weighıs of 1 g, one weight of 2 g, one weight of 5 g, two weights of 10 g, etc. We then speak of a standard set of weights. This provides weights with which we can make all combinations from 1 g up to and including 1999 g.

This leads us to pose the problem: For a given number of weights, determine the set of weights which enable us to weigh the largest possible number of objects whose weights increase by single grams (starting with 1 g). We call this the weights puzzle of Bachet (1587–1638).

With two weights we can weigh up to and including 3 g if we use weights of 1 g and 2 g. To weigh an object of 4 g, we need another weight. The best we can do is to take a weight of 4 g; we can then weigh up to and including $4 + 3 = 7$ g. To continue, we add another weight, one of $7 + 1 = 8$ g, because this gives a maximum extension, up to and including $8 + 7 = 15$ g. Continuing in this way, we find that, for the case of thirteen weights (the same number of weights as in a standard set), we obtain the maximum range from a set composed as follows: 1–2–4–8–16–32–64–128–256–512–1024–2048–4096. We can then weigh all objects by steps of 1 g up to and including 8191 g.

This shows we should make the set contain weights for which the numbers of grams are 1, 2, and the successive powers of 2; we then speak of a binary set of weights. With such a set we can weigh up to (but not including) the first weight that does not occur in the set. If the set contains no weights heavier than 1024 g, we can weigh up to and including 2047 g. So with two weights fewer than in a standard set, we can get further by 48 g.

With a binary set of weights we can make up each weight (provided it is not too high) in just one way. If the weights of the set have been placed in ascending order, and we make a weighing using the weights from the positions 1, 2, 4, 5, 6, 8, 9, 11, 12, and 13, then the weight of the object, written in the binary system, is 1110110111011 g, which is 7611 g in decimal notation. So if we did our computations in the binary system (like the ancient Chinese), the use of a binary set of weights would be particularly convenient.

The standard set of weights has been adapted to the decimal system, and apart from this it has been constructed in such a way as to provide a large number of consecutive weights from a small number of weights. In the absence of a requirement for maximum range, there are 976 weights (out of 1999) than can be made up in more than one way. These weights are those for numbers of grams which require one or more of the digits 2 and 7 when expressed in the scale of ten, as in the cases 7, 12, 20, 22, 27, 29, 102, 122, 200, 222, 227, 228, for instance. For in these weighings we can then use a weight of 2 g (which can be replaced by two weights of 1 g each), or a weight of 20 g (which can be replaced by two weights of 10 g each), or the like. If the number of grams has only a single digit 2 or 7, then the object can be weighed in two ways. If each of two digits is either a 2 or a 7, then it can be done in $2^2 = 4$ ways, for example: $27 = 20 + 5 + 2 = 20 + 5 + 1 + 1 = 10 + 10 + 5 + 2 = 10 + 10 + 5 + 1 + 1$.

If the weight of the object is a number of grams in which each of three digits is either a 2 or a 7, then the weight can be made up from the set of weights in $2^3 = 8$ ways, for example:

$$\begin{aligned}272 &= 200 + 50 + 20 + 2 = 200 + 50 + 20 + 1 + 1 = 200 \\ &\quad + 50 + 10 + 10 + 2 = 200 + 50 + 10 + 10 + 1 + 1 \\ &= 100 + 100 + 50 + 20 + 2 = 100 + 100 + 50 + 20 + 1 + 1 \\ &= 100 + 100 + 50 + 10 + 10 + 2 = 100 + 100 + 50 + 10 \\ &\quad + 10 + 1 + 1.\end{aligned}$$

90. Weights puzzles with weights on both pans. In §89 it was tacitly assumed that the weighing took place with the object situated on one pan, and the weights on the other pan. More weighings can be made with a given number of weights if we also allow weights on the pan which contains the object to be weighed. With this new assumption we again ask how we should construct the set of weights so that the maximum number of objects (weighing 1 g, 2 g, 3 g, and so on) can be weighed with the help of a given number of weights.

With two weights, one of 1 g and one of 3 g, we can weigh objects of 1, 2, 3, and 4 g; with an object weighing 2 g, we then put a weight of 1 g along with the object on one pan, and a weight of 3 g on the other pan. If we want to be able to weigh objects of 5 g, too, then we must include another weight in the set. The heaviest weight that permits this weighing is one of $4 + 5 = 9$ g; the weighing of the object of 5 g then involves putting the weights of 1 g and of 3 g along with the object, and putting the weight of 9 g on the other pan. By choosing these three weights, of 1 g, 3 g, and 9 g, we make it possible to weigh objects up to and including 13 g. This is evident from the relations: $6 = 9 - 3$, $7 = 9 + 1 - 3$, $8 = 9 - 1$, $10 = 9 + 1$, $11 = 9 + 3 - 1$, $12 = 9 + 3$, $13 = 9 + 3 + 1$. For example, from $7 = 9 + 1 - 3$ we see that an object of 7 g can be weighed by putting a weight of 3 g along with the object and having a weight of 9 g and a weight of 1 g on the other pan.

To be able to weigh an object of 14 g, we must make the next weight in the set no heavier than $13 + 14 = 27$ g. If in fact we add a weight of 27 g, we can weigh up to and including 40 g, as is evident from the relations:

$$14 = 27 - 9 - 3 - 1$$
$$15 = 27 - 9 - 3$$
$$16 = 27 + 1 - 9 - 3$$
$$17 = 27 - 9 - 1$$
$$18 = 27 - 9$$
$$19 = 27 + 1 - 9$$
$$20 = 27 + 3 - 9 - 1$$
$$21 = 27 + 3 - 9$$
$$22 = 27 + 3 + 1 - 9$$
$$23 = 27 - 3 - 1$$
$$24 = 27 - 3$$
$$25 = 27 + 1 - 3$$
$$26 = 27 - 1$$
$$27 = 27$$
$$28 = 27 + 1$$
$$29 = 27 + 3 - 1$$
$$30 = 27 + 3$$
$$31 = 27 + 3 + 1$$
$$32 = 27 + 9 - 3 - 1$$
$$33 = 27 + 9 - 3$$
$$34 = 27 + 9 + 1 - 3$$
$$35 = 27 + 9 - 1$$
$$36 = 27 + 9$$
$$37 = 27 + 9 + 1$$
$$38 = 27 + 9 + 3 - 1$$
$$39 = 27 + 9 + 3$$
$$40 = 27 + 9 + 3 + 1$$

If the set also contains a weight of $40 + 41 = 81$ g, then we can weigh up to and including $81 + 27 + 9 + 3 + 1 = 81 + 40 = 121$ g. If additionally we have a weight of $121 + 122 = 243$ g, then we can weigh up to and including $243 + 121 = 364$ g, and so on.

From this it is evident that for weighings with weights on both scales, we obtain the maximum range by constructing the set of weights as follows: 1–3–9–27–81–243–729–2187.

As the numbers of grams for these weights are 1, 3, and the successive powers of 3, we speak of a ternary set of weights. If (as in the last case) it contains eight weights, so that the largest weight is 3^7 g, then we can weigh numbers of grams up to and including $\frac{1}{2}(3^8 - 1) = 3280$. With a ternary set of weights that contains thirteen weights we can weigh numbers of grams up to and including $\frac{1}{2}(3^{13}-1) =$ 797,161. Each weighing can be performed in only one way.

In the light of this, the following problem will no longer offer any difficulty: A shopkeeper has a bar of lead which weighs 40 kg. He wants to cut the bar into four pieces, in order to use these pieces as weights in weighing. How should he do this so that he can weigh all objects from 1 kg up to and including 40 kg?

91. Relation to the ternary system. If we have made a weighing with a ternary set of weights, and have the object in one pan together with (for example) a weight of 243 g $= 3^5$ g, a weight of 9 g $= 3^2$ g, and a weight of 1 g, while in the other pan we have weights of 2187 g $= 3^7$ g, of 729 g $= 3^6$ g (1000000 g in the ternary system), and of 27 g $= 3^3$ g (1000 g in the ternary system), then the weight in grams of the object, written in the ternary system is: 11001000 − 100101 = 10200122 (equivalent to 2690 g in the decimal system). For this type of weighing, it would be convenient if we did our computations in the ternary system.

Consideration of the ternary system also shows clearly that a ternary set of weights will allow us to weigh any object (provided it is not too heavy) if we can place weights in both pans, and that this can in fact be done in only one way. To determine this, we use the ternary system to write the weight of the object (expressed in grams). If the number thus obtained contains no digits other than 0 and 1, then we should not put any weights along with the object; digits 1 in the number then indicate which weights have to be put in the other pan. If the number contains one or more digits 2, then we eliminate these by adding a number (also in ternary notation) in which no digits other than 0 and 1 occur. The latter number is to be chosen in such a way that the sum is another number which contains no digits other than 0 and 1, such that the two numbers with digits 0 and 1 nowhere have a digit 1 in the same place; this can always be done, and in one way only as the following examples clearly show:

$$\begin{array}{r} 10200122 \\ 100101 \\ \hline 11001000 \end{array} \qquad \begin{array}{r} 20012021220221 \\ 10011001010010 \\ \hline 100100100001001 \end{array}$$

The sum indicates which weights have to be placed in the pan that does not contain the object. The number that is added to the weight of the object indicates which weights have to be placed along with the object.

II. EXAMPLE OF A BINARY PUZZLE

92. Disks puzzle. A horizontal board is a support for three vertical pegs which we shall call *a*, *b*, and *c*. Peg *a* passes through holes in ten disks of increasing size, the largest disk being at the bottom, on top of this a smaller disk, on top of that one a still smaller disk, and so on. We may take disks off a peg, always one at a time, and put each disk on another peg, but never in such a way that a disk is put on top of a smaller disk. The problem is to transfer all the disks from peg *a* to peg *b*. We ask: How should this be done, and furthermore, what is the least number of times we must make a transfer to a disk from one peg to another in order to do it?

Figure 76, which represents the starting position, is a puzzle in its

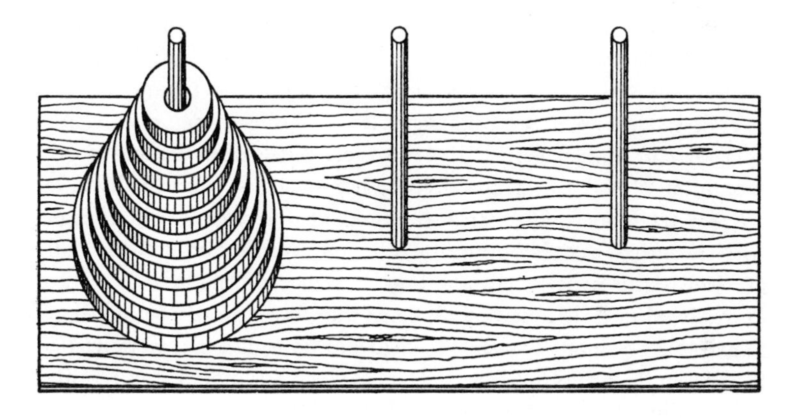

Fig. 76

own right: How is the figure to be viewed in order to have a clear picture of what it is intended to convey? A direct view shows a set of circles with vertical displacements. The secret of the optical illusion is that the disks have been drawn by projecting them onto the base at an angle of 45°, so that they remain circles in the figure, instead of becoming elongated in a horizontal direction. If we now look at the figure from this same angle of 45°, the result is the clear picture which we want. To obtain it, stand the book upright and look at the picture from above.

We number the disks 1, 2, . . . , 10, downwards, so that 1 is the smallest disk and 10 the largest. We write 5*b* to mean that disk 5 is transferred to peg *b*, and similarly in other cases. Disks 1 and 2 are first transferred to peg *b* in the following way: 1*c*–2*b*–1*b*. Then disk 3 can be moved, and we continue by 3*c*. Disks 1 and 2 are next transferred to peg *c* in exactly the same way, so that they end up on top of disk 3. This therefore is done as follows: 1*a*–2*c*–1*c*. The number of times a disk has been transferred is now equal to $7 = 2^3 - 1$. The result is to transfer disks 1, 2, and 3 to peg *c*. This is the only way to release the disk 4, apart from the possibility that the disks could instead have been transferred to peg *b* (unprofitably, as will appear later). We move disk 4 to peg *b*, and then proceed to transfer disks 1, 2, and 3 to peg *b* in the same way as above, which puts them on top of disk 4. This requires the transfers: 4*b*–1*b*–2*a*–1*a*–3*b*–1*c*–2*b*–1*b*.

So far there have been $15 = 2^4 - 1$ transfers in all, and disks 1, 2, 3, and 4 have been transferred to peg *b*. Next, disk 5 is transferred to peg *c* (in the same way as indicated above); this requires another fifteen transfers. In all, then, $2^4 + 2^4 - 1 = 2^5 - 1$ transfers have been made. We continue in the same way. After $2^6 - 1$ transfers, disks 1, 2, 3, 4, 5, and 6 are around peg *b*; $2^7 - 1$ transfers put disks 1, 2, 3, 4, 5, 6, 7 around peg *c*; $2^8 - 1$ transfers put 8 disks around peg *b*; then after $2^9 - 1$ transfers 9 disks have moved to *c*, and finally, after $2^{10} - 1$ transfers (that is, 1023 transfers) all 10 disks are around peg *b*, with the largest at the bottom, and other requirements met. Admittedly the job requires perseverance, but it cannot be made shorter. For example, after disks 1, 2, 3, 4, 5, 6 are transferred to peg *b*, and disk 7 to *c*, it takes as many transfers to clear peg *b* for disk 8 as were needed to transfer 1, 2, 3, 4, 5, 6 to *b*. Hence every additional disk requires the number of transfers to be doubled and then augmented by 1.

The previous discussion shows that it is always an even-number pile which is transferred to peg *b*, and an odd-number pile to peg *c*. To achieve this we must begin by transferring disk 1 (a single disk) to peg *c*, because 1 is an odd number. Since our problem is to transfer all the disks to peg *b*, we decide that we should begin by transferring disk 1 to *c* because the total number of disks is even. If we had started out with nine disks or eleven disks, then we should have begun by transferring disk 1 to peg *b*.

The like has to be kept in mind continually, for every transfer of a pile of disks to another peg. If the pile is odd, we begin by transferring its top disk to the bar to which the whole pile has to be moved; with an even pile we begin by transferring the top disk not to this peg but to the other peg.

The total number of transfers can be written immediately in binary notation. This makes it a number with all digits equal to 1, and a number of digits equal to the number of disks.

93. Origin of the disks puzzle. The disks puzzle seems to be of a venerable age. There is a legend that three pillars were erected in an Oriental temple long ago, two of silver and one of gold. Around one of the silver pillars there were 100 perforated alabaster disks decreasing in dimensions from the bottom upwards. Every believer who visited the temple was allowed to transfer a disk from one pillar to another; however, a larger disk was never allowed to be placed on a smaller disk. If by this means all 100 disks could be transferred to the gold pillar, the end of the world would be at hand.

Assuming that the transfers were performed in the most efficient manner, this would require $2^{100} - 1$ transfers of disks from one pillar to another. Now 2^{10} amounts to 1024, which exceeds 1000. Consequently, 2^{100}, which is the tenth power of 2^{10}, exceeds the tenth power of 1000, which makes it larger than the fifth power of a million. So the number exceeds 10^{30}. Even if a disk were transferred every second, it would still take over 300×10^{18} centuries before the entire job was complete. In a century there are $100 \times 365.24 \times 24 \times 60 \times 60$ seconds, something under 3.2×10^{9} seconds.

III. ROBUSE AND RELATED BINARY PUZZLES

94. Robuse. In the interesting card game called robuse, which is played by two persons, we encounter a problem similar to the disk puzzle (but somewhat simpler); the game consists of getting rid of cards efficiently and requires close attention. As the game is not very well known, we shall give a description of how it is played.

Two players, John and Peter, each pick up a well-shuffled bridge deck (52 cards) in such a way that they can see only the backs (red in one deck, and blue in the other). Each player deals out a pile of twelve cards, face down, and puts the thirteenth card on top, face up. The one who starts is the player whose card is the higher according to the rules of bridge (3 above 2, diamonds above clubs, etc.). In the case of identical cards (which of course is rare) the fourteenth card decides. The cards on the table are called the "dirty" cards. At the beginning, each of the players has thirteen of these. Each player then deals four cards from the cards in his hand, putting them face up in a row on the table, where they provide fixed locations called "houses."

Between the two rows of houses some vacant space is left, which is called the "ace space"; this has to be large enough to allow room for eight cards in four rows of two. The player who starts, let us say John, is allowed to move a card from one of the eight houses onto a card in another house, provided the latter is the next higher card of the same suit; this gives John an empty house, or, briefly, a house. If there is an ace in one of the eight houses, John is allowed to move it to the ace space; this also provides John with a house. Furthermore, John is allowed to put one of the eight cards on top of the pile of Peter's dirty cards, provided that the top card of that pile is either the next higher or the next lower card of the same suit. Peter is then said to receive a "setback," because the winner of the game is the player who first gets rid of all his cards. John is allowed to turn a single card from the top of the ones he keeps in his hand, and with that card he may do the same things as with a card from one of the eight houses: he may move it to an empty house; he may move it onto a card in one of the houses, if the new card is a matching card, in the sense of providing the next lower card of the same suit; he may put it on Peter's dirty card if that is either the next higher or the next lower card of the same suit; or, if it is an ace, he may move it to the ace space. John may do the same things with the top card of his pile of dirty cards, after which he turns the card that has then become the topmost dirty card, and puts this face up on top of his pile of dirty cards. If the top card of Peter's dirty cards is an ace, then John can put a 2 of the same suit on it. If Peter's topmost dirty card is a 2, then John is allowed to put an ace of the same suit on it, instead of putting the ace in the ace space. However, once John has put the ace in the ace space, or given it to Peter, he cannot take it back again; on the other hand, an ace that remains in one of the houses can be taken up and moved, onto a 2 of the same suit which is in another house, for example. John may also put a 2 from a house, or from his hand (the top card), or from his dirty cards (again the top card), onto an ace of the same suit that is in the ace space; once he has done this, he may not take away that 2 to use it elsewhere; this transfer is allowed only into the ace space and not away from it (in fact, once a card has been put into the ace space it cannot be taken back).

John is allowed to continue in this way as long as he can find something he can do. For instance, if he has put the jack of diamonds onto the queen of diamonds in one of the houses, and if he also has the 10 of diamonds at his disposal, then he is allowed to put this onto the jack of diamonds in its turn, and so on, and the result can be a sequence continuing as far as the 2 of diamonds. If John then also turns up the ace of diamonds, he may put it in the ace space, put the 2 of diamonds on top of it, and then the others, to make a pile (which is called an ace pile) with the queen of diamonds on top; by this procedure John obtains an empty house, and he can therefore continue with the disposal of further cards. It should be noted that when a card is transferred onto a matching card, it is irrelevant whether or not both cards belong to the same deck of 52 cards. If a card is lying face up (for example, in a house), you should not be able to tell the deck to which it belongs. You can of course see this when the card is put down for the first time, but if you want to make use of this fact, you must remember it; you are not allowed to turn over a card which is face up, to discover from the back whether it is from your own deck, or from your opponent's deck. So if Peter has a 10 of hearts, say, on top of his dirty cards, and if there is a 9 of hearts in one of the houses, then John can put it on top of Peter's 10 of hearts, with no need to recall whether the 9 of hearts came from his own deck or from Peter's. If John also has a 10 of hearts available (this came from his own deck, of course), he may put this in its turn on top of Peter's dirty cards. On top of his dirty cards, Peter can in this way acquire a set of cards, all of the same suit, which are partly in descending order and partly in ascending order, for example: 7 6 5 6 7 8 9 10 9 8. This is called a "bump"; and we shall see that it can be very unpleasant for Peter.

If John can no longer dispose of any further cards, he must discard the top card from the pile in his hand onto the table, face up, next to his pile of dirty cards. This is called "snipping." John is allowed to snip at an earlier stage, but in general this will not be advantageous. However, it can easily occur that John will snip too early because he has overlooked something. In this event he is not allowed to take back the snipped card (once he has laid it down), and his turn is over. John is not here considered to have infringed the rules of the game, which he would be, if he put a card behind or on top of another card that did not match it. After John's snipping gives Peter his turn, Peter may dispose of cards in the same way as John could when it was his turn. Additionally, Peter is allowed to put the next higher or lower card of the same suit on top of John's snipped card, so that here, too, a bump can arise (if this procedure is repeated). When Peter in his turn can dispose of no further cards (or if he only thinks so because he overlooks something), then he too snips, and it is John's turn again. Now John has the additional choice of disposing of his topmost snip card (the one with which he last snipped, or which Peter may have put on this), provided it is a matching card for a card in a house, for an ace or a card on top of one, or for Peter's topmost dirty card or snip card, always according to the previous rules (same suit, descending in houses, ascending from aces, either ascending or descending for the dirty card or snipped card).

With repeated snipping, a second pile of cards, called the snip pile, is built up next to the pile of dirty cards. If a player has disposed of all the cards which he holds in his hand, he then takes up the pile of snip cards face down (by turning over the pile). He then can turn the top card of that pile, and may dispose of it if possible. Whenever a player has exhausted the cards in his hand and can no longer dispose of any other cards, he must take up the snip pile face down and turn the top card. As long as he can do something else, he may leave the snip pile on the table, in order to be able to dispose of its top card (if this should become possible). If a player has exhausted all the cards from his hand and also the whole of the snip pile, when he still has dirty cards remaining, the top one of which cannot be placed (for the pile of dirty cards is not to be turned over), this too finishes his turn. He then makes it known, by a knock on the table, that he cannot do anything (or rather that he does not see anything he can do), which has the effect of snipping; once he has knocked, his turn is over, even if he then sees that he could have put his topmost dirty card on the ace pile, on a card belonging to his opponent or onto a card in a house, or that he could have moved a card from a house onto the ace pile, onto a card in another house or onto a card belonging to his opponent. If a player has no cards remaining in his hand, and neither dirty cards nor snip cards additionally, he then has won the game.

If one of the players has got to the stage of picking up and turning over his snip pile, he then can know the exact order of these cards and use this fact to advantage when choosing cards to put into empty houses. However, it takes some doing to remember the identity and correct order of all the cards you snipped and all the cards your opponent put on top of them. You are not allowed to examine the order of the snip cards. So the snip cards should not be put down carelessly, but in such a way that the other cards are covered by the topmost card. The same applies to the cards in the pile of dirty cards. However, cards which lie on top of other cards in the houses should be placed in such a way that you can clearly see how many cards there are, one on top of another. If they have been stacked too closely, a player is allowed to check the composition of the sequence by moving the cards slightly apart.

It should be noted explicitly that a player is not allowed to use the next higher card of the same suit when he puts another card onto a card in a house, also that he is allowed to take away only the lowest ranking card from a sequence in a house, and this only if he can find a place for it elsewhere. However, when there are one or more empty houses, these restrictions can in effect be overcome by making repeated displacements of the cards; this is called "transposing." For example, if a house contains only the 7 of spades and Peter turns up (either from his dirty cards or from his hand) an 8 of spades with which he can do nothing, then this gives him a setback, because he has to snip and John gets an empty house since he can give the 7 of spades to Peter; however, if Peter has all this and an empty house as well, he can put the 8 of spades into this, and then put the 7 of spades onto the 8, so that the result is the same as if Peter had placed the 8 of spades under the 7; for Peter gets his empty house back again.

The annoying thing about a bump on the dirty cards is that you can get rid of the cards concerned only when you have a sufficient number of houses available (this is also true without a bump, if the cards lie in decreasing sequence downwards), except when you can give these cards to your opponent, or transfer them, wholly or partly, to an ace pile. If John starts to get rid of dirty cards before he has a sufficient number of houses, he will be forced to snip in a situation in which the topmost card of the dirty cards forms a setback; Peter can then give John all his dirty cards back, often with some additional cards, and thus he gets John's empty houses, possibly with one or more other houses also. If you cannot get rid of dirty cards, there will be cards below these (often including cards which you could use with advantage to secure empty houses, so-called matching cards) which will remain out of circulation for a shorter or a longer time, and which will hinder the disposal of your cards. A bump in the snip cards has a much less damaging effect, because cards are placed on top of the snip cards continually, and so there is little danger that the bump will increase; at a later stage, the pile of snip cards is turned over, with the result that snip cards do not remain out of circulation. If you turn over the snip cards and happen upon a bump, then this may at times become dangerous, but often the situation has meanwhile altered to such an extent that you can put the cards of the bump, in whole or in part, into the ace space.

It may happen that neither of the two players can get rid of his dirty cards. The game then results in a draw. However, this rarely occurs when there is correct play.

However simple it may seem to be, to dispose of cards according to these rules, there is much more to it than you might think. First, efficient transposition of cards (to obtain the largest possible number of open houses) requires some practice: but even this is not the greatest difficulty. You can increase your chances by attention to various other points of good play, including some affecting the choice of the card (or cards) to be disposed of when you have one or more empty houses, and this is precisely where the charm of the game is to be found.

It would take us too far afield to list all the considerations here which require attention in good play of robuse. We shall therefore restrict ourselves to giving the more important hints:

Take care to get rid of dirty cards as soon as possible, by giving priority to dirty cards for the occupation of houses, unless, of course, you see that this would lead to your receiving a setback.

Remember the cards you have received on your pile of dirty cards, and in their correct order. This is necessary if you are to be able to judge how many houses you need to get rid of these cards.

Do not put a card in a house too soon, that is, without first checking whether you need this house for transposing, possibly to obtain more houses or to get rid of other cards without reducing the number of houses. For you are not allowed to take back a card once you have got rid of it (however, you can move it to another house, or put it onto an ace pile, or give it to your opponent). Also, do not put a card on an ace pile too soon, because this card, too, cannot be taken back.

Remember the card (or preferably the first two or three cards) which lie below the card which you snipped last, so that you can judge whether it is appropriate to dispose of the snip card (usually this is not the case). It is only towards the end of the game that you need to know these cards, because then there is a greater chance of the speedy acquisition of a house.

Also, you need to know what you can and what you cannot achieve by transposition. We shall discuss this in the form of a few puzzles.

95. Transposition puzzles. In robuse you frequently encounter situations that present the following puzzles, which we shall call transposition puzzles.

I. In one house there is a sequence of cards which have a 9 as their card of highest rank. In another house there is a sequence of the same suit, with the 10 as the card of lowest rank, or just a single 10 (which comes to the same thing here). There are two empty houses. What is the maximum number of cards that the first-mentioned sequence can contain, if it is to be possible to transpose the whole of it onto the 10?

The answer is: $2^2 = 4$ cards (hence 9–8–7–6). First you put the 6 and the 7 into the empty houses, next you transfer the 6 onto the 7, and then you put the 8 into the house which you have just made empty. Then you can place the 9 onto the 10 and then the 8 onto the 9, after which (by first putting the 6 in an empty house) you can put the 7 onto the 8, and then the 6 onto the 7.

II. One house contains a sequence of cards which have a 9 as their card of highest rank. You turn up the 10 of the same suit, at a time when there are two empty houses. What is the maximum number of cards that the sequence can contain, if it is to be possible to transpose and put the 10 under the 9?

Answer: $2^2 - 1 = 3$ cards (hence 9–8–7). You should not start by putting the 10 into the empty house; then you would have only one house left, and your efforts would fail. First you put the 7 into one house, and the 8 into the other house, then you put the 7 onto the 8. This gives you an empty house once again, and it is only now that you should put the 10 into one, after which you can put the 9 onto the 10 and follow with the others.

I′. The same puzzle as I, with the difference that there are three empty houses.

The sequence can now consist of $2^3 = 8$ cards (hence 9–8–7–6–5–4–3–2); even if the 10, the jack, or the queen were the card of highest rank in the sequence (and then of course the card of lowest rank in the other house would be the jack, the queen, or the king, respectively), this would not let the first sequence become any longer. First you transpose 5–4–3–2 to one of the empty houses in the manner of puzzle I (first 3–2 to one open house, then 4 and 5 to the other two empty houses, next the 4 onto the 5, and so on). In the manner of puzzle I, you can then transpose 9–8–7–6 onto the 10, and then 5–4–3–2 onto the 6.

II′. The same puzzle as II, but with three empty houses.

The sequence can now consist of $2^3 - 1 = 7$ cards (hence

9–8–7–6–5–4–3). First you put 6–5–4–3 into one of the empty houses (as for puzzle I), after which you can transpose the 10 under the 9–8–7 in the manner indicated in puzzle II. After that you put 6–5–4–3 onto the 7 (as for puzzle I). So you should not put down the 10 until you have put 6–5–4–3 into one empty house and 8–7 into another. If you put the 10 into an empty house before this, your efforts will fail.

From the foregoing, it is easy to see what the answers to the puzzles would be if the sequences could be longer, that is, if you were to use a deck of cards in which each suit had more than thirteen cards. With four empty houses you could transpose a sequence of $2^4 = 16$ cards, running upwards to a 19, say, onto a 20 in another house. If you turn up the 20 when you have four empty houses, you can transpose a sequence of $2^4 - 1 = 15$ cards (upwards to a 19 as the highest) onto the 20. The 20 should not be disposed of until you have put 12–11–10–9–8–7–6–5 in one empty house, 16–15–14–13 in another, and 18–17 in a third; then you put the 20 into the fourth empty house.

We now return to straightforward robuse. Our puzzles will also serve to provide an answer to the following question:

How many houses are required if you are to remove a given bump from the top of the dirty cards?

As an example we take the case where John has acquired the worst possible bump on his pile of dirty cards, namely:

k–*q*–*k*–*q*–*j*–10–*j*–10–9–8–9–8–7–6–7–6–5–4–5–4–3–2–3–2,

with the king as the base, on the assumption, of course, that no ace of this suit is as yet upon the table. With six houses available, John can succeed in arranging the 24 cards in two sequences of twelve, and still leave himself with four empty houses. If John tries to get rid of his dirty cards when he has only five empty houses (even this is an unusual number to have), he will get stuck. He can manage to transpose a sequence 9–8–7–6–5–4–3–2 into each of two empty houses. He then has three empty houses remaining, and this is not enough to let him transpose a 10 under a 9. John can fill the three houses with *j*–10, *j*–10 again, and a queen. This gives him a king as his topmost dirty card when he has to snip, and Peter can then return all the dirty cards which John has just tried to put away. If John has some other cards as well as his dirty cards, he can fill two of the five houses each by 9–8–7–6–5–4–3–2 from the dirty cards, and the other three houses by cards other than dirty cards, hoping either to have four empty houses again later, and thus dispose of the other cards at the top of his dirty cards, or else for an early appearance of the corresponding ace. If John has the bump in question and five empty houses, but no cards other than dirty cards, he knows that his corresponding ace must be among these dirty cards. If Peter has already gone right through the cards in his hand without the corresponding ace appearing, then Peter knows that his ace, too, is among his dirty cards. To make things difficult for John, Peter will not get rid of any dirty cards after he follows John's knock on the table by returning John's dirty cards, thus obtaining the five empty houses. He will start upon his dirty cards only when he has got rid of all his other cards, and then it is virtually certain that he will win.

This example clearly shows how dangerous it is to be left with dirty cards. Admittedly it will rarely be as catastrophic as has been assumed above, but often you can ascribe the loss of a game to undue delay in getting rid of your dirty cards (though maybe, of course, you would have lost anyway).

***96. Other transposition puzzles.** The question of how few moves will suffice for success in puzzles I, II, I′ and II′ of §95 is hardly of any importance in practical play of robuse. In puzzles II and II′, putting the card from your hand into a house is counted as a move. The relevant numbers are then 9, 8, 27, and 26 respectively (when the sequence is of maximum length).

Additionally to this, Tables 3 and 4 give indications of numbers of moves for cases where the sequences are shorter than the number of

TABLE 3

		1	2	3	4	5	6	7	8	9	10	11
	1	1	3	—	—	—	—	—	—	—	—	—
	2	1	3	5	9	—	—	—	—	—	—	—
	3	1	3	5	7	11	15	19	27	—	—	—
I	4	1	3	5	7	9	13	17	21	25	29	33
	5	1	3	5	7	9	11	15	19	23	27	31
	6	1	3	5	7	9	11	13	17	21	25	29
	7	1	3	5	7	9	11	13	15	19	23	27
	8	1	3	5	7	9	11	13	15	17	21	25

available houses would allow, for various lengths of sequence (up to and including 11, the greatest possible length) and for all numbers of houses (up to and including the largest number, 8).

TABLE 4

		1	2	3	4	5	6	7	8	9	10	11
	1	2	—	—	—	—	—	—	—	—	—	—
	2	2	4	8	—	—	—	—	—	—	—	—
	3	2	4	6	10	14	18	26	—	—	—	—
II	4	2	4	6	8	12	16	20	24	28	32	40
	5	2	4	6	8	10	14	18	22	26	30	34
	6	2	4	6	8	10	12	16	20	24	28	32
	7	2	4	6	8	10	12	14	18	22	26	30
	8	2	4	6	8	10	12	14	16	20	24	28

Table 3 relates to puzzles I and I′ of §95, and applies to the case where the sequence has to be placed on top of a card which is already in a house. Table 4 relates to puzzles II and II′ of §95 and applies to the case where the sequence has to be placed on top of a card which is not yet in a house. In the tables, each row (horizontally) is for some number of houses; the numbers have been indicated to the left of the rows. Each column (vertically) is for some number of cards in the sequence that has to be transposed; the numbers have been indicated at the top of the columns. The different compartments display the minimum numbers of moves of separate cards by which the transposition of the sequence can be achieved. A dash has been put in the compartment in cases where the number of houses is too small to permit the required transposition. The tables show, among other things, that with four houses it will take at least 33 moves to transpose a sequence of eleven cards onto a king which is already in a house, and that it will take at least 40 moves to transpose this sequence onto a king which is not yet in a house (that is, onto a king that you have in your hand, or that is your topmost dirty card). We leave the reader to look up appropriate numbers for other cases.

Chapter VI:
GAMES WITH PILES OF MATCHES

I. GENERAL OBSERVATIONS

97. General remarks. One or more piles of matches are provided. John and Peter take turns removing one or more matches according to definite rules which constitute the rules of the game concerned. The one who removes the last match is the winner.

The rules of the game may be such that a situation can arise in which not all the matches have been removed, but the player whose turn it is has no move left, in the sense that he cannot remove a number of matches which is allowed by the rules of the game, because (for example) the rules say that he should take either two or three matches, and there is only one match left on the table. The obvious interpretation to make here is that the player whose move it is has then lost. This interpretation ensures that removing the last match means a win. If John has taken all the matches that were left on the table, then Peter has no move left because the rules of the game demand, among other things, that he has to remove one or more matches. Hence to decide the result of the game, it is enough to agree that a player who has no move becomes the loser.

Instead of this, of course, one can have the opposite rule, that the player who has no move is the winner. This includes an implication that a player who takes the last match (because he cannot otherwise comply with the rules of the game) then becomes the loser.

98. Winning situations. There are certain situations (specified by piles with certain numbers of matches) in which the player who is not to move will win, provided he continues correctly. These situations we shall call winning situations. Hence, a player should try to move in such a way that a winning situation arises. The remaining situations will be called losing situations. A player who has to reply to a losing situation, or who replies with a winning situation, is in the winning position, that is to say, he can win if he continues correctly. The other player is then in the losing position.

The characteristics of winning and losing positions are:

(1) Each move must change a winning position into a losing one, while in every losing position there is at least one move that will change it into a winning position.

(2) If it has been agreed that certain final situations imply a win (or loss), then these final positions belong to the winning (or losing) positions.

With this as an aid, we can track down the winning situations by starting from the simplest situation and proceeding to consider situations of increasing complexity (characterized by larger piles of matches). To be more concrete, we take the most common case, where removal of the last match means a win, with a move always possible as long as there is at least one match on the table. "No match on the table" is then a winning situation.

Next we assume that there is one pile of matches left on the table. Piles which may be taken away in their entirety (according to the rules of the game) represent losing situations. Each pile that always passes into a losing situation in one move is then a winning situation. We then move on to two piles of matches, starting with the simplest cases, and so on.

When we have discovered (or conjectured) a pattern in the winning situations found, we can prove that the result is indeed correct by showing that the situations concerned have the properties (1) and (2). Often this proof is much shorter and much clearer than the procedure that led to discovering the winning situations. Without showing how we have obtained the result, we can convince someone of its correctness in this way, that is, in terms of the properties (1) and (2). Yet, such an isolated proof is often unsatisfactory in the sense that it gives no idea of the way in which the result has been obtained. This comes out very clearly in the game of nim (about which we will say more in §§113 *ff.*); in such a case one can only imagine that the solution was the result of some lucky and apposite discovery.

II. GAMES WITH ONE PILE OF MATCHES

99. Simplest match game. One of the simplest imaginable games with matches is the following. There is one pile of matches. John and Peter take turns removing either one match or two matches, as they choose. The one who takes the last match wins.

A pile of 0 matches is a winning situation. Both a pile of 1 and a pile of 2 matches are losing situations, because they can be changed into a pile of 0 matches in one move. With a pile of 3 matches this is not possible; such a pile produces a losing situation after any move and is thus a winning situation. A pile of 4 or 5 matches is a losing situation again, because it can be changed into a pile of 3 matches. Continuing in this way, we find that piles with a number of matches divisible by 3 (hence 3, 6, 9, 12, 15, etc.) are the winning situations.

If this result is made to appear out of the blue, its correctness can be proved in terms of the properties (1) and (2) from §98. When a number that is divisible by 3 is diminished by 1 or 2, it turns into a number that is not divisible by 3; a number that is not divisible by 3 turns into one that is so divisible, when diminished either by 1 (as for 4, 7, 10, 13, etc.) or by 2 (as for 5, 8, 11, 14, etc.). Furthermore, it should be noted that 0 belongs to the numbers divisible by 3.

We have discussed this simple, one may indeed say childish, game, to illustrate what has been said in general about the winning situations, and also to use it as a preliminary to considerably more difficult games. The game is so simple that many will notice the solution at once, and will be able to indicate the correct way of playing. This runs as follows: Remove one match when your opponent has removed two matches, and two when he has removed one, provided you have earlier managed to make the number of matches a multiple of 3; if your opponent has left a number of matches that is not a multiple of 3, you should change it into a multiple of 3 immediately.

100. Extension of the simplest match game. The match game in §99 is not altered essentially when John and Peter are allowed to take turns removing as many matches as they please up to a certain maximum, but at least one match, as (for example) 1, 2, 3, 4, or 5 matches. In this new case the winning situations are the piles in which the number of matches is a multiple of $5 + 1 = 6$. If John has been able to reply with such a winning situation (or if the game started off with such a winning situation and Peter had to make the first move), then the correct way for John to play is to remove the number of matches which added to the number of matches last taken by Peter will give a total of 6; hence he takes 3 matches if Peter took 3, he takes 4 matches if Peter took 2, and so on. This lets John's next reply again produce a winning situation, namely a multiple of 6, so that finally he reaches 0 (when all matches are removed).

If we introduce the further modification that the one who takes the last match loses, then we have to make a slight and obvious modification in the method of playing, to aim not at 0 matches, but at 1 match. If the player is allowed to take, say, 1, 2, 3, 4, 5, 6, 7, 8, or 9 matches, then the winning situations are not the multiples of 10 themselves, but multiples of 10 increased by 1, hence 1, 11, 21, 31, etc. After a winning position is secured, the correct way of playing is still to supplement the number of matches last taken, to produce a joint total of 10. The only difference is that the numbers of matches of the winning situations have been increased by 1.

The situation can be made seemingly more difficult by a form of camouflage, namely by starting with several piles of matches, and by stating the rule that 1, 2, 3, 4, 5, or 6 matches, say, may be taken, which need not all belong to the same pile. Obviously, this comes to entirely the same thing as when all piles have been combined into a single pile. If John knows this trick and plays the game against Peter who does not know it, then John will choose his matches from different piles; in doing so he will pretend to check carefully from which piles he will take matches, and how many matches from each, although in reality all that matters is the total number of matches which he removes. If Peter is not too bright, he will be misdirected by this procedure.

101. More difficult game with one pile of matches. The game can become considerably more difficult when the number of matches to be removed with each move (a minimum of one) ranges over nonconsecutive values. We assume that 1 occurs among these values, so that a move can always be made as long as not all the matches have been taken. Once again, it is immaterial whether the last match is associated with a win or a loss; in the latter case the player has to aim for 1 in entirely the same way as he aims for 0 in the first case, so that in the case in which the taker of the last match loses, the winning situations always contain one additional match. We may as well assume that the taker of the last match is the winner.

The game is still very simple when only an odd number of matches can be taken, for instance 1, 5, or 7 matches. The winning situations are found in the well-known way (by starting with the smallest number); they turn out to be the even piles. Once this has been surmised, its correctness can be easily confirmed in terms of the properties (1) and (2) from §98. An even pile can be immediately followed by an odd pile only, and conversely. Starting in a winning position, John stays in a winning position however he plays; hence, John cannot lose, even if he wanted to. So the game has no interest except when a player has to try to decide whether he will begin himself or let his opponent begin.

The game becomes more interesting when the player is allowed to take either one match or else a fixed even number of matches, for example either 1 or 8. Here the winning situations are the multiples of 9, the multiples of 9 plus 2, the multiples of 9 plus 4, and the multiples of 9 plus 6. Then a losing situation is converted to a winning situation by either reply, except for a multiple of 9 plus 7 (from which one should take 1 match) and for a multiple of 9 plus 8 (in which case one can secure a winning position only by taking 8 matches). The simplest way to stay in a winning position, once you have got there, is by bringing the number last taken up to 9. This receives an obvious modification when you have the choice of taking, say, 1 or 10 matches. In that case the winning positions are the multiples of 11, and the multiples of 11 increased by 2, 4, 6, or 8.

The only essential, of course, is to know the winning situations. Below, we present some simple examples of these. If the numbers of matches that may be removed are 1, 2 and one or more numbers not divisible by 3, then the winning situations are the multiples of 3 (hence the same as when you are allowed only to take 1 or 2 matches). If the numbers to be taken are 1, 2, 3 and one or more numbers not divisible by 4, then the winning situations are the multiples of 4. If the numbers to be taken are 1, 2, 3, 4 and one or more numbers not divisible by 5, then the winning situations are the multiples of 5 (as for 1, 2, 3, 4 above), and so on.

We shall now discuss some more complicated examples. In Table 5 the numbers of matches that can be taken from the pile are given in the left-hand column (see the numbers not in parentheses), while the winning situations, hence the situations at which one should aim, are given in the right-hand column.

The numbers in parentheses in the left-hand column indicate other numbers of matches that can be allowed to be taken without changing the winning situations. These numbers are numbers by which one cannot get from one winning situation to another, and so they can be determined directly from the winning situations; we could have added more numbers in the parentheses, but we imposed a maximum of 10. One or more of the numbers in parentheses can be added as desired. For example, not only 1–3–6, but 1–3–6–8 and 1–3–6–10 and 1–3–6–8–10 as well, all give the multiples of 9, the multiples of 9 plus 2, and the multiples of 9 plus 4 as winning situations.

TABLE 5

m. = multiple

1–4–6(9)	m. of 5 m. of 5 + 2
1–3–4(6–8–10)	m. of 7 m. of 7 + 2
1–2–6(5–8–9)	m. of 7 m. of 7 + 3
1–4–5(3–7–9) or 1–3–4–7(5–9)	m. of 8 m. of 8 + 2
1–2–4–6(7–9–10) or 1–2–6–7(4–9–10)	m. of 8 m. of 8 + 3
1–4–7(9)	m. of 8 m. of 8 + 2 m. of 8 + 5
1–3–4–6–7(5–9) or 1–4–5–6–7(3–9)	m. of 10 m. of 10 + 2
1–2–4–5–6(8–9)	m. of 10 m. of 10 + 3

1–4–5–6(3–8–10)	m. of 9 m. of 9 + 2
1–3–6(8–10)	m. of 9 m. of 9 + 2 m. of 9 + 4
1–2–5–6–7(4–9–10)	m. of 11 m. of 11 + 3
1–5–6(3–8–10)	m. of 11 m. of 11 + 2 m. of 11 + 4
1–6–7(3–5–9)	m. of 12 m. of 12 + 2 m. of 12 + 4
1–4–6–7(9)	m. of 13 m. of 13 + 2 m. of 13 + 5 m. of 13 + 10

The games given in Table 5 have the drawback that they are somewhat artificial, in that they can be varied endlessly. The most interesting of these games is surely the one in which you are allowed to take 1, 3, or 4 matches. This rule is simple, while the correct way of playing, once you know it, can be easily applied (make a multiple of 7 or a multiple of 7 plus 2). However, it is less simple to find the correct way of playing (this is not immediately obvious), and in particular it is not so easy to learn it by just watching how your opponent plays. You do have the latter possibility in a game where the winning situations are multiples of some fixed number. A player who does not know the game will of course understand at once that the remaining number of matches is what counts. Hence, if he notices that his opponent appears to know the game, and always makes that number a multiple of 5, say, then he in turn will try to do the same thing if he gets the chance.

III. GAMES WITH SEVERAL PILES OF MATCHES

102. Case of two piles. We assume that John and Peter start out with more than one pile of matches, and that they take turns removing a number of matches (each time with a free choice, from a minimum of one match up to a certain maximum), but all from one pile (also to be freely chosen every time). Taking the last match means a win. This game is not so easy to play, and becomes more difficult as the fixed maximum and the number of piles gets larger.

It is still rather simple in the case of two piles. To be more concrete, we assume that the fixed maximum is 5. The number of matches in a pile can be a multiple of 6, a multiple of 6 plus 1 (which may be 1 itself), a multiple of 6 plus 2, a multiple of 6 plus 3, a multiple of 6 plus 4, or a multiple of 6 plus 5. We then speak of a 0-pile, a 1-pile, a 2-pile, a 3-pile, a 4-pile, and a 5-pile, respectively. Two 3-piles (for example, one of 3 and one of 15 matches) will be called similar, and so forth; a 2-pile and (for example) a 4-pile are dissimilar; hence, "similar" means that the numbers of matches in the two piles differ by a multiple of 6 (which may be 0).

The winning situations all consist of two similar piles, for if you have to reply to two similar piles, you must make them dissimilar. However, if it is your turn and you are faced with two dissimilar piles, for instance a 3-pile and a 5-pile, you make them similar by taking 2 matches from the 5-pile (or 4 matches from the 3-pile if it is sufficiently large). Thus you finally reach two piles of 0 matches each (all matches taken), because these 0-piles are similar.

The rule of §100 is included in the above, for you can consider a single pile as equivalent to two piles, one of which contains 0 matches and therefore is (and remains) a 0-pile. Hence you should try to turn the other pile into a 0-pile, that is, make it contain a number of matches that is divisible by 6.

It is clear how this has to be modified when the maximum is different, for instance 7. "Similar" then means that the difference between the numbers of matches in the two piles is divisible by $7 + 1 = 8$. In a 3-pile the number of matches is now a multiple of 8 plus 3, and so on.

103. Case of more than two piles and a maximum of 2. We consider the game of §102 with a maximum of 2 (the smallest possible maximum), for an arbitrary number of piles. "Similar" then means that the difference between the numbers of matches is a multiple of 3. Now the winning situations are those in which the number of 1-piles (piles in which the number of matches is a multiple of 3 plus 1) is even, as well as the number of 2-piles; the number of 0-piles is irrelevant.

To see this, we note that every move turns a pile into a dissimilar pile. If the number of 1-piles and the number of 2-piles are both even, then a move causes at least one of these numbers to become odd. If the number of 1-piles is even and the number of 2-piles is odd (or conversely), then you can make both numbers even by taking two matches from a 2-pile (or one match from a 1-pile), so that a 0-pile arises. If the number of 1-piles and the number of 2-piles are both odd, then both numbers become even if one match is taken from a 2-pile, to change it into a 1-pile; if there is still a 1-pile with at least four matches, there is another possibility, to change this into a 2-pile by taking two matches. Note also that the final position (two empty piles, with the numbers of 1-piles and 2-piles both zero) satisfies the above-mentioned characteristic of the winning situation (because zero is an even number).

This last rule also holds, of course, when there are only two piles. The situation is then winning when there are two 0-piles, two 1-piles, or two 2-piles, in accordance with §102.

In the foregoing, we made the winning situations appear out of the blue. It is more laborious to produce them in the well-known manner we described, starting from small numbers. This is all the more true when the maximum number of matches that can be taken is larger.

104. Case of more than two piles and a maximum of 3. We begin by explaining an expression which we shall need in what follows. We shall say that two numbers have the same parity when they are both even or both odd.

If we are allowed to take 1, 2, or 3 matches from one pile, where we have a choice from an arbitrary number of piles each time, then "similar" means differing by a multiple of 4, with a related meaning for a 0-, 1-, 2-, or 3-pile (multiple of 4, multiple of 4 plus 1, and so on). The winning situations are those for which the numbers of 1-, 2-, and 3-piles have the same parity. Obviously, the final position (in which these three numbers are all zero) satisfies this condition.

Again the proof rests on the fact that a move transforms a pile into a dissimilar pile. For example, if the three numbers are all odd and the move changes a 2-pile into a 3-pile (or a 0-pile, or changes a 0-pile into a 2-pile), then the number of 2-piles becomes even, while the number of 1-piles remains odd. If a situation does not satisfy the condition in question, for instance if the number of 1-piles is odd, while the numbers of 2- and 3-piles are even, then we can change this situation into three even numbers in one move, by taking one match from a 1-pile. This is always possible, because there is at least one 1-pile. If there is a non-empty 0-pile, we have the alternative of taking two matches from it, so that it becomes a 1-pile. If there is a 3-pile (which need not be the case, since the even number of 3-piles can be 0), then we could take one match from that pile, so that it becomes a 2-pile, and the numbers of 2- and 3-piles would both become odd. Finally, if there is a 2-pile with at least six matches, we could change it into a 3-pile by taking three matches from it.

Once more, of course, the result for the case of two piles is included in the general result.

***105. Case of more than two piles and a maximum of 4 or 5.** If we increase the maximum to 4, then "similar" means that the difference is a multiple of 5. Now the winning situations are those for which the number of 4-piles is even, and the numbers of 1-, 2-, and 3-piles have the same parity, for we can verify easily that this characteristic (which is satisfied by the final position) is lost after any single move, whatever this is.

Now we have to prove also that when the property in question is not present, there is always at least one move after which it will appear. If the numbers of 1-, 2-, and 3-piles have the same parity, while the number of 4-piles is odd, then the required characteristic arises by changing a 4-pile into a 0-pile (by taking 4 matches from it). If the number of 4-piles is even, as well as the numbers of 1- and 3-piles, while the number of 2-piles is odd, then we change a 2-pile into a 0-pile. If the number of 4-piles is even, as well as the number of 3-piles, while the number of 1- and 2-piles are both odd, then we change a 2-pile into a 1-pile. If the number of 4-piles is odd, while the numbers of 1-, 2-, and 3-piles do not all have the same parity, then we change a 4-pile into a 1-pile, say, when the number of 1-piles differs in parity from that of the numbers of 2- and 3-piles.

In the foregoing we have given a move that is always feasible. However, in various cases there are other ways to bring about the desired characteristic of a winning situation; this, of course, is of no importance for the proof. But these other moves are not always feasible, because (for example) we can change a 4-pile into a 2-pile only when there is a 4-pile, and we can change a 2-pile into a 4-pile (by taking 3 matches from it) only when there is a 2-pile with at least 7 matches.

If we further increase the maximum to 5 (in which case "multiple of 5" has to be replaced by "multiple of 6"), then the characteristic of the winning situations is the following: The number of 2-piles has the same parity as the number of 3-piles; the number of 4-piles has the same parity as the number of 5-piles; the number of 1-piles has the same parity as the sum of the numbers of 2- and 4-piles. The last requirement can also be expressed by saying that the number of 1-piles must be even or odd, according as the numbers of 2- and 4-piles have the same or different parities.

We leave it to the reader to prove this. It requires detailed consideration of various possibilities, but in any case this is much easier than finding the winning situations (by starting with the smallest numbers of matches).

***106. As before, but the last match loses.** We now introduce into the previous rules the modification that taking the last match means losing. In the case of 1 pile this gave only an obvious difference (aim for 1 instead of for 0). However, with several piles the modification in question makes the game entirely different and much more of a complication.

We show this in the simple case where 1 or 2 matches may be taken from some single pile. If the last match wins, we have a winning situation when the numbers of 1- and 2-piles are both even (see §103). If the last match loses, the winning situations are: an odd number of 1-piles and no 2-piles; an even number (at least two) of 2-piles, and an even number (possibly 0) of 1-piles.

Even with two piles this makes a considerable difference. If the last match wins, there is a winning situation when both piles are similar. If the last match loses, then there is a winning situation when one pile is a 0-pile and the other a 1-pile (for example, a pile of 3 and one of 4), and also when both piles are 2-piles (for example, a pile of 2 and one of 5). As in the previous sections, the proofs of the stated results are left to the reader.

IV. SOME OTHER MATCH GAMES

107. Game with two piles of matches. A simple case is obtained when John and Peter have to take turns removing one or two matches (as they like), but not 2 matches from the same pile, where the last match wins. The winning situations are those in which both piles are even.

When the last match loses, we have the same winning situations as long as there are two (non-empty) piles. With a single pile the situation is winning if it is odd. Hence, once John is in a winning position, he has to play the game in the same way, regardless of whether the last match wins or loses. However, as soon as Peter makes a pile of 1 match, which he will eventually be forced to do, John has to take that match and make or leave the other pile odd. Hence, being in a winning position, John can leave to Peter the choice as to the last match winning or losing, which gives a very generous impression; it is only when Peter reduces one of the piles (or both piles) to a single match that John has to know where he stands.

108. Game with three piles of matches. The game of §107 (taking away one or two matches, but not two from the same pile), in the case where taking the last match wins, becomes more interesting when you begin with three piles. Then the winning situations are those in which the three piles have the same parity, and hence are either all even or all odd. Assuming that John is in a winning position, he has a choice of two correct moves every time, because he can make all piles even, but he can also make them all odd, and he can use this to conceal the correct way of playing from Peter. However, as soon as Peter makes an empty pile, John has only one correct reply, to make the two other piles even.

If John is in a winning position, he need not continually count the piles to stay in a winning position, for he can imitate Peter's move, that is, take a match from the same pile or the same 2 piles as Peter did, or he can reply with the complementary move; by this we mean that John diminishes the pile or the 2 piles from which Peter did not take a match on his last move. When Peter has made an empty pile, John can reply only with the complementary move, and afterwards he imitates Peter's last move every time.

***109. Extension to four or five piles.** We now extend the game of §§107 and 108 to five piles. If we list the numbers of matches in the piles in increasing order (or rather in non-decreasing order, since there can be equal piles), then there are 4 kinds of winning situations: even-even-even-even-even, even-even-odd-odd-odd, odd-odd-even-even-odd, odd-odd-odd-odd-even.

This also applies to four piles if we consider the fifth pile as empty (and therefore even). This is then the smallest pile, and it is even, so that the winning situations are: even-even-even-even, even-odd-odd-odd; that is, the smallest pile (or one of the smallest piles) is even, while the other three piles have the same parity. The result for three piles is again included in that for four piles, and the result for two piles in that for three piles.

When there are five piles, John, provided he is in a winning position, usually has two correct replies, and sometimes three. Thus, John can change even-even-odd-odd-even into even-even-even-even-even or even-even-odd-odd-odd or odd-odd-odd-odd-even. It may also occur, however, that John has only one correct reply; thus to 4–6–6–7–9 he can only reply with 4–6–6–6–8.

***110. Modification of the game with three piles of matches.** We modify the game of §108 to the effect that taking the last match loses. As a result, the correct way of playing becomes completely different, and much more complicated. The winning situations are now: (*a*) the two smallest piles are equal and the third pile has the other parity; (*b*) the three piles have the same parity and the two smallest piles are unequal.

This rule also applies to two piles or one pile, if we consider these as equivalent to three piles, one or two of which contain a zero number of matches. The smallest pile is then even, while the two smallest piles are unequal or equal, according as the number of non-empty piles is 2 or 1; hence, with two piles the situation is winning when they are both even, and with one pile, when that pile is odd.

If John is in a winning position, with three piles, he can easily stay in a winning position without continually counting the piles, by the procedure of replying to each of Peter's moves with the complementary move (see §108); for the effect of both moves together is that each of the three numbers is diminished by 1, so that the situation has remained a winning one. If the complementary move is no longer possible, because John would have to take a match from an empty pile, then John makes two even piles or one odd pile.

111. Match game with an arbitrary number of piles. The game of §105 can be extended, after some modification, to an arbitrary number of piles: John and Peter are allowed to take matches from as many piles as they please, with a total of at least one match, and a restriction to a certain maximum—5, say—in any single pile. If the last match wins, the situation is winning when in every pile the number of matches is a multiple of (5 + 1). The game does not differ essentially from the game in §100.

However, a difference arises when the last match loses. Then you should not aim for piles in which the number of matches is a multiple of 6 plus 1, for the winning situations are unaltered, provided the number of non-empty piles is at least two. But instead of eventually making one pile in which the number of matches is a multiple of 6, you should make that number equal to a multiple of 6 plus 1.

***112. Case in which loss with the last match is the simpler game.** In the foregoing, the case in which the last match wins is always simpler than the case in which the last match loses. We now give an example of a case in which the opposite is true. Given three or more piles of matches, John and Peter have to take turns removing a total of 1, 2, or 3 matches from at most two of the piles. Hence, at every move John or Peter is given the choice from which pile or from which two piles he will take matches, and how many he takes in all (1, 2, or 3).

When there are three piles and the last match means a loss, then the winning situations are those for which the total number of matches is a multiple of 4 plus 1. The case in which the last match wins (for three piles) is slightly more complicated. Then the winning situations are those in which the total number of matches is a multiple of 4, with the exception of the case 1–1–2 (that is, two piles of 1 and a pile of 2), and with the addition of the case 1–1–1.

In the case of four piles, where the last match wins, the winning situations are those for which the total number of matches is a multiple of 4, with the exception of 1–1–1–(multiple of 4 plus 1) and 1–1–2–(multiple of 4), and with the addition of 1–1–1–(multiple of 4). The case in which the last match loses (for four piles) is simpler. Then the winning situations are those for which the total number of matches is a multiple of 4 plus 1, with the exception of 1–1–1–2, and with the addition of 1–1–1–1.

With five piles and a loss for the last match, the winning situations are those in which the total number of matches is a multiple of 4 plus 1, with the exception of 1–1–1–1–(multiple of 4 plus 1), and 1–1–1–2–(multiple of 4), and with the addition of 1–1–1–1–(multiple of 4). The case in which the last match wins (for five piles) is more complicated.

The difficulty of the game lies especially in the fact that towards the end of the game one has to deviate from the rule (make a multiple of 4, or a multiple of 4 plus 1).

V. GAME OF NIM

113. General remarks. The finest of all match games is without doubt the game of nim, which is said to be of ancient Chinese origin. It is usually played with three piles of matches. John and Peter are allowed to take turns removing as many matches as they please (but at least one), from one pile only (which can be freely chosen anew every time). Prior agreement decides whether taking the last match means a win or a loss.

It is very difficult to resolve this game, in the sense of determining all the winning situations. If neither of the two players knows all the winning situations, then the player who knows the larger number of them wins, provided the piles are sufficiently large. Of course, the players will note some winning situations while playing, for example, the situation 1–2–3, both when the last match wins and when it loses. If Peter has to reply to 1–2–3, and if he changes it into 0–2–3 or 1–2–2, then John replies 0–2–2. After Peter's reply (0–1–2 or 0–0–2) John can still leave to Peter the choice whether the last match will win or lose; in the first case John replies to 0–1–2 with 0–1–1 and to 0–0–2 with 0–0–0; in the second case John replies with 0–0–1. Even when Peter plays differently in the situation 1–2–3, John also wins; when the last match means a win, then John replies with 0–1–1, while otherwise he replies with 1–1–1 or with 0–0–1.

114. Game of nim with two piles. The game of nim with two piles is as easy as that with three piles is difficult. In the first case, the winning situations are two equal piles as long as they each contain more than one match. If John is in a winning position, it is only when Peter makes a pile of 0 or of 1 that John has to know whether the last match will win or lose; in the first case John wins by 1–1 or 0–0; in the second case by 0–1.

When John has managed to make two equal piles of at least 2 matches, he can continue almost up to the end of the game and leave Peter the choice whether taking the last match means a win or a loss. The correct way of playing for John is perfectly simple: Peter is forced to make the piles unequal, John makes them equal again.

115. Some winning situations. Even if the player knows no winning situations in a 3-pile game other than those already mentioned, he still has quite a great advantage over someone who does not know a single one, and he will win a vast majority of the games. He takes care not to make two equal piles, because otherwise he runs the risk that his opponent will take the third pile entirely. He will make this move himself if his opponent makes two equal piles. For the rest, he will lie in wait for the opportunity to create the situation 1–2–3, or at any rate he will try to move in such a way that his opponent cannot create that situation.

Proceeding from the winning situations discussed (1–2–3, and those in which one pile has been taken entirely and the other two are equal), the player can easily form winning situations with larger piles. In order of simplicity, 1–2–3 is followed by 1–4–5. If Peter does not take a pile and does not make two equal piles, John can reply with 1–2–3. In this way we find the following winning situations:

1–2–3,	1–4–5,	1–6–7,	1–8–9,	1–10–11,	1–12–13, etc.,
2–4–6,	2–5–7,	2–8–10,	2–9–11,	2–12–14,	2–13–15, etc.,
3–4–7,	3–5–6,	3–8–11,	3–9–10,	3–12–15,	3–13–14, etc.,
4–8–12,	4–9–13,	4–10–14,	4–11–15,	4–16–20,	4–17–21, etc.

If you know these four rather simple sequences of winning situations, you will be practically invincible, until you meet your master—someone who has a knowledge of the complete system of winning situations. Against such a person, you cannot fail to lose, provided, of course, the piles in the initial position are not too small, for example a smallest pile of at least six matches, and no two equal piles.

VI. GAME OF NIM AND THE BINARY SYSTEM

116. Relation to the binary system. In order to describe the winning situations in their entirety, the numbers of matches for the three piles have to be written in the binary system of notation. Thus the ancient Chinese (many centuries B.C.), who seem to have developed a more or less consistent binary system, were virtually predestined to discover the game of nim.

To find the winning situations, we proceed as follows. We write the three numbers in the binary system one below the other (that is, in such a way that the units digits become aligned vertically, as well as those for the twos, and for the other places). The sum of the digits in the same column will be called a digit-sum. If all digit-sums are even, we speak of an even situation, otherwise of an odd situation. In an odd situation not all digit-sums need be odd; but, at least one of the digit-sums must be odd. Examples of an even and of an odd situation are:

1010001	1101100
1001101	1011000
11100	101110
2022202 even situation,	2213210 odd situation.

In decimal notation these examples are 81–77–28 (even situation) and 108–88–46 (odd situation).

Now, if taking the last match means a win, we have this simple rule: the even situations are winning, the odd situations are losing.

The rule for the case of two piles is included in this, for then in one of the three numbers (corresponding to the empty pile) no digit 1 occurs. Hence, in the other two numbers the digits below one another have to be equal if the sum of the numbers is to be even; thus these numbers are equal when the situation is even.

117. Proof of the rule for the winning situations. To prove the rule, we have to show three things: (1) The final situation, which yields a win, is even; (2) an even situation cannot be changed into another even situation in one move; (3) an odd situation can always be changed into an even situation in one move.

The correctness of (1) follows from the fact that the final situation concerned is 0–0–0 (all matches taken). There is only one digit-sum in that case, and it is 0, hence even. A move diminishes one of the three numbers and hence changes at least one of the digits of that number from 1 to 0. Hence, the corresponding digit-sum changes from even to odd or conversely (giving a change in parity). So, if the situation is even, it becomes odd, and (2) has been proved. We now assume that the situation is odd. In the column farthest to the left which has an odd digit-sum, at least one digit 1 must occur. We diminish the number which has this digit 1 by changing this digit into 0; digits to the right of it in the same number are changed (if necessary) in such a way that all the digit-sums become even. Thereby we have created an even situation, and (3) has been proved.

This proof is quite simple, but the great difficulty lies in hitting on the idea of writing the three numbers in the binary system. It is true that some pattern can be detected in the four sequences given in §115, but there is no obvious connection with the binary system. This is still the case when one forms several such sequences (with 5 or 6 as the smallest number, and so on). The notation of the decimal system just does not lend itself to discovering this connection. It is not known who was the first to see the relation with the binary system. The unknown individual who had such a brilliant idea must have been no ordinary person.[1]

118. Remarks on the correct way of playing. If we assume arbitrary values for two of the three numbers of matches, the third number can be chosen in only one way, to give an even situation; for the requirement that all digit-sums are to be even determines every digit of the third number. For instance, of 10111010011 and 10110010 are two of the numbers, then only 10101100001 for the third number makes the situation even, as we see immediately by writing the two first-mentioned numbers one below the other. The foregoing shows that it would be extremely coincidental if three piles put down arbitrarily formed an even situation. When it is necessary to reply to an odd situation, and the odd digit-sum farthest to the left is equal to 1, then exactly one of the three numbers has the digit 1 in the corresponding column. In that case there is only one correct reply, and it consists of diminishing the number in question. Thus, in the second example of §116, the only correct reply is to reduce 1011000 to 1000010 (hence 88 to 66) by taking 22 matches from the middle pile).

If the odd digit-sum farthest to the left is equal to 3, then there are three correct replies. It is then immaterial from which pile you remove matches; once the pile has been chosen, you can make a winning (even) situation in only one way. With correct play, the case in question can occur only on the first move, for after the reply to an even situation the odd digit-sum farthest to the left must be equal to 1.

From the foregoing we conclude that it is highly improbable, indeed practically impossible, that someone who does not know the game will find the correct moves by accident, and win. Hence, John, who knows the game, can confidently leave the choice as to who will begin to Peter, who knows nothing or little of the game, provided the three piles put down are not too small and all three are different. Should Peter happen to make a winning situation on his first move, then John takes only a few matches, taking care not to make two equal piles and not to make the smallest pile too small.

---

[1] [Priority of publication can be assigned to C. L. Bouton, *Annals of Mathematics* (2), **3** (1901–02), 35–39—T. H. O'B.]

When John and Peter both know the game, and Peter can choose who will begin, then he is certain to win. He will then begin himself if the initial situation is odd (which will be so in the vast majority of cases when the piles have been laid down arbitrarily), otherwise he will let John begin.

119. Case in which the last match loses. When taking the last match means losing, the winning situations are still the even situations, with the exception of 0–1–1, and also of 0–0–0, of course. On the other hand, 1–1–1 and 0–0–1 are now winning situations. When John is in a winning position, he can continue almost to the end of the game, leaving to Peter the choice whether the last match will win or lose. It is only when John has to reply to a situation in which there is only one pile with more than a single match, combined or not with one or two single-match piles, that Peter has to decide whether the last match wins or loses. If Peter says that the last match wins, then John creates the situation 0–1–1 or 0–0–0, depending on whether he has to reply to more than one pile or to only one pile of matches. If Peter decides that the last match loses, then John creates the situation 1–1–1 or 0–0–1, depending on whether he has to reply to three piles or fewer than three piles.

120. Simplest way to play. From the connection with the binary system it may appear that it is difficult to achieve the correct way of playing. One might get the impression that John, who knows the game, has to count the three piles of matches and do paper-work to convert the resulting number to the binary system, a thing which would, of course, make no pleasant impression upon his opponent.

However, John can proceed differently, and very simply, by arranging the matches of each of the piles as inconspicuously as possible, in the way described in §88. For convenience we assume that each of the piles contains fewer than 30 matches, something which will practically always be the case. If possible, John makes a group of 16 from the matches in a certain pile, then similarly a group of 8, of 4, or of 2. For instance, if there are 23 matches, then he makes a group of 16, and from the remaining matches a group of 4, a group of 2, and a group of 1. For this it is not necessary to count the matches. By grouping the matches, John has counted them in the binary system, as it were, and has found the representation 10111 (see *Figure 77*). In the same way he groups the two other piles.

The situation is winning (even) when the number of groups of 16 is none or two, and similarly for groups of 8, of 4, of 2, and of 1. When

John has to reply to an odd situation, and there is only one pile in which a group of 16 occurs, then he first takes that group and (if necessary) puts back as many matches as may be required to make the number of groups of 8, of 4, of 2, and of 1, all even; to do this, it

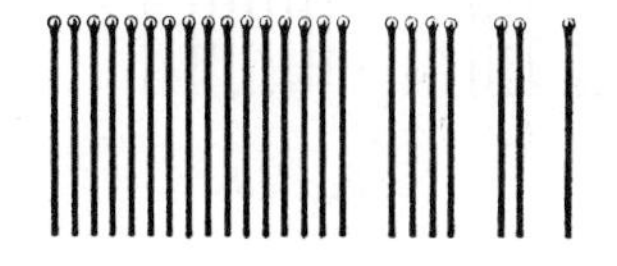

Fig. 77

may instead be necessary to take more than 16 matches from the pile, and indeed he may have to take exactly 16 (if the number of groups of 8 is already even, as well as the number of groups of 4, of 2, and of 1).

An example of this is presented in *Figure 78*. John has to take, from

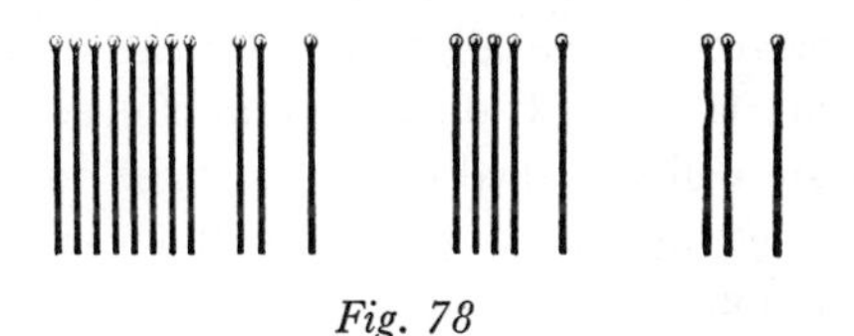

Fig. 78

the largest pile, the group of 8 and the group of 1, and put back 4 of these 9 matches to get two groups of 4. Hence, he should take 5 matches from the largest pile. It is better if John first finds out mentally how many matches he has to take (something which can certainly be done after some practice), and then takes the correct number at once, at the same time disturbing the division into groups. After Peter's reply, John sets out to restore the division into groups again as inconspicuously as possible, pretending that he is just counting the piles.

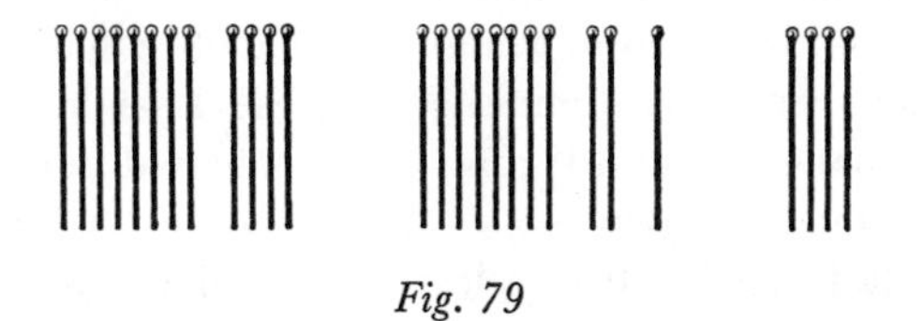

Fig. 79

As a second example we take the situation in *Figure 79*, where three matches should be taken from the middle pile, and as a third example the situation in *Figure 80*, where one match has to be taken from the smallest pile.

Fig. 80

VII. EXTENSION OR MODIFICATION OF THE GAME OF NIM

121. Extension of the game of nim to more than three piles. The game of nim undergoes no essential change when we begin with more than three piles of matches. To find the winning situations, we use the binary system to write the numbers of matches in the various piles in appropriate vertical alignment (or else we can arrange the matches in binary fashion). If the digit-sum of every column is even, then the situation will be called even, otherwise odd. When the last match wins, the winning situations are without exception the same as the even situations.

Since the final situation (all matches taken) is even, and every move changes an even situation into an odd situation, we only have to show that for every odd situation there is at least one move that causes an even situation; in this process, the correct way of playing will also be determined. To do this, we find the odd digit-sum farthest to the left. From one of the piles (uneven in number) which contribute 1 to this digit-sum we take as many matches as is necessary to make all digit-sums even; hence we change the digit 1 in question into a 0, and also modify, if necessary, the digits to the right of it in the same number.

When Peter has replied to a winning situation, John, to stay in a winning position, has then to diminish a pile of matches other than the one which Peter has diminished, for the 1 farthest to the left which Peter changed into a 0 cannot be converted into a 1 by any removal of matches by John.

When the last match means a loss, the winning situations are the even situations in which at least one of the piles contains more than one match, and the odd situations in which no pile contains more than a single match. As long as there are at least two piles that contain more than one match each, it is not necessary that John, when replying in a winning position, should know whether the last match will win or lose.

***122. Further extension of the game of nim.** We modify the game of nim to the effect that we start with several piles of matches and that we are allowed at each turn to take as many matches as we like, but from at most two piles (or from at most three piles, or from at most four piles, and so on), while—in the simpler case—the last match wins.

As an example we take the case in which we are allowed to take matches from at most two piles. The game makes sense only when we start with at least three piles. With three piles, the winning situations are three equal piles. To such a situation, we can reply only with three piles that are not all equal, and with the following reply the three piles can be made equal again. For an arbitrary number of piles of one match each, the winning situations are those in which the number of piles is a multiple of 3, for then the game is nothing but the simple game of §99.

The foregoing suggests the idea that the winning situations are those for which the digit-sums are all a multiple of $(2 + 1)$. As before, the numbers of matches in the various piles are written one below the other in the binary system. The final situation (all matches taken) is among those mentioned, and it is not possible to get from such a position to another such position in only one move.

When the digit-sums are not all a multiple of 3, they can all be changed into a multiple of 3 in one move. To do this, we proceed from the first digit-sum from the left that is not a multiple of 3. Of the 1's in that column we change one or two of them into 0, depending on whether the digit-sum in question is a multiple of 3 plus 1, or a multiple of 3 plus 2. By this, either one or two piles are indicated from which we have to take matches. If two piles are involved (because the digit-sum is a multiple of 3 plus 2), then we first take so many matches from these piles that the two digits 1 from the column in question each change into a 0, and hence, the corresponding digit-sums become multiples of 3. We have then taken enough matches to ensure (by putting back some if necessary, or by taking still more matches from these two piles) that all the other digit-sums also become a multiple of 3.

When only one pile (A, say) from which matches have to be taken, has been indicated so far (because the digit-sum in question is a multiple of 3 plus 1), then there is the possibility that we can make all digit-sums equal to a multiple of 3 by taking matches from pile A only. This case presents itself when pile A contributes 1 to each digit-sum that is a multiple of 3 plus 1, and 0 to each digit-sum that is a multiple of 3 plus 2; by changing all those 1's into 0's, and all these 0's into 1's, each digit-sum becomes a multiple of 3. If this cannot be achieved with pile A by itself, we consider the first digit-sum from the left that is not a multiple of 3 and that cannot be changed into a multiple of 3 by altering the corresponding digit in A. Then there exists a second pile, B, in which the digit that corresponds to that digit-sum is equal to 1. We change this digit into 0 by taking matches from pile B, too. When the last-mentioned digit-sum is a multiple of 3 plus 1, A contributes 0 to it; this digit 0 of the number of matches in A we leave unchanged. When the last-mentioned digit-sum is a multiple of 3 plus 2, A contributes 1 to it; this digit 1 of the number in A we change to 0, so that two groups of matches are then taken from A. Now we have taken sufficient matches from A and B to be able to bring it about, by putting back some of them (if necessary), that the remaining digit-sums also become multiples of 3; to secure this, it may also be necessary to take still more matches from the piles A and B. By playing the game a couple of times, starting with four or five piles, say, we can convince ourselves that we can always create an even situation in the manner indicated, when we have to reply to an odd situation.

When it is agreed that the last match loses, this produces a difference in the winning situations only towards the end of the game, when all piles of more than 1 match have disappeared. Then the winning situations are those in which the number of single-match piles is a multiple of 3 plus 1.

The treatment of the case in which we are allowed to take from at most three, or from at most four piles, is left to the readers who take pleasure in it. The winning situations then are (when the last match wins) those in which the digit-sums of the numbers in the binary system are all a multiple of $(3 + 1)$ or $(4 + 1)$, respectively. The proof of this is completely analogous to the above proof, but slightly more complicated.

***123. Special case of the game of §122.** A special case of the game of §122 is obtained when we are allowed to take as many matches from as many piles as we please, but not from all piles; this is to be understood in the sense that when one or more piles have been entirely depleted, we are allowed to take matches from all remaining piles at the same time at the next move. The winning situations of §122 now become (when the last match wins) the situations in which all piles contain the same number of matches.

That these are the winning situations can be seen more easily without the binary system. When Peter has to reply to equal piles, he cannot leave them all equal, after which John can make them equal again, and thus achieve the final position (all matches taken). When the last match loses, the correct way for John to play is the same, as long as all piles contain more than one match; when Peter makes an empty pile or a pile of one, John takes all matches but one.

***124. Modification of the game of nim.** We modify the game of nim in the following way. From four piles of matches John and Peter take turns removing as many matches as they please, but from exactly two piles (no more and no fewer). The player who can no longer make a move loses; this may mean that there is not a single match on the table, but also that only one pile is left.

The winning situations are now those in which three of the piles are equal and the fourth contains the same number of matches or more. When John has created such a situation, he can create it again after any reply from Peter and thus achieve the situation in which he leaves either no single pile or only one pile; both situations satisfy the characteristic in question, because then there are three equal piles of 0 matches each, and a pile that is not smaller.

When we make the agreement that the player who can no longer make a move wins, then the winning situations are the same, as long as each of the four piles still contains more than one match; there are, besides, the winning situations 0–1–1–1 and 0–0–1–(at least 1).

Hence, towards the end of the game, John (who is assumed to be in a winning position) has to take care not to make the situation 1–1–1–(at least 1), because Peter can reply to this with 0–0–1–1, after which John has to make the situation 0–0–0–0.

Also, we could agree (but this is less natural) that a player is compelled to take from two piles only when there are at least two piles left, so that it is still necessary to take one or more matches when the number of piles has dropped to one. When the last match means a win, the winning situations are still the same, as long as each of the four piles contains more than one match; besides these, 1–1–1–1 and 0–1–1–(at least 1) are now winning situations. The case in which the last match loses becomes more complicated. The winning situations remain the same (three equal piles, and one equal or larger pile), as long as each of the four piles contains at least four matches; in addition, the following are winning situations: 3–3–3–3, 2–2–2–2, 1–3–3–(at least 3), 1–2–2–2, 0–2–2–(at least 2), 0–0–1–1, and 0–0–0–1. Hence, the player who is in a winning position has to be very careful towards the end of the game.

Chapter VII: ENUMERATION OF POSSIBILITIES AND THE DETERMINATION OF PROBABILITIES

I. NUMBERS OF POSSIBILITIES

125. Multiplication. As we have seen, solving a puzzle largely reduces to distinguishing among various possibilities, each of which then has to be investigated separately. It often happens in examining a certain possibility, that a new distinction among different cases has to be made, while it is no rarity that the consideration of one of these cases necessitates a further division, and so on.

In many cases the number of the various possibilities can be determined without stating these possibilities separately. This is of importance especially when the number of possibilities is somewhat large. If it appears to be so large that examining the various possibilities one by one is not feasible, this negative result may still contain an important indication, for it suggests that the different possibilities have not been grouped efficiently, and so it often emerges that a better classification of the various cases can be made, and that the number of these cases can be reduced considerably by simple arguments.

For other purposes, too, a count of possibilities is of importance. I only have to mention the computation of probabilities, which is often based on such a count. Usually this is not a count in the ordinary sense of the word, counting one by one, because this would take far too much time, but a shortened way of counting by a suitable classification of the possibilities into groups.

A frequently occurring case of this nature is one in which there is a distinction into (say) six cases, and a different sort of division into (say) five cases. Here we assume that each of these six cases can occur in combination with each of the five cases. It is evident that then there are in all 6×5, hence 30 possibilities, with no need to state the 30 possibilities explicitly. Moreover, such explicitness is by no means desirable, since complete detailing would make the argument much too long-winded. This applies with greater force in a case in which there is a division into possibilities in (say) three ways: first a division into 11 possibilities, then one into 9 possibilities, and finally one into 2 possibilities; one then obtains $11 \times 9 \times 2 = 198$ cases; as an example we take population returns of the Netherlands in which the inhabitants are classified (1) by provinces (11 possibilities); (2) by age (0–10, 10–20, 20–30, 30–40, 40–50, 50–60, 60–70, 70–80, and over 80, hence 9 possibilities); and (3) according to sex (2 possibilities).

We shall refer to the rule which we have discussed as the rule of multiplication of numbers of possibilities. Each multiplication of numbers can be considered as a multiplication of numbers of possibilities. For example, if John, Peter, and Charles possess five objects each, then there are fifteen objects in all. If we think of John's objects as being numbered 1, 2, 3, 4, 5, and the same for Peter's and Charles's objects, and if we point out some object, there are three possibilities, because it may be an object belonging to John, to Peter, or to Charles. Considered from another point of view, there are five possibilities because the number assigned to the object may be 1, 2, 3, 4, or 5. The combination of these two divisions leads to 3×5 cases.

As an example we take the problem of the number of possible throws with three colored dice, a red one, a blue one, and a green one, where two throws are considered to be different even when the three spot-numbers agree, if these are distributed differently among red-blue-green. Each of the three dice leads to 6 cases, so that we get $6 \times 6 \times 6 = 6^3 = 216$ possible throws in all.

126. Number of complete permutations. We imagine the situation in which a certain number of objects (6, say) are arranged side by side in a row, and we require the number of ways in which this can be done. Such an arrangement in a row is called a (complete) permutation of the objects. Two arrangements for which the order is the same, but in one case from left to right, and in the other from right to left, are considered to be different.

We can imagine the 6 objects to be numbered in sequence. The problem of finding the number of permutations is then the problem of finding the number of ways in which the numbers 1, 2, 3, 4, 5, 6 can be placed in a row. We can start with the number farthest to the left; this gives 6 possibilities, because it may be 1, 2, 3, 4, 5, or 6. When a choice, 4 say, has been made, there are still 5 possibilities for the number next to the right; since this cannot be a 4, it is one of the numbers 1, 2, 3, 5, 6. When a choice has been made, 1 say, for this number too, there are still 4 possibilities for the third number from the left, because it can be 2, 3, 5, or 6. Continuing in this way we find, according to the rule of multiplication of numbers of possibilities discussed in §125, a total of $6 \times 5 \times 4 \times 3 \times 2 \times 1 = 720$ permutations. The product $6 \times 5 \times 4 \times 3 \times 2 \times 1$ is written more shortly as $6!$, which is read as "six factorial."

Likewise, the number of permutations of 8 objects is, of course, equal to $8 \times 7 \times 6 \times 5 \times 4 \times 3 \times 2 \times 1 = 8! = 40{,}320$. Since the number of permutations increases very rapidly, writing out all permutations is practicable only with a small number of objects. Thus the possible permutations of the numbers 1, 2, 3 are:

1–2–3, 1–3–2, 2–1–3, 2–3–1, 3–1–2, 3–2–1.

Below we give the numbers of permutations for objects up to 20 in number:

$$1! = 1,\ 2! = 2,\ 3! = 6,\ 4! = 24,\ 5! = 120,\ 6! = 720,\ 7! = 5040,$$
$$8! = 40320,\quad 9! = 362880,\quad 10! = 3\,628800,\quad 11! = 39\,916800,$$
$$12! = 479\,001600,\quad 13! = 6227\,020800,\quad 14! = 87178\,291200,$$
$$15! = 1\,307674\,368000,\qquad 16! = 20\,922789\,888000,$$
$$17! = 355\,687428\,096000,\qquad 18! = 6402\,373705\,728000,$$
$$19! = 121645\,100408\,832000,\qquad 20! = 2\,432902\,008176\,640000.$$

To give an idea of the size of the number $20!$, we note that it is equal to the number of seconds in over 770 million centuries.

Various questions can be reduced to permutations. As an example we take the number of ways in which 7 persons can be seated at a round table when two arrangements in which everyone has the same neighbors are considered to be the same (here we pay no attention to left or right). Indicating the persons by 1, 2, 3, 4, 5, 6, 7, it is irrelevant where 1 is seated. The remaining 6 persons can now be seated in $6!$ ways. In this we do take left and right into account, so that the desired number is half of $6!$, hence 360.

127. Number of restricted permutations. The number of permutations of 9 objects, say, can be interpreted as follows. We imagine a box with 9 numbered balls. These balls are successively drawn from the box until it is empty. The numbers are written down in the order in which they are drawn. The number of ways in which this can happen is evidently $9!$.

We now introduce the modification that the box is not emptied entirely, but 6 balls, say, are drawn successively. The first drawing can occur in 9 ways. If, for example, the number 3 has been drawn, the second number drawn can only be one of the 8 numbers 1, 2, 4, 5, 6, 7, 8, 9. At the third drawing only 7 possibilities are left, and so on. Hence, the number of ways in which the drawing of the 6 balls can occur (taking into account the order of the numbers drawn), is equal to $9 \times 8 \times 7 \times 6 \times 5 \times 4 = 60{,}480$. This number, which can also be written as $9!/3!$, is called the number of permutations of the 9 objects in groups of 6. Likewise, the number of permutations of 12 objects in groups of 5 is equal to $12 \times 11 \times 10 \times 9 \times 8 \times 7 \times 6 = 12!/7! = 95{,}040$.

In very simple cases only, that is, for very small numbers, is it feasible to state the individual cases. Thus, the permutations of 5 numbered objects in groups of 3 are:

123, 124, 125, 132, 134, 135, 142, 143, 145, 152, 153, 154,
213, 214, 215, 231, 234, 235, 241, 243, 245, 251, 253, 254,
312, 314, 315, 321, 324, 325, 341, 342, 345, 351, 352, 354,
412, 413, 415, 421, 423, 425, 431, 432, 435, 451, 452, 453,
512, 513, 514, 521, 523, 524, 531, 532, 534, 541, 542, 543,

here the principle according to which the permutations have been ordered can be clearly seen. Without taking the trouble to write down all the permutations, we see that their number is equal to

$$5 \times 4 \times 3 = 60.$$

128. Number of combinations. Again we take a box with 9 numbered balls from which 6 balls are drawn, but now we introduce the modification that the 6 balls are not drawn one after another, but simultaneously; hence, after a single drawing we have a handful of 6 balls out of the 9. The number of ways in which this can be done is called the number of combinations of 9 objects in groups of 6. This shows that the difference from permutations lies in the fact that with permutations we do take the order into account, whereas with combinations we do not.

The combinations of 5 numbered objects in groups of 3 are:

123, 124, 125, 134, 135, 145, 234, 235, 245, 345.

From these combinations we can construct permutations by permuting the 3 numbers of a group. Each of the 10 combinations then leads to $3! = 6$ permutations; thus the combination 235 leads to the permutations 235, 253, 325, 352, 523, 532. In this way the 10 combinations lead to $6 \times 10 = 60$ permutations of 5 objects in groups of 3, as written out above.

The foregoing also provides the means to determine the number of combinations without first writing down all these combinations; this indeed would be a hopeless task with somewhat large numbers. In the example (5 objects and groups of 3) the number of permutations is $5 \times 4 \times 3$. Of these permutations, groups of $3!$ belong to the same combination, in various cases, so that the number of combinations of 5 objects in groups of 3 is equal to $\frac{5 \times 4 \times 3}{3!} = 10$. Since we can also write the number of permutations as $5!/2!$, the number of combinations of 5 objects in groups of 3 can also be represented in the form

$$\frac{5!}{3! \times 2!}.$$

We reason in entirely the same way when the total number of objects and the number of objects per group are varied. Thus, the number of permutations of 9 objects in groups of 6 is equal to $9 \times 8 \times 7 \times 6 \times 5 \times 4 = \frac{9!}{3!}$. Every combination of the 9 objects in a group of 6 gives rise to a number of permutations equal to the number of permutations of 6 objects, thus to $6!$ permutations. For the number of combinations this gives:

$$\frac{9 \times 8 \times 7 \times 6 \times 5 \times 4}{6!} = \frac{9!}{6! \times 3!} = \frac{9 \times 8 \times 7}{3!} = 84.$$

The number of combinations of 12 objects in groups of 5 is the same as for groups of $12 - 5 = 7$, namely, $\frac{12!}{5! \times 7!}$. We can see directly that both numbers of combinations are equal, even without knowing the method of determining the number of combinations. For groups of 5, we seek the number of ways in which we can draw a group of 5 balls out of a box with 12 numbered balls. When we draw, we leave 7 balls in the box, so that we are also seeking the number of ways in which we can leave 7 balls in the box—which is the same as considering the drawing of a group of 7 balls.

129. Number of permutations of objects, not all different. We require the number of permutations of the letters *abcdeee*, to give the number of words that can be formed with these 7 letters. Here each order is considered as a "word," regardless of whether it has a meaning and whether it can be pronounced. If we first imagine the three letters *e* to be different—for example, by using different founts of type—we obtain $7!$ permutations. Since the three letters *e* which occur in every permutation can be permuted in $3!$ ways (including the original arrangement), there are groups of $3!$ permutations that become identical when we neglect the distinction between the types of the *e*'s. Hence, the number of permutations of *abcdeee* is $7!/3!$.

Next, we require the number of permutations of the letters *aaaabcdeee*. If we first consider the four *a*'s as being different, and also the three *e*'s, we obtain $10!$ permutations. By disregarding the difference between the four types of *a*, groups of $4!$ permutations become identical, and the number of permutations reduces to $10!/4!$. Here the three *e*'s are still considered as different. By also disregarding the distinction between the types of *e*, groups of $3!$ permutations again coalesce into single permutations, so that the number of permutations is divided once more, this time by $3!$. Hence, the desired number of permutations is $\frac{10!}{4! \times 3!}$.

It is clear how this can be extended to other cases. For example, the number of permutations of the 17 letters *aaaabcdeeefghiiii*, among which there are 4 equivalent *a*'s, 3 equivalent *e*'s, and 4 equivalent *i*'s, is equal to:

$$\frac{17!}{4! \times 3! \times 4!} = 102{,}918{,}816{,}000.$$

At the end of §20 we had to make 16 moves each of a different piece, 8 times for an odd piece, and 8 times for an even piece. According to the foregoing this can be done, so far as the alternation of even and odd is concerned, in $\frac{16!}{8! \times 8!} = 12{,}870$ ways.

130. Number of divisions into piles. We take a bridge pack (with 52 cards), and we want to divide it into 4 piles, a pile A of 17 cards, a pile B of 15, a pile C of 12, and a pile D of 8 cards. To determine the number of ways in which this can be done, we imagine the cards to be numbered consecutively from 1 to 52. By associating each number with the letter of the pile in which the corresponding card is located, every division into piles (containing numbers of cards as above) leads to a permutation of 52 letters, namely 17 A's, 15 B's, 12 C's, and 8 D's. Hence, the number of ways in which the division into piles can be made is, according to §129, equal to

$$\frac{52!}{17! \times 15! \times 12! \times 8!}.$$

The number of ways in which the pack of cards can be divided into 4 piles A, B, C, D, of 13 cards each, is, as appears from the foregoing,

$$\frac{52!}{13! \times 13! \times 13! \times 13!} = 2^9 \times 3^3 \times 5^4 \times 7^4 \times 17^3 \times 19^2 \times 23^2 \times 29 \times 31 \times 37 \times 41 \times 43 \times 47$$
$$= 53{,}644{,}737{,}765{,}488{,}792{,}839{,}237{,}440{,}000 = \text{more than } 5 \times 10^{28}.$$

Here it is assumed that a distinction is made between one pile of 13 cards and another, hence that a new division into piles is obtained when we interchange the piles. When we make no distinction between one pile and another, we still have to divide the number found by 4! (the number of ways in which we can arrange the 4 piles), so that the number of ways in which the 52 cards can be divided into four piles of 13 cards is then equal to $\frac{52!}{(13!)^4 \times 4!}$.

As an illustration we take smaller numbers, namely 4 cards which are divided into two piles of 2 cards each. The number of ways in which this can be done is equal to $\frac{4!}{2! \times 2!} = 6$, when we distinguish between one pile and the other; otherwise the number of ways becomes $\frac{4!}{2! \times 2! \times 2!} = 3$. We can also interpret this result as follows: 4 persons go for a walk, and walk in twos. This can be done in 3 ways, as can also be seen immediately (without the previous considerations). However, if the 4 persons walk on a narrow road, so that they have to walk one pair behind the other, and if we distinguish between the first pair and the second pair, then the number of ways becomes twice as large, hence equal to 6.

We now return to the division of the cards in bridge. Here, not only are four piles of 13 cards involved, but it is equally important to know to which player a given pile corresponds. Hence, a distinction should here be made between one pile and another, so that the number of possible divisions amounts to more than 5×10^{28}.

From this enormous number, which we just cannot imagine, we see clearly how utterly incredible the story is that has turned up now and then in the newspapers, according to which the following rare sequence of bids in bridge was made somewhere or other (without the possibility of a joke or cheating): "grand slam clubs," "grand slam diamonds," "grand slam hearts," "grand slam spades." The dealer here held 13 clubs, his left-hand neighbor 13 diamonds, and so on. According to our calculation, such a distribution of the cards occurs not even once in 5×10^{28} deals, on the average. If four persons play night and day, and if we assume that each hand takes five minutes, then the cards are dealt 288 times per 24 hours, that is, approximately 100,000 times per year. If we now assume that the entire adult population of the world, say 10^8 tables of 4, do nothing but play bridge, day in, day out, then the cards are dealt 10^{15} times per century. So even then it would still take millions and millions of centuries before we eventually met with a distribution of the cards like that mentioned in the newspaper report. Hence we can safely say that such a distribution is impossible, and that the whole story is a mystification.[1]

II. DETERMINING PROBABILITIES FROM EQUALLY LIKELY CASES

131. Notion of probability. In a game of dice we can usually distinguish various cases which we feel are entirely equal in degree of uncertainty. By this we mean that we have no reason at all to expect the occurrence of one case rather than another, and that we cannot imagine that such a reason could be suggested. We then speak of equally likely cases. Thus, when we throw an "unloaded" die, throws of 1, 2, 3, 4, 5, 6 are equally likely cases. When we draw a card from a bridge pack, the 2 of clubs, the 3 of clubs, and so on, are equally likely cases.

However, it is not always so easy to judge whether cases can be considered as equally likely. Often this is not a matter of calculation, but of common sense. Only when we have succeeded in finding a satisfactory division into equally likely cases can the calculation be started. The French mathematician Laplace (1749–1827) expressed this by saying that, in the end, the theory of probability is nothing but common sense reduced to calculation. When someone is convinced that round numbers offer a worse chance in a lottery than

[1] [The joke or cheating type of explanation seems the likeliest, when the facts are indisputable: but the author's argument assumes a perfection of shuffling which may not apply in actual practice—T. H. O'B.]

other numbers, then it is not possible to talk him out of this superstition by argument or by calculations.

We shall assume now that the division into equally likely cases has been made in such a way that each of these cases means either a win or a loss. This means that we assume that among these cases there is not a single one in which the choice between a win and a loss depends on still other circumstances. The cases that entail a win are called favorable cases. By the probability or chance of a win, or, more generally, the probability of a certain event, we then understand the value of the following quotient:

$$\frac{\text{number of favorable cases}}{\text{total number of equally likely cases}}.$$

The probability of a loss is the number of unfavorable cases divided by the total number of cases. As the sum of the number of favorable cases and that of the unfavorable cases is equal to the total number of cases, the sum of the probability of a win and the probability of a loss is equal to 1. The probability of a loss is also called the complementary probability. Hence we can also say that probability + complementary probability = 1.

As appears from the given definition of probability, a probability of 1 means certainty, and a probability of 0 means impossibility.

Since the calculation of probabilities reduces to counting numbers of possibilities, it is clear that the considerations of §§125–130 will stand us in good stead.

132. Origin of the theory of probability. The main import of the theory of probability does not lie in its application to the so-called games of chance, but rather in its applications to life insurance, to statistics, to the improvement of inaccurate measurements (the so-called theory of errors), and particularly to developments in the modern natural sciences. One might say that wherever certain knowledge ceases, the theory of probability enters.

However, it was games of chance which provided the motive for developing the theory of probability. No doubt, more or less correct calculations must have been made as long as people have indulged in gambling; however, the oldest writings do not date from earlier than the beginning of the seventeenth century, although a few isolated cases of calculation of probabilities can be found as early as the fifteenth century. The well-known Italian mathematician Cardano (1501–1576), who was himself a great enthusiast for dicing, had an accurate knowledge of the chances of the various throws. In addition, not inconsiderable contributions were made by gamblers who were not mathematicians but who devoted themselves to calculations of probabilities. They often obtained results which turned out to be contradictory to experience, usually as a consequence of misjudging the equal likelihood of cases. One player had noted that when three dice are thrown, throws of 10 and 11 are more frequent than 9 and 12, in spite of the fact that each of the four numbers can arise in six ways as a sum of three numbers:

$$9=1+2+6=1+3+5=1+4+4=2+2+5=2+3+4=3+3+3,$$
$$10=1+3+6=1+4+5=2+2+6=2+3+5=2+4+4=3+3+4.$$

With this objection he came to Galileo (1564–1642), the famous Italian mathematician, physicist, and astronomer, who soon solved the difficulty by observing that these 6 ways are not equally likely cases: the throw 3–3–3 can arise in only one way, namely when each of the dice shows a 3. However, the throw 1–4–4 can arise in three ways, because the 1 can come from any of the three dice. A throw like 1–2–6 can arise in $3! = 6$ ways. Hence, we should distinguish not 56, but $6 \times 6 \times 6 = 216$ equally likely cases. Of the 56 cases, there are 20 with three different spot numbers, 30 with two different spot numbers (with the additional implication that two are equal), and 6 with three equal spot numbers; these represent $6 \times 20 + 3 \times 30 + 1 \times 6$, hence, as is proper, 216 equally likely cases. Hence, the probabilities of throwing a 9 and of throwing a 10 are not equal to 6/56, but 9 has a probability of 25/216, and 10 the slightly larger probability of $27/216 = 1/8$. The probability of 12 is also 25/216, and that of 11 is 1/8.

The error made 50 years later by another gambler, the Chevalier de Méré, was of a different nature, but he, too, obtained a result that turned out to be contradictory to experience. He applied to the great French philosopher Blaise Pascal (1623–1662), who managed to solve the apparent paradox without much trouble. De Méré, who, according to a letter from Pascal to Fermat (1601–1665), was very intelligent but no mathematician, had observed that it is advantageous to bet on at least one 6 in 4 throws with one die, but that it is disadvantageous to bet on at least one occurrence of 6–6 in 24 throws with two dice. Since there are 6 different throws with one die, and $6 \times 6 = 36$ throws with two dice, de Méré considered this to be in contradiction to arithmetic, which shows that 4 and 6 are in the proportion of 24 and 36. Hence, de Méré started out with a correct division into equally likely cases, but he assumed wrongly that a simple proportion had to exist between the various numbers. With 4 throws with one die there are $6^4 = 1296$ equally likely cases, of which $5^4 = 625$ are unfavorable; hence 671 are favorable for throwing at least one 6. Thus, the probability of this is $671/1296 = 0.5177$. With 24 throws of two dice the probability of at least one double 6 is equal to

$$\frac{36^{24} - 35^{24}}{36^{24}} = 1 - \left(\frac{35}{36}\right)^{24} = 1 - 0.5086 = 0.4914.$$

Hence, the first probability is actually slightly larger than 0.5 and the second probability slightly smaller than 0.5, so that de Méré had given evidence of very good observation.

From such questions put to Pascal, who communicated them in correspondence with Fermat and Huygens, the theory of probability arose. Christiaan Huygens (1629–1690), the great son of the Dutch poet Constantijn Huygens, played an especially important part in the development of the theory of probability through his *Van Rekeningh in Spelen van Geluck* (On Calculations in Games of Chance), which remained the only book on probability for half a century. As Bertrand (1822–1900) puts it so strikingly in his beautifully written *Calcul des probabilités*, the great names of Pascal, Fermat, and Huygens adorn the cradle of the theory of probability.

Even great mathematicians have made errors with regard to equal likelihood. For example, the French mathematician d'Alembert (1717–1783) discussed the probability of throwing heads at least once when tossing a coin twice. Here he distinguished three cases: tails appears on the first toss and on the second toss; tails appears on the first toss and heads on the second toss; heads appears on the first toss. In the last case there is no second toss, so that no more than three cases have to be distinguished. Of these, two are favorable, and for this reason he determined the desired probability as 2/3. However, the three cases are by no means equally likely. To obtain equally likely cases, it is necessary to toss the coin again even when the first toss results in heads, so that the third case splits into two: heads–tails and heads–heads. Thus, we obtain four cases, and these are equally likely. Of these, three are favorable, so that the probability of heads at least once is really 3/4. We can also recognize this fact by imagining that two coins are tossed simultaneously, and that we are seeking the probability that at least one of these coins will come up heads.

A misconception of a more general nature, which can, however, also be considered as an incorrect judgment of equal likelihood, is common even now among gamblers. This is the belief (in the coin-tossing example) that when heads has appeared a number of times in succession, the probability that tails will appear the next time is increased. Even d'Alembert held this view, which, however, was characterized as absurd by his contemporary Euler (1707–1783), one of the greatest mathematicians who ever lived. And indeed it is impossible to see how past results could have an influence on tosses which have yet to be made. However those previous tosses have turned out, heads and tails are equally likely cases again for each succeeding toss.

133. Misleading example of an incorrect judgment of equal likelihood. An interesting case, in which one is easily inclined to misjudge matters of equal likelihood, can be found in the above-mentioned book by Bertrand. Three similar chests contain two drawers each. Both drawers of chest A contain a gold coin, chest B has a silver coin in both drawers, but chest C has a gold coin in one drawer and a silver coin in the other. Someone who knows this, but who is unaware of which chest contains the gold coins, which the silver, and so on, opens a drawer and finds a silver coin in it. What is the probability that the other drawer of the same chest contains a gold coin?

A drawer of chest B or of chest C has been opened. We are inclined to consider both cases as equally likely, and hence to evaluate the required probability as $1/2$. However, the probability is $1/3$, because the three drawers with the silver coins represent equally likely cases, and only one of these is favorable, namely, the drawer with the silver coin in chest C. That the probability is $1/3$ is particularly evident from the fact that the problem reduces to the question: What is the probability that chest C (the one with a gold and a silver coin) has been opened? Initially, this probability is $1/3$, and it remains $1/3$ when a drawer has been opened and a silver coin has been found in it, because gold and silver coins are distributed in exactly the same way over drawers and chests. Hence, someone who made a bet on chest C will not rate his chances either better or worse after a silver coin has been found in the drawer which was opened first.

At the same time, this example of Bertrand's shows us some of the peculiarities of the concept of probability. In the first place, probability is subjective, that is, it is different for one person from what it is for another. A probability depends on our knowledge about factors that exert influence on the occurrence of the event. When our knowledge of these factors is increased, this may change the probability. When we have not yet seen the contents of the opened drawer, the probability that the chest is A is equal to 1/3, as is the probability that it is B; however, when we have found a silver coin in the drawer, the probability that it is A drops to 0, while the probability for B rises to 2/3.

In the second place, the example shows that probability need not be related to an event in the future, but that it may also refer to the past. In the case discussed, it is already certain whether or not the event (in the example, the placing of a gold coin in the other drawer of the box) has occurred. However, as long as we are in the dark concerning its occurrence, we can speak of the probability of the event in question. The possibility of referring to the past is closely related to the subjective nature of probability, for two persons provided with different data about what has occurred will assign different values to the probability of some particular result. This amounts to saying that cases that are equally likely for one person need not be so for the other.

We shall illustrate this once more by considering the question of the probability of throwing 10 with three dice; this probability is 1/8 (see §132). The probability is still 1/8 when the throw has taken place, but no one has seen any of the dice. When someone has seen one of the dice, and found it to show a spot number of 2, then for him the probability of 10 from the three dice together is the same as the probability of 8 from throwing two dice, which is 5/36, as can be easily verified. When he sees another die, his estimate of the probability will either drop to 0 or increase to 1/6, depending on whether that second die shows less than 2, or at least 2.

III. RULES FOR CALCULATING PROBABILITIES

134. Probability of either this or that; the addition rule. It often happens that we seek the probability of occurrence of a certain event which can come about in more than one way. To be more concrete, we take two ways, which we shall call A and B, so that we require the probability of either A or B. Suppose that a division has been made into, say, 15 equally likely cases, 11 of which are favorable. We further assume that of these 11 favorable cases, 4 belong to A and 7 to B. The probability for A is then 4/15, and that for B 7/15. The probability for the event which consists of the occurrence of A or of B

is 11/15, or 4/15 + 7/15. Hence, the probability of an event which can come about in two ways, A and B, is equal to the sum of the probabilities A and B. This is called the rule of total probability, or of "either-or" probability. We also speak of the addition rule.

In the foregoing, there must be explicit assurance that the ways A and B are mutually exclusive, so that it is not possible for the event to come about in both ways simultaneously. Further, it is clear that the rule can be extended to three and more ways. In that case, too, these ways have to be mutually exclusive, since otherwise we would count some favorable cases more than once, and thus find too large a value for the required probability.

This shows that the probability of drawing from a pack of 52 cards a heart or an honor (jack, queen, king, or ace) is not equal to 1/4 (the probability of a heart) + 4/13 (the probability of an honor), which would make it equal to 29/52, but smaller than this by 4/52 (the probability of an honor of hearts), and hence 25/52. This is a consequence of the fact that the four cases in which the card drawn is both a heart and an honor have been counted twice in the incorrect calculation. A correct application of the rule of total probability is obtained by writing for the required probability:

1/4 (probability of a heart) + 1/13 (probability of club honor) + 1/13 (probability of a diamond honor) + 1/13 (probability of a spade honor) = 25/52,

or, alternatively, by writing this probability as:

4/13 (honor) + 9/52 (heart, but not an honor) = 25/52.

As a second example, we take the case in which John and Peter draw a card from a pack of 52, when hearts are trumps. First John draws a card; it is the seven of diamonds. Then he replaces it, and Peter draws a card. The probability that Peter will win is:

7/52 (probability of a higher diamond) + 1/4 (probability of a trump) = 5/13.

The rule of total probability in itself is of little direct advantage, as it is immaterial in practice whether (to stick to the last example) we determine the number of favorable cases by adding the numbers for both ways (hence as 7 + 13 = 20) before dividing the sum by 52, or by dividing both numbers separately by 52, before adding the quotients (which are the probabilities). However, in combination with other rules of the theory of probability, the rule for the probability of "either this or that" is very useful.

135. Probability of both this and that; the product rule. When an event consists of the occurrence of two events A and B, we speak of a compound event. The probability of this is called a compound probability, or a "both-and" probability. For the compound event, we assume that 30 equally likely cases are to be distinguished, 11 of which are favorable to the event A. We further assume that among these 11 cases favorable to A (which, after A has taken place, represent the total number of possible cases for the event B), there are 5 cases that are also favorable to B. The probability of A is then 11/30, whereas, if A has taken place, the probability of B is equal to 5/11. The probability of the compound event (A and B) is 5/30, equal to $11/30 \times 5/11$. Hence, the probability of an event which consists of the occurrence of two events A and B is equal to the product of the probabilities of A and B separately.

In applying this rule of compound probability, which is also called the product rule, it may happen that the two events do not influence each other. As an example of this we take two boxes, one with 3 white and 2 black balls, and one with 5 white and 4 black balls. From each of these boxes a ball is drawn. The probability that both balls are white is $3/5 \times 5/9 = 1/3$, since the probability that a white ball will be drawn from the first box (event A) is 3/5, while the probability that a white one will appear from the second box (event B) is 5/9. When successive drawings are made from the first and the second box, the probability of event B is not influenced by the occurrence or non-occurrence of event A.

When event A does influence event B, the value which we must take for the probability of event B (as appears from the derivation of the rule we found) must be the value that it assumes after the occurrence of event A. As an example we again take the two above-mentioned boxes with white and black balls. From the first box, the one containing 3 white and 2 black balls, we draw a ball; we put it in the second box, shake up the balls, and draw a ball from the second box. The probability that both balls drawn will be white is $3/5 \times 6/10 = 9/25$, since the second box, which initially contained 5 white and 4 black balls, will contain 6 white and 4 black balls after receiving from the first box a ball that is known to be white. Hence, the probability that white will be drawn from the second box, if the first ball was white, has become equal to 6/10 as a result of the addition of the first ball to the second box.

The probability of "both this and that" can also be applied to the puzzle of the seven coins (see §§14 and 15) when we make someone move coins along free lines at random. However, the result depends on what cases are considered to be equally likely. When we take these to be the still unoccupied vertices of the star, then the probability that someone will succeed in placing the seven coins by accident is equal to $\frac{2}{7} \times \frac{2}{6} \times \frac{2}{5} \times \frac{2}{4} \times \frac{2}{3} \times \frac{2}{2} = \frac{4}{315}$, since at the placing of the second coin, 7 free vertices are left, 2 of which are correct; at the placing of the third coin, 2 of the 6 free vertices then remaining are correct, and so on. On the other hand, if we consider the remaining free lines as equally likely cases (counting each free line for two cases, since the line can be traversed in either direction), the probability becomes much smaller, namely, $\frac{2}{12} \times \frac{2}{10} \times \frac{2}{8} \times \frac{2}{6} \times \frac{2}{4} \times \frac{2}{2} = \frac{1}{720}$. The decision as to which of these two probabilities (both small) is the better representation of the truth will depend on whether the problem solver directs most of his attention to the vertices or the lines; he may, perhaps, even pay attention sometimes to the vertices, and sometimes to the lines, which would make the calculation of a probability impossible.

136. Examples of dependent events. In the following example we obtain an incorrect result by not paying attention to the dependence of the events that are to occur simultaneously. We draw a ball from a box containing sixteen balls numbered in sequence from 1 to 16. The probability of an even number is $8/16 = 1/2$, the probability of a multiple of 3 is 5/16. However, the probability of a multiple of 6 (both even and a multiple of 3) is not $\frac{1}{2} \times \frac{5}{16} = \frac{5}{32}$, but $\frac{2}{16} = \frac{1}{8}$, namely the probability of drawing the number 6 or the number 12 from the sixteen balls. When we know that the number drawn is even, then the probability of a multiple of 3 is no longer 5/16, but $2/8 = 1/4$ (because fewer multiples of 3 occur among the even numbers than among the odd numbers); thus we find that the required probability is $\frac{1}{2} \times \frac{1}{4}$.

An example in which we are even more inclined to overlook the dependence of the events is the following. Two equally skillful marksmen, John and Peter, shoot at a target. John is allowed to fire one shot, and Peter two shots. The one who fires the best shot wins. How great is John's probability of winning? We can safely assume that John fires the first shot. The probability that Peter's first shot is worse is 1/2. The probability that Peter's second shot is worse than John's shot is also 1/2. Hence, the probability that Peter's shots are both worse than John's shot, in other words, the probability that John wins, is $\frac{1}{2} \times \frac{1}{2} = \frac{1}{4}$. However, this argument is incorrect, because the two events are not mutually independent. The fact that Peter's first shot is worse than John's makes it probable that John has fired a good shot and that therefore Peter's second shot will also be worse than John's. Hence, in reality, John's probability is greater. It is 1/3, since each of the three shots has the same probability of being the best, and only one of these three equally likely cases is favorable to John. From the probability 1/3 it appears further that, if Peter's first shot is worse than John's, there is a probability of 2/3 that Peter's second shot will also be worse than John's.

137. Maxima and minima of sequences of numbers. Mistakes like the one mentioned in the problem of the marksmen of §136 are made repeatedly, and only recently this happened in connection with a question of entirely the same kind. Suppose that a large number of dice (10, say) are thrown a number of times in succession, and that the various totals thrown are noted down; this gives a sequence of numbers, for example: 41, 13, 37, 51, 15, and so on. What is the probability that the fifth number of the sequence is a maximum, that is, larger than the fourth number and larger than the sixth? Here we disregard the possibility that two of these three numbers are equal. Then we can argue as follows. The probability that the fourth number is smaller than the fifth is 1/2; the probability that the sixth number is smaller than the fifth is also 1/2. Hence, the probability that both the fourth and sixth numbers are smaller than the fifth number is $\frac{1}{2} \times \frac{1}{2} = \frac{1}{4}$. Here, the same error is made as was made above, with the marksmen. The actual probability that the fifth number is a maximum, hence larger than the two adjacent numbers, is 1/3, since the fourth, the fifth, and the sixth number each have the same probability of being the largest of the three.

From the foregoing it appears that the 1/3 probability of a maximum also applies to other sequences of numbers that have arisen by chance, such as (for example) the numbers which win the first prize in successive drawings of a national lottery, or successive numbers that win in roulette. The probability we have determined leads us to expect that about 1/3 of the numbers are local maxima, and also, of course, that 1/3 of the numbers are local minima. In 6000 numbers it will occur about 1000 times that 3 successive numbers are in ascending order, and also about 1000 times that the three numbers are in descending order, hence (for example) that numbers 27, 23, 9 appear successively.

These considerations are also of importance for sequences of numbers that are governed partly by regularity, partly by chance, such as (for example) the observed temperatures from a weather bureau; here they help us decide whether observed fluctuations can be ascribed to chance, or indicate a periodical regularity.

138. Extension to several events. As is the case with the sum rule, the product rule can be extended to more than two events: A, B, C, and D, say. If the probabilities of A, B, C, and D are 1/3, 2/5, 3/4, and 2/3, respectively, then the probability of the coincidence of A, B, C, and D is equal to $\frac{1}{3} \times \frac{2}{5} \times \frac{3}{4} \times \frac{2}{3} = \frac{1}{15}$. If the events occur one after the other in the order indicated, and if they exert influence on one another, then for the probability of B we have to take the value it assumes after A has taken place, for the probability of C the value it assumes after A and B have occurred, and for the probability of D the value it assumes after the occurrence of A, B, and C.

A frequently arising case is that in which the events that are to coincide are mutually independent and all have the same probability. The probability of the compound event is then a product of equal factors, and hence a power of the simple probability.

As an example we take the case in which a die is thrown four times in succession. The probability of four sixes is then $(1/6)^4 = 1/1296$. The probability of no six is $(5/6)^4 = 625/1296$, so that the probability of at least one six is $1 - \frac{625}{1296} = \frac{671}{1296}$; in §132 this probability was found by directly counting the equally likely cases and the favorable cases.

However, in somewhat more complex examples the use of the product rule is a little more convenient. By these means we can often avoid having to consider numbers of combinations, too. As an example we take a box containing 12 white and 13 black balls. From it we draw four balls, and seek the probability that they are all white. The equally likely cases are the combinations of $12 + 13 = 25$ balls in groups of four; the number of these is $\frac{25!}{4! \times 21!} = \frac{25 \times 24 \times 23 \times 22}{4!}$ (see §128). The favorable cases are the combinations of 12 white balls in groups of four; the number of these is

$$\frac{12!}{4! \times 8!} = \frac{12 \times 11 \times 10 \times 9}{4!}.$$

Hence, the required probability is $\frac{12 \times 11 \times 10 \times 9}{25 \times 24 \times 23 \times 22} = \frac{9}{230}$. The calculation becomes easier if we imagine that the four balls are drawn one after the other, which obviously does not change the required probability. Hence, the probability that the first ball is white is 12/25. If it is white, then there are 24 balls left in the box, among which are 11 white balls. Hence, the probability that after drawing a white ball the second ball is white, too, is 11/24. If this one is also white, the probability that the third ball is again white is 10/23. After three white balls, the probability of the fourth ball being white is 9/22. Hence, the required probability is $\frac{12}{25} \times \frac{11}{24} \times \frac{10}{23} \times \frac{9}{22} = \frac{9}{230}$.

139. Combination of the sum rule and product rule. In the calculation of a probability, we usually have to apply both the sum rule and the product rule. As an example we take two similar boxes *A* and *B*. Box *A* contains 3 white and 4 black balls; box *B* contains 4 white and 5 black balls. Someone draws a ball at random from one of the boxes. What is the probability that it is white? The probability that he chooses box *A* is 1/2. Hence, the probability that he draws a white ball that originates from box *A* is $\frac{1}{2} \times \frac{3}{7} = \frac{3}{14}$. Likewise, the probability that he draws a white ball that originates from box *B* is $\frac{1}{2} \times \frac{4}{9} = \frac{2}{9}$. So, according to the rule of the probability of "either this or that," the probability of his drawing a white ball is $\frac{3}{14} + \frac{2}{9} = \frac{55}{126}$.

In this problem, it is not possible to indicate equally likely cases. The 16 balls in the two boxes do not represent equally likely cases, because a ball from the box containing 7 balls has a larger probability of being drawn than one from the box containing 9 balls. We can obtain equally likely cases by modifying the problem to the effect that both boxes contain the same number of balls, with 27 white and 36 black balls in *A*, and 28 white and 35 black balls in *B*. The proportion of white to black balls has not changed in either box, hence the probability of drawing a white ball has not changed either. In the modified problem, there are $2 \times 63 = 126$ equally likely cases (the balls in the two boxes), of which $27 + 28 = 55$ are favorable (the 55 white balls in the two boxes). This, too, leads to the probability 55/126.

The problem discussed in §132 (heads or tails), of which an incorrect solution was given by d'Alembert, can also be treated in a convincing manner by using the sum rule and the product rule. The probability of tossing heads at least once in two tries is $\frac{1}{2} + \left(\frac{1}{2} \times \frac{1}{2}\right) = \frac{3}{4}$, since there is a probability 1/2 of tossing heads the first time, and a probability $\frac{1}{2} \times \frac{1}{2}$ of doing this on the second try only. The desired probability can also be found from the fact that the complementary chance (the probability of tails twice) is $\frac{1}{2} \times \frac{1}{2} = \frac{1}{4}$.

The foregoing can be extended to the case in which an event can happen in several mutually exclusive ways. We then have to multiply the probabilities of each of these ways by the probability that the contingency in question (if operative) will bring about the event; to get the required probability, we must add these products. For example, the event may again be the drawing of a white ball, while the ways in which this can occur can be from different boxes containing white and black balls; the probability that a certain way (or box) brings about the event will depend on the number of white and black balls in that box.

140. More about maxima and minima in a sequence of numbers. The considerations of §137 were based on the assumption that we can neglect the occurrence of equal numbers among three successive numbers of a sequence that has been brought about by chance. It is only under that assumption that there is a probability 1/3 of an arbitrary number of the sequence being a maximum, that is, larger than the two adjacent numbers. The probability is no longer equal to 1/3 when the numbers of the sequence can take on only a few different values.

As an example we take the case in which the numbers of the sequence are the spot numbers of successive throws with one die. We then have to agree upon what is to be understood by a maximum, so that, for example, we have to know whether the middle number of 4–4–3 should or should not be considered as a maximum.

The probability that three successive throws with the die are equal, hence that the second and the third throw are equal to the first, is $(1/6)^2 = 1/36$. The probability that the three throws are different, hence that the second throw is different from the first, and the third different from the first and the second throws, is $\frac{5}{6} \times \frac{4}{6} = \frac{5}{9}$. Hence, the probability that two of the three throws are equal, without all three being equal, is $1 - \frac{1}{36} - \frac{5}{9} = \frac{5}{12}$.

If we speak of a maximum in the sense that the middle throw is not exceeded by either of the two adjacent throws, so that (for example) the middle throw of 4–4–2 or 4–4–4 is to be reckoned a maximum, then with three equal throws the probability that the middle throw is a maximum is equal to 1. With three different throws that probability is 1/3, as has been shown in §137. If two of the throws, but not all three, are equal, the middle throw is certainly either a maximum or a minimum; hence, the probability that it is a maximum is 1/2, since obviously the probability of the middle throw being a maximum is as large as that of its being a minimum. Hence, the probability that a certain throw is a maximum, in the sense here attached to this, is $\frac{1}{36} \times 1 + \frac{5}{9} \times \frac{1}{3} + \frac{5}{12} \times \frac{1}{2} = \frac{91}{216}$, or approximately 0.4213, that is, something greater than 1/3. The probability that the middle throw is a minimum is, of course, also equal to 91/216.

We now take the case in which there are successive throws with one die and a maximum is understood to be present only when the middle throw is larger than the two adjacent throws. When two throws are equal but not all three, then the probability that the odd throw is the largest is 1/2, and the probability that the odd throw is the middle one is 1/3; hence there is a probability $\frac{1}{2} \times \frac{1}{3} = \frac{1}{6}$ that the middle throw is largest. Hence, the probability that a certain throw is a maximum according to the last definition, is

$$\left(\frac{1}{36} \times 0\right) + \left(\frac{5}{9} \times \frac{1}{3}\right) + \left(\frac{5}{12} \times \frac{1}{6}\right) = \frac{55}{216},$$

or approximately 0.2546, that is, something less than 1/3.

IV. PROBABILITIES OF CAUSES

141. A posteriori probability: the quotient rule. As in §139, we assume that an event can be brought about in several mutually exclusive ways. These ways are also called causes; hence this word does not quite have its usual meaning.

We assume that the event has occurred, but that we do not know from which of the possible causes. In the example this means that a white ball has been drawn, but that we do not know from which box it came. We then might seek the probability that a certain cause has been operating, or, in the example, the probability that the white ball came from some definite box. This probability is called the a posteriori probability of that cause. The a posteriori probability can be found by using two ways to express the probability that the event is brought about by the particular cause. In the first place, this probability is the product of the a priori probability that the cause will be operative, and the probability that the cause (if operative) will give rise to the event. In the second place, the probability that the event will be brought about by the particular cause is equal to the product of the probability that the event will occur and the a posteriori probability that it has been brought about by the particular cause. This shows that after the occurrence of the event in question we have:

$$\text{a posteriori probability of a certain cause} = \frac{\text{a priori probability of the event (as a result of that cause)}}{\text{total a priori probability of the event}}.$$

This rule, which we shall call the quotient rule, is due to the English mathematician Bayes (d. 1763), and so is called Bayes' rule. We also speak of the rule of probabilities of causes.

As an illustration we take the two boxes of §139, box A containing 3 white and 4 black balls, and box B containing 4 white and 5 black balls. The a priori probability that a white ball is drawn from box A is $\frac{1}{2} \times \frac{3}{7} = \frac{3}{14}$. The probability that a white ball is drawn irrespective of the box is (as we found in §139) equal to 55/126. When a white ball has been drawn, and we do not know from which box it came, then the probability that the ball comes from box A is, according to the quotient rule, equal to $\frac{3}{14} \div \frac{55}{126} = \frac{27}{55}$.

Bertrand's problem (§133) about the three chests A, B, and C, with two drawers each, is also a problem of an a posteriori probability. When we open a drawer of one of the chests and find a silver coin in it, then the causes through which the event can have been brought about are the choice of chest B and the choice of chest C. The a priori probabilities of B and C are both equal to 1/3. The probability that cause B, if operative, brings about the event is 1; the probability that cause C, if operative, leads to finding a silver coin, is 1/2. Hence, the a priori probability of finding a silver coin is $\frac{1}{3} \times 1 + \frac{1}{3} \times \frac{1}{2} = \frac{1}{2}$; this probability can also be found from the fact that the gold and silver coins are distributed in the same way over both chests and drawers, and that therefore there is as much chance of finding a silver coin as of finding a gold coin. When a silver coin has been found in the drawer opened first, then, according to the quotient rule, the probability that the other drawer of the same chest contains a gold coin (hence, the a posteriori probability that chest C has been chosen) is equal to $\frac{1}{6} \div \frac{1}{2} = \frac{1}{3}$.

142. Application of the quotient rule. A bridge pack (52 cards) is distributed between John and Peter. From his pile of 26 cards, Peter draws a card at random. Then John does the same with his own pile. John wins if he draws a higher card of the same suit, and also when he can trump, because (for example) Peter draws a diamond and John draws a spade (when spades are trumps). Charles sees that John takes the trick, but he did not see either of the two cards. What (to Charles) is the probability that John has trumped a plain card of Peter's?

The a priori probability that John draws a card of the same suit as Peter's is 12/51, because after Peter has drawn there are 12 cards of the same suit as Peter's card among the 51 remaining cards; the probability that it is higher is as large as the probability that it is lower, hence the a priori probability that John draws a higher card of the same suit is $\frac{12}{51} \times \frac{1}{2} = \frac{2}{17}$. The probability that Peter does not draw a trump card but John does is $\frac{3}{4} \times \frac{13}{51} = \frac{13}{68}$. Hence, the total probability that John will win is $\frac{2}{17} + \frac{13}{68} = \frac{21}{68}$. Hence, when John has won, the a posteriori probability that this has occurred through

John trumping a plain card is $\frac{13}{68} \div \frac{21}{68} = \frac{13}{21}$.

143. Another application. Another application is the following. A box contains 100 balls, some white, some black, possibly all white or all black. We assume that all assortments have the same probability, namely 1/101, because a number has been drawn at random from the 101 numbers 0, 1, 2, . . . , 100 to determine the number of white balls to be put in the box. When we draw a ball from the box, we get a box containing 99 balls, with the same probability for all assortments, namely 1/100, since a box with, say, 37 white and 62 black balls can have arisen from drawing a white ball from a box containing 38 white and 62 black balls, but also by drawing a black ball from a box containing 37 white and 63 black balls. So the probability that 37 white and 62 black balls remain in the box is

$$\left(\frac{1}{101} \times \frac{38}{100}\right) + \left(\frac{1}{101} \times \frac{63}{100}\right) = \frac{1}{100};$$

any other assortment obviously leads to the same probability.

The result found can be applied repeatedly. If we draw another ball, 98 balls remain in the box, and all assortments have the same probability, namely 1/99. If we draw 20 balls from the box, all assortments of the 80 remaining balls still have the same probability; this is then equal to 1/81. Since the balls drawn and the balls remaining in the box play the same role, the assortments of the 20 balls drawn also all have the same probability, namely 1/21.

We draw 20 balls at random from the box. They turn out to be all white. Now we draw one more ball, without having put back the 20 balls in the box, and seek the probability that the twenty-first ball is white again.

The situation can be interpreted like this: a draw of 21 balls has been made. As long as we have not seen the color of the balls, all assortments have the same probability; hence there is a probability of 1/22 that they are all white. The probability that the 20 balls drawn first are all white is 1/21. This can have 2 causes, namely: there is one black ball among the 21 balls, and there is no black ball among the 21 balls. The a priori probability of the latter event is 1/22, hence the a posteriori probability of 21 white balls is $\frac{1}{22} \div \frac{1}{21} = \frac{21}{22} = 1 - \frac{1}{22}$.

Hence the sought-for probability is $1 - \frac{1}{22}$. It turns out to be independent of the number of balls originally in the box, provided all assortments have the same probability, and the balls drawn are not replaced. When we draw 33 balls, and they are all white, then the probability that the thirty-fourth ball is also white is $\frac{34}{35} = 1 - \frac{1}{35}$; with 47 balls drawn the probability becomes $1 - \frac{1}{49}$, and so on. These results also hold, of course, but then approximately only, when we do replace the balls drawn, provided the number of balls in the box is very large compared with the number of balls drawn.

These results have been applied in situations like the following: On the way to work, you meet the same person 20 days in succession. The probability that when going to work the next day you will again meet that person is 21/22. However, it remains very doubtful whether the assumptions that have led to this probability are to any extent justified here.

More doubtful still is the application that was once made to the question: What is the probability that the sun will rise tomorrow? One can certainly say that the sun has risen for millions of successive days (about 27 centuries), since an event as striking as the sun not rising would certainly have been recorded. Application of our result would lead to a probability $1/10^6$ of the sun not rising tomorrow. In reality, however, this probability is very much smaller, because here we are not concerned with an event that depends on chance, but with a phenomenon which occurs according to fixed laws that are accurately known by us. On the other hand, in this way we may perhaps find too large a value for the probability that the sun will rise tomorrow. If a traveller visits the polar regions, and observes that every day the sun attains a lower point above the horizon, then on a day when the sun has just been barely visible he will set the probability that he will see the sun the next day at a value much smaller than 31/32 if he has seen the sun rise in the polar region for 30 successive days, say. A pattern is observed here, and it makes no sense at all to apply formulae that have been derived from cases where such a pattern is absent and where everything depends on chance.

Chapter VIII: SOME APPLICATIONS OF THE THEORY OF PROBABILITY

I. VARIOUS QUESTIONS ON PROBABILITIES

144. Shrewd prisoner. Long ago a prisoner was to be executed. In response to his supplications, he was promised that he would be released if he drew a white ball from one of two similar urns. The provisions were that he had to distribute 50 white and 50 black balls between the two urns, in any way he liked, after which he had to draw a ball at random from one of these urns. The story goes that he drew a white ball and was released.

How did the prisoner arrange to make his chance of success as great as possible? If he had put equal numbers of white balls and black balls into one of the urns, so that the other urn also contained as many white balls as black balls, then his probability of drawing white would obviously have been 1/2. Therefore, he put more white balls than black balls in urn A, and more black balls than white balls in urn B. If he were unlucky and chose urn B, then the probability of drawing white from it would be less than 1/2. He made this probability less than 1/2 by as little as possible by making the number of black balls in B only one more than the number of white balls, and also by putting as many balls as possible in urn B, hence 49 white balls and 50 black balls. He put the remaining white ball in A, and by so doing he increased his probability to certainty if he had the luck to choose urn A. In this manner his probability of success became

$$\frac{1}{2} \times 1 + \frac{1}{2} \times \frac{49}{99} = \frac{74}{99},$$

hence a little less than 3/4.

His prospects would have been still better if he had been allowed to distribute 100 balls among four urns, say. He would then have put a single white ball in each of three of the urns, and the remaining 97 balls in the fourth urn. His chance would then have been

$$\frac{1}{4} \times 1 + \frac{1}{4} \times 1 + \frac{1}{4} \times 1 + \frac{1}{4} \times \frac{47}{97} = \frac{169}{194},$$

which is a little less than 7/8. With four urns and a larger number of balls (1000 white and 1000 black balls, say) his chance would have been still closer to 7/8.

145. Game of kasje. This dice game, which is very popular, is played with three dice by two players, John and Peter. Winning throws are 4–1–1, 4–1–2, 4–1–3, 4–1–4, 4–1–5, and 4–1–6, called kasje 1, kasje 2, and so on; kasje 5 is a better throw than kasje 3, and so on. A kasje need not be made in 1 throw: the player is also allowed to do it in two or in three throws. If he does not throw a 4 the first time, he continues to throw with three dice. If he gets a 4 on the first or second throw, but not a 1, he leaves the 4 (or one of the fours) and throws once more with the other two dice. If he has obtained a 4 and a 1 on the first or second throw, he is allowed to throw once more with the third die, in the hope of thus improving his throw; however, he is also allowed to abandon the next throw, and leave the position as it is. In addition, there is the stipulation that the throw 4–4–4, provided it is obtained when the dice are first thrown, is higher than all other throws, hence even higher than kasje 6.

First, John makes one, two, or three throws, after which it is Peter's turn. The player who has managed to attain the higher throw wins. In the case of two equal throws the whole procedure is repeated. The throws are also equal when both players have nothing. By "nothing" we mean that there was no first throw of 4–4–4 and that three throws have given neither player a combination 4–1.

In view of the choice of stopping or continuing after obtaining 4–1 at an early stage, a player can exercise some degree of judgement in this game. If John has obtained 4–1–1 before his third throw, then of course he throws once more (with one die), because by so doing he cannot lower his throw, whereas he does have the possibility of improving it. Also with 4–1–2 or 4–1–3 (from the first or second throw) John is well-advised to throw once more with the third die, since with 4–1–3 he has the probability 1/2 of improving his throw, the probability 1/3 of lowering it, and the probability 1/6 of leaving his throw unchanged. With 4–1–4 from the second throw, and always with 4–1–5, John should stop.

With 4–1–4 from the first throw, John is well-advised to continue to throw. For, if he throws 1, 2, or 3 with his odd die, he is allowed to throw that die again. Therefore, John's probability of a final result lower than 4–1–4 (which arises only when he throws 3 or lower twice in succession, because on 4–1–4 in two throws he stops) is equal to

$\left(\frac{1}{2}\right)^2 = \frac{1}{4}$. Hence, John has a probability of $\frac{3}{4}$ for a final result 4–1–4 or higher, since he has a probability of $\frac{1}{4}$ for each of the results 4–1–4, 4–1–5, and 4–1–6 separately. Hence, by continuing to throw when he has 4–1–4 from his first throw, John has a probability of $\frac{1}{2}$ to improve his holding and a probability of only $\frac{1}{4}$ to lower it. Hence, it is quite advantageous in this case to continue to throw.

Peter, who throws last, has an easier game, because he is in a better position to judge whether or not he should continue with some particular kasje. Peter stops when he is higher than John, and continues (if this is still possible) as long as his result is lower. If Peter is level with John before his third throw, then he continues to throw with 4–1–3 or less, and also with 4–1–4 on the first throw; in the remaining cases he stops, because then he has a draw already, and a larger probability of losing than of winning if he were to continue. In view of all this, Peter has a slightly larger probability of winning than John has. Therefore John and Peter should take turns to begin when the game is played a number of times in succession.

***146. Simplification of the game of kasje.** In calculating the probability of winning, we shall assume for the sake of simplicity that each of the players is allowed to throw only twice at the most. As before, the player who throws first is assumed to be John.

John's probabilities on the first throw (as can readily be shown by counting favorable cases) are 4–4–4: probability **1/216**; 4–1–1 and 4–1–4: probability **1/72** each; 4–1–2, 4–1–3, 4–1–5, and 4–1–6: probability **1/36** each; no 4: probability **125/216**; a 4, no 1, and not 4–4–4: probability **5/18**.

The simplest way to find the probability of no 4 is to derive it in the form $(5/6)^3$, and for the final case the last probability (5/18) is found most simply from the fact that the sum of all probabilities is equal to 1. We assume that with 4–1–3 or less, John continues to throw, but that he refrains when he has 4–1–4 or higher. When he throws again after 4–1–1, 4–1–2, or 4–1–3, his probabilities of 4–1–1, 4–1–2, 4–1–3, 4–1–4, 4–1–5, and 4–1–6 all become equal to **1/6**. With a 4, no 1, and not 4–4–4, John throws once more with two dice. By so doing, his probabilities become: for 4–1–1, probability **1/36**; for 4–1–2, 4–1–3, 4–1–4, 4–1–5 and 4–1–6, **1/18** each; for a "nothing," **25/36**.

With no 4, John throws once more with the three dice, and so his probabilities become: for 4–1–1 and 4–1–4, probability **1/72** each;

for 4–1–2, 4–1–3, 4–1–5, 4–1–6, **1/36** each; for nothing, **31/36**.

Hence, the total probability that John will get 4–1–1 is:

$$\frac{1}{72}\times\frac{1}{6}+\frac{1}{36}\times\frac{1}{6}+\frac{1}{36}\times\frac{1}{6}+\frac{5}{18}\times\frac{1}{36}+\frac{125}{216}\times\frac{1}{72}$$
$$=\frac{425}{1552}=0.02733.$$

John's total probability of 4–1–2, equally that of 4–1–3, is:

$$\frac{1}{72}\times\frac{1}{6}+\frac{1}{36}\times\frac{1}{6}+\frac{1}{36}\times\frac{1}{6}+\frac{5}{18}\times\frac{1}{18}+\frac{125}{216}\times\frac{1}{36}$$
$$=\frac{335}{7776}=0.04308.$$

John's total probability of 4–1–4 is:

$$\frac{1}{72}\times\frac{1}{6}+\frac{1}{72}+\frac{1}{36}\times\frac{1}{6}+\frac{1}{36}\times\frac{1}{6}+\frac{5}{18}\times\frac{1}{18}+\frac{125}{216}\times\frac{1}{72}$$
$$=\frac{761}{15552}=0.04893.$$

John's total probability of 4–1–5, and equally of 4–1–6, is:

$$\frac{1}{72}\times\frac{1}{6}+\frac{1}{36}\times\frac{1}{6}+\frac{1}{36}\times\frac{1}{6}+\frac{1}{36}+\frac{5}{18}\times\frac{1}{18}+\frac{125}{216}\times\frac{1}{36}$$
$$=\frac{551}{7776}=0.07086.$$

In addition, John still has a probability $1/216 = 0.00463$ of obtaining 4–4–4. His probability of nothing is

$$\frac{5}{18}\times\frac{25}{36}+\frac{125}{216}\times\frac{31}{36}=\frac{5375}{7776}=\mathbf{0.69123}.$$

Naturally the sum of these probabilities is equal to 1.

Now it is Peter's turn. The probabilities printed above in **boldface** apply for his throws also. If John has obtained nothing, Peter has the probability 0.69123 to draw, the probability 0 for a loss, and hence the probability 0.30877 of a win.

If John has obtained 4–1–1, Peter does not throw again with 4–1–2, and his probability of drawing is

$$\frac{1}{72}\times\frac{1}{6}+\frac{5}{18}\times\frac{1}{36}+\frac{125}{216}\times\frac{1}{72}=\frac{281}{15552}=0.01807,$$

his probability of losing is 0.69123, hence his probability of winning is 0.29070.

If John has obtained 4–1–2, Peter continues to throw with 4–1–2, but does not with 4–1–3 and his chances are:

to draw:

$$\frac{1}{72} \times \frac{1}{6} + \frac{1}{36} \times \frac{1}{6} + \frac{5}{18} \times \frac{1}{18} + \frac{125}{216} \times \frac{1}{36} = \frac{299}{7776} = 0.03845,$$

to lose:

$$\frac{1}{72} \times \frac{1}{6} + \frac{1}{36} \times \frac{1}{6} + \frac{5}{18} \times \frac{13}{18} + \frac{125}{216} \times \frac{63}{72} = \frac{3701}{5184} = 0.71393,$$

hence his chance of winning is 0.24762.

If John has obtained 4–1–3, then Peter, who throws again with 4–1–3, but not with 4–1–4, has a probability 0.04308 of a draw and a probability

$$\frac{1}{72} \times \frac{2}{6} + \frac{1}{36} \times \frac{2}{6} + \frac{1}{36} \times \frac{2}{6} + \frac{5}{18} \times \frac{14}{18} + \frac{125}{216} \times \frac{65}{72}$$
$$= \frac{11845}{15552} = 0.76164$$

of losing, hence a probability 0.19528 of winning.

If John has obtained 4–1–4, Peter, who does not continue with 4–1–4, has a probability 0.04893 of drawing and a probability

$$\frac{1}{72} \times \frac{3}{6} + \frac{1}{36} \times \frac{3}{6} + \frac{1}{36} \times \frac{3}{6} + \frac{5}{18} \times \frac{15}{18} + \frac{125}{216} \times \frac{67}{72}$$
$$= \frac{12515}{15552} = 0.80472$$

of losing, hence a probability 0.14635 of winning.

If John has obtained 4–1–5, then Peter, who continues with 4–1–4, but not with 4–1–5, has a probability

$$\frac{1}{72} \times \frac{1}{6} + \frac{1}{72} \times \frac{1}{6} + \frac{1}{36} \times \frac{1}{6} + \frac{1}{36} \times \frac{1}{6} + \frac{1}{36} + \frac{5}{18} \times \frac{1}{18} +$$
$$\frac{125}{216} \times \frac{1}{36} = \frac{569}{7776} = 0.07317$$

of drawing, a probability

$$\frac{1}{72} \times \frac{4}{6} + \frac{1}{72} \times \frac{4}{6} + \frac{1}{36} \times \frac{4}{6} + \frac{1}{36} \times \frac{4}{6} + \frac{5}{18} \times \frac{16}{18} + \frac{125}{216} \times \frac{34}{36}$$
$$= \frac{3301}{3888} = 0.84902$$

of losing, and a probability 0.07780 of winning.

If John has obtained 4–1–6, Peter has a probability

$$2 \times \frac{1}{72} \times \frac{5}{6} + 3 \times \frac{1}{36} \times \frac{5}{6} + \frac{5}{18} \times \frac{17}{18} + \frac{125}{216} \times \frac{35}{36}$$
$$= \frac{7135}{7776} = 0.91757$$

of losing, a probability 1/216 = 0.00463 of winning, and hence a probability 0.07780 of drawing.

If John has obtained 4–4–4, Peter has a probability 0.00463 of drawing, a probability 0.99537 of losing, and a probability 0 of winning.

Hence the total probability that Peter will win is

$$0.69123 \times 0.30877 + 0.02733 \times 0.29070 + 0.04308 \times 0.24762$$
$$+ 0.04308 \times 0.19528 + 0.04893 \times 0.14635$$
$$+ 0.07086 \times 0.07780 + 0.07086 \times 0.00463 = 0.25346.$$

The total probability that Peter will lose is:

$$0.02733 \times 0.69123 + 0.04308 \times 0.71393 + 0.04308 \times 0.76164$$
$$+ 0.04893 \times 0.80472 + 0.07086 \times 0.84902$$
$$+ 0.07086 \times 0.91757 + 0.00463 \times 0.99537 = 0.25162.$$

Hence, the total probability that Peter will win is in fact slightly larger than the probability that he will lose. The probability of a draw is 0.49492.

In view of the fact that a high throw by John compels Peter to aim for a high result when he continues to throw, the question arises whether it is perhaps more advantageous for John (who begins) to continue to throw when he obtains 4–1–4 from his first throw. To settle this point, we have to repeat the foregoing computation assuming this new strategy is used by John. However, the various probabilities undergo slight changes only, so that it is not necessary to do the whole computation over again. The total probability that John will obtain 4–1–1 now becomes larger by $\frac{1}{72} \times \frac{1}{6} = \frac{1}{432} = 0.00231$, as do John's probabilities for obtaining 4–1–2, 4–1–3, 4–1–5, and 4–1–6. However, the probability that John will obtain 4–1–4 is diminished by $5/432 = 0.01157$. As a result, Peter's probability of winning is increased by

$$0.00231 \times (0.29070 + 0.24762 + 0.19528 + 0.07780 + 0.00463 - 5 \times 0.14635) = 0.00231 \times 0.08428 = 0.000195.$$

On the other hand, the probability that Peter will lose is diminished by

$$0.00231 \times (5 \times 0.80472 - 0.69123 - 0.71393 - 0.76164 - 0.84902 - 0.91757) = 0.00231 \times 0.09021 = 0.000209,$$

so that the probability of a draw is increased by 0.000014. Hence, it is slightly better for John to stop with 4–1–4 on his first throw (with our assumption that only two throws are allowed), although the difference in probability is insignificant.

A similar computation shows that when John stops with 4–1–3 on the first throw, his probability of winning decreases by 0.00078, and his probability of losing increases by 0.00097, so that the probability of a draw decreases by 0.00019. Here the disadvantage to John is larger than when he continues to throw with 4–1–4, as was to be expected without computation. Considered in isolation, stopping at 4–1–3 seems to be disadvantageous in the same way as continuing to throw with 4–1–4. However, because Peter takes note of John's throw and of the possible need to aim for a high result, the disadvantage to John of stopping at 4–1–3 is increased, and the disadvantage of continuing with 4–1–4 is diminished.

147. Poker dice. This is another popular game with dice, somewhat similar to the game of kasje but rather more complicated, which offers a smaller probability of a draw. This game requires five dice, which have faces bearing 9, 10, jack, queen, king, ace (derived from the pack of cards). The five dice are thrown successively by two players, John and Peter; the winner is the one who makes the higher throw. In the case of equal throws (a draw), there is a fresh start. The winning throws, in descending order are: five of a kind, four of a kind, a full house (three of one kind plus two of another kind), a big straight (10, j, q, k, a), a little straight (9, 10, j, q, k), three of a kind, two pairs, one pair (or two of a kind).

If no such combination is thrown (the throw being 9, 10, j, q, a or 9, 10, j, k, a or 9, 10, q, k, a or 9, j, q, k, a), the player has nothing; the throw (called a bust, or a broken straight) then ranks as the lowest possible, and can lead to a draw at the best. As for five of a kind, five aces beats five kings, five kings beats five queens, and so on. With four of a kind, four aces is the highest; here the fifth die plays no role, nor do the other two dice with three of a kind, the fifth die with two pairs, and the other three dice with one pair. If both players have a full house, the one with the higher three of a kind wins; if these are the same, the higher pair decides. If John and Peter both have two pairs, the one with the highest pair wins; if they both have the same highest pair, then the lowest pair decides, so that two kings plus two jacks beats two kings plus two tens. With three of a kind or a pair the higher value wins.

As in the game of kasje, it is not necessary to achieve the result in one throw. If John throws first, he is allowed to make three throws at most, and at each throw he can leave as many dice on the table as he chooses; he is also free to choose which dice he leaves, and he is also allowed to make a fresh throw with all five dice. The same is true for Peter, with the additional condition that he is not allowed to make more throws than John. Hence, if John obtains a somewhat high throw the first time, and stops, then Peter in his turn is allowed to make only one throw; so if John's first throw is on the high side, his wise course is not to try to improve it by continuing to throw, but to stop, and thus deny Peter the chance of making more throws.

In contrast to the situation for the game of kasje, which has no restriction on the number of Peter's throws, it is difficult to tell whether poker dice gives an advantage to John, who begins, or to Peter. Peter again has the advantage that he is in a better position to decide whether to continue to throw or not, while John now has the advantage that he can prevent Peter from continuing to throw. An extremely laborious computation would be required to determine which of these advantages tips the scale. It seems likely that John's chances and Peter's are more or less equal.

148. Probabilities in poker dice. In the game of §147, the probabilities for the result of the first throw can be easily calculated. Since six different throws are possible with one die, the number of possible throws with five dice is $6^5 = 7776$. To compute the probability of a certain throw, all that remains to be done is to determine the number of favorable cases. Thus, two aces, two kings, and a jack counts as $\frac{5!}{2! \times 2!} = 30$ cases, the number of permutations of five objects among which are two pairs of equivalent objects (see §129). Further, three aces and two kings counts as $\frac{5!}{3! \times 2!} = 10$ cases, the number of permutations of five objects among which there are three equivalent objects of one type and two equivalent objects of another type.

The probability of five aces is 1/7776. This is also the probability of five kings, five queens, and so on. The total probability of five of a kind is six times this value, hence $1/1296 = 0.00077$.

The probability of four aces (and also that of four kings, and so on) is 25/7776, since the fifth die can have 9, 10, jack, queen, or king, and can also be any of the five dice, so that there are $5 \times 5 = 25$ favorable cases. The total probability of four of a kind is again six times as large, hence $25/1296 = 0.01929$.

The probability of three aces and two kings is $10/7776 = 5/3888$. The total probability of a full house is 6×5 times as large, hence $25/648 = 0.03858$.

The probability of a big straight is $5!/6^5 = 5/324$, which is also the probability of a little straight. Hence the probability of a straight is $5/162 = 0.03086$.

With three aces, the remaining two dice can still fall in $\frac{5!}{2! \times 3!} = 10$ ways, while the three aces and the other two values can also be distributed in $5!/3! = 20$ ways among the five dice. Hence, the number of favorable cases is $10 \times 20 = 200$, so that the probability of three aces is $200/7776 = 25/972$. The total probability of three of a kind is six times as large, hence $25/162 = 0.15432$.

The probability of two aces and two kings is $120/7776 = 5/324$. The total probability of two pairs is 15 times as large, hence $25/108 = 0.23148$.

The probability of two aces is $\frac{10 \times 60}{7776} = \frac{25}{324}$, hence the total probability of one pair is $25/54 = 0.46296$.

The probability of a bust is $\frac{4 \times 5!}{7776} = \frac{5}{81} = 0.06172$.

The fact that the sum of all these probabilities is equal to 1 can serve as a check.

The probability of a throw which allows a possible win is $1 - 0.06172 = 0.93828$. Since even at the first throw there is little probability of getting a bust, and since there are 71 different throws by which one can win, the probability of a draw is extremely small in this game.

From the probabilities stated it appears that at the first throw there is a probability 0.44572 of throwing lower than two aces and a probability 0.47532 of throwing higher than two aces. From this it appears that for John, who begins, it is not advantageous to stop at two aces. Hence, after two aces at the first throw, he throws the other three dice again, as a result of which he has a probability 13/18 = 0.72222 to improve this throw, made up of a 5/18 probability of two pairs, a 5/18 probability of three aces, a 5/216 probability of two aces plus three of a kind, a 5/72 probability of three aces plus two of a kind, a 5/72 probability of four aces, and a 1/216 probability of five aces. If he obtains three aces or better, he stops; otherwise he throws once more. If John throws higher than two aces on his first throw, he is well-advised to stop, because then Peter is allowed to throw only once, and therefore the probability that he will throw lower than John is greater than the probability that he will throw higher.

If John has a bust on his first throw, he will have thrown a single ace. The best thing he can then do is to leave the ace and to throw the other four dice again. With three aces or better he should stop, whereas otherwise he does better to make the third throw.

Peter, who has seen John's throw, has of course a different basis for deciding whether to continue to throw or not, and for choosing which dice he leaves. Here he takes note of John's result. If this is (say) four tens (in two throws), and if Peter throws two nines on his first throw, he does not leave the nines (as he would indeed do, if John had, for instance, two tens) because if he leaves the nines he can avert a loss only by throwing five nines, when he must throw three nines with the other three dice, the probability of which is 1/216. If Peter throws all five dice on his second throw, he has the larger probability 53/3888 of better than four tens and, besides, the probability 25/7776 of a draw. It is still more advantageous for Peter to leave one die that shows higher than a 10; he then has the probability 1/54 of better than four tens, and besides, the probability 1/1296 of a draw.

II. PROBABILITIES IN BRIDGE

149. Probability of a given distribution of the cards. In playing bridge, South, who plays the hand, sees 26 of the 52 cards: the 13 cards in his hand, and 13 cards on the table. For any given suit (trumps, say) he knows how many cards East and West have between them. Supposing there are 5 such cards, he may wonder: What is the probability that these 5 cards are distributed 2–3, what is the probability of 1–4, and what is the probability of 0–5? It is assumed that the bidding has not provided South with any information on this point, and that no card has as yet been played.

The number of ways in which the 26 cards which are concealed from South can be distributed between East and West is, according to §129, equal to $\frac{26!}{13! \times 13!}$ (the number of equally likely cases). If East and West between them have 5 cards of a certain suit, then these can be distributed between East and West in $\frac{5!}{3! \times 2!}$ ways; the remaining 21 cards of East and West, of which East has 11 cards and West 10, can be distributed in $\frac{21!}{11! \times 10!}$ ways. Hence, the total number of ways in which the cards can be distributed between East and West in such a way that East has 2 of these 5 cards, and West 3, is equal to $\frac{5!}{2! \times 3!} \times \frac{21!}{11! \times 10!}$. This is the number of cases favorable to the assumption that East has 2 and West has 3. Hence the probability of this is

$$\frac{5!}{2! \times 3!} \times \frac{21!}{11! \times 10!} \div \frac{26!}{13! \times 13!}$$
$$= \frac{5 \times 4}{2} \times \frac{13 \times 12 \times 13 \times 12 \times 11}{26 \times 25 \times 24 \times 23 \times 22} = \frac{39}{115} = 0.33913$$

This is also the probability of 3 with East and 2 with West.

In a similar fashion we find for the probability of 1 with East and 4 with West (also the probability of 4 with East and 1 with West):

$$\frac{5!}{1! \times 4!} \times \frac{21!}{12! \times 9!} \div \frac{26!}{13! \times 13!}$$
$$= 5 \times \frac{13 \times 13 \times 12 \times 11 \times 10}{26 \times 25 \times 24 \times 23 \times 22} = \frac{13}{92} = 0.14130.$$

The probability that the 5 cards are all with West (also the probability that East has them all) is:

$$\frac{21!}{13! \times 8!} \div \frac{26!}{13! \times 13!} = \frac{13 \times 12 \times 11 \times 10 \times 9}{26 \times 25 \times 24 \times 23 \times 22}$$
$$= \frac{9}{460} = 0.01957.$$

The fact that the sum of these probabilities is equal to 1 can serve as a check.

Likewise we find a value for the probability that, of 8 given cards, East has 4 and West 4 cards:

$$\frac{8!}{4! \times 4!} \times \frac{18!}{9! \times 9!} \div \frac{26!}{13! \times 13!}$$
$$= \frac{8 \times 7 \times 6 \times 5}{4 \times 3 \times 2} \times \frac{13 \times 12 \times 11 \times 10 \times 13 \times 12 \times 11 \times 10}{26 \times 25 \times 24 \times 23 \times 22 \times 21 \times 20 \times 19}$$
$$= \frac{143}{437} = 0.32723.$$

In this way we compute Table 6 which gives the probabilities of the various possible distributions of known totals 2, 3, 4, . . . , 13 cards (of one suit) between East and West. Not all these probabilities are of importance in bridge, but since it is difficult to draw a dividing line between cases of frequent occurrence and cases which hardly ever arise, we have included all cases for the sake of completeness.

Each (horizontal) row of this table corresponds to some number of cards that East and West hold between them; this number has been

TABLE 6

2	0.24	0.52	0.24				
3	0.11	0.39	0.39	0.11			
4	0.04783	0.24870	0.40696	0.24870	0.04783		
5	0.01957	0.14130	0.33913	0.33913	0.14130	0.01957	
6	0.00745	0.07267	0.24224	0.35528	0.24224	0.07267	0.00745
7	0.00261	0.03391	0.15261	0.31087	0.31087	0.15261	0.03391
8	0.00082	0.01428	0.08568	0.23561	0.32723	0.23561	0.08568
9	0.00023	0.00535	0.04284	0.15707	0.29451	0.29451	0.15707
10	0.00005	0.00175	0.01890	0.09239	0.23098	0.31183	0.23098
11	0.00001	0.00048	0.00722	0.04764	0.15880	0.28585	0.28585
12	0.00000	0.00010	0.00231	0.02117	0.09528	0.22868	0.30490
13	0.00000	0.00002	0.00058	0.00786	0.04915	0.15926	0.28312
	0	1	2	3	4	5	6

indicated to the left of the row. Each (vertical) column refers to the case in which a certain number of these cards (which has been indicated at the bottom of the column) is held by East. In the compartment corresponding to a given row and a given column, the probability of the associated distribution is given.

Thus we see from the table that when East and West have between them 8 cards of a certain suit, the probability of East having 3 and West 5 of these cards is then 0.23561; this is also the probability that 5 of these cards are held by East and the remaining 3 by West, so that the probability of a 3–5 distribution is $2 \times 0.23561 = 0.47122$. Further, we read from the table (for example) that the probability of a 3–3 distribution for 6 outstanding cards is equal to 0.35528. With 2, 4, 8, 10, or 12 outstanding cards, the probability of an equal division between East and West is equal to 0.52, 0.40696, 0.32723, 0.31183, 0.30490, respectively; as was to be expected, the probability of an equal division decreases as the number of outstanding cards increases.

150. A posteriori probability of a certain distribution of the cards. We take the case that East and West have 7 low diamonds (some other suit being trumps), and that West begins by leading a diamond. Further we assume that no conclusion can be drawn from this lead (to avoid the consideration of details peculiar to the game, which cannot well be brought into a calculation of probabilities). If East has followed suit, the probability for South that the 7 cards are distributed 3–4 is no longer 0.62174, which it was before West's lead. The a priori probability that East and West can both follow suit in diamonds is $0.06782 + 0.30522 + 0.62174 = 0.99478$ (which can also be found as $1 - 2 \times 0.00261$). When, from the course of the first trick, this has become certainty, then for South (according to the quotient rule on the a posteriori probability) the probability that East originally held 3 and West 4 of the cards has become:

$$\frac{0.31087}{0.99478} = 0.31250;$$

for South, the total probability of a 3–4 distribution after the first trick is thus $2 \times 0.3125 = 0.625$. This result implies that the probability of a 0–7 distribution has dropped to 0, and that the ratios of the other probabilities remain unchanged.

If diamonds are played to the second trick also, and if East and West follow suit (the a priori probability of which is 0.30522 +

0.62174 = 0.92696), then for South the probability of an initial 3–4 distribution with East holding 3 cards, and West 4, say, increases to $\frac{0.31087}{0.92696} = 0.33537$. Again, some probabilities drop to 0, and the ratios of the other probabilities remain unchanged.

If East and West follow suit in the third trick of diamonds, the total probability of an initial 3–4 distribution increases to 1, that is, to certainty.

In Table 7 various probabilities of this kind have been assembled.

TABLE 7

	1	2	3	4	
4	0.27500	0.45000	0.27500	0	after 1 trick
5	0.14706	0.35294	0.35294	0.14706	after 1 trick
6	0.07377	0.24590	0.36025	0.24590	after 1 trick
	0	0.28846	0.42307	0.28846	after 2 tricks
7	0.03409	0.15340	0.31250	0.31250	after 1 trick
	0	0.16463	0.33537	0.33537	after 2 tricks
8	0.01430	0.08582	0.23600	0.32777	after 1 trick
	0	0.08835	0.24295	0.33742	after 2 tricks
	0	0	0.29509	0.40984	after 3 tricks
9	0.00535	0.04286	0.15714	0.29465	after 1 trick
	0	0.04332	0.15884	0.29783	after 2 tricks
	0	0	0.17391	0.32609	after 3 tricks

Each (horizontal) row refers to the number of cards that East and West held between them originally; this number has been indicated to the left of the row. Each (vertical) column refers to the number of cards held by East originally; this has been indicated at the bottom of the column. For the rest, the construction of the table speaks for itself.

In Table 8 the various probabilities have been grouped differently. The numbers at the left indicate the numbers of cards of the suit in question that remain with East and West after East and West have both followed suit in the first 0, 1, 2, etc. tricks (of that suit). For example, if originally 12 diamonds are held by East and West, and if there have been four tricks of diamonds in which both East and West followed suit, then for South the probability that East holds none of the four diamonds then outstanding is equal to 0.1, and the probability that East has one of these four diamonds is equal to 0.24.

These probabilities can also be applied to the distribution of the trump cards held by East and West after the leader has won one of the first tricks in another suit. If East and West have followed suit in these tricks, this will have an extremely small influence on the probabilities of the various distributions of the trumps, so small that it can safely be left out of consideration. Indeed, we have to do this if we want to avoid a mass of complications, not all of which can be treated by the theory of probability.

151. Probabilities in finessing. If a player wants to extract trumps, and has at his disposal all the high trumps except the king, he has the choice either of leading the ace in the hope that the king will drop, or of finessing the king.

Suppose that the dummy has won one of the first tricks and that the declarer (South) has the ace and queen of trumps, while the dummy has only small trumps. From the dummy South now plays a small trump, which East will cover with the king only if he has the king as a singleton. We assume that East follows suit but does not play the king, and we seek the probabilities that East has the king, and that West has the singleton king.

If East and West originally held two trumps between them (the king and a small one), it becomes evident from East's playing that he originally held either the king and a small trump (the a priori probability of which, according to Table 6, is equal to 0.24), or just a single small trump (a priori probability $1/2 \times 0.52 = 0.26$), since with an equal division of the outstanding trumps between East and West, there is the same probability that the king is with East or with West; after East plays a small trump, both probabilities are increased in the same proportion (doubled, actually), so that the probability that West has the king remains the larger one, although the difference in probability is small.

Things are different when the opponents hold the king and two small trumps. After East plays a small trump the following possibilities remain: East originally held the king of trumps and two small trumps (probability 0.11); East originally held two small trumps (probability $1/3 \times 0.39 = 0.13$); East originally held the king of trumps and a small trump (probability $2/3 \times 0.39 = 0.26$); East originally held one small trump only (probability $2/3 \times 0.39 = 0.26$).

The stated probabilities are to be taken as a priori probabilities;

TABLE 8

	0	1	2	3	
2	0.24	0.52	0.24		after 0 tricks
	0.275	0.45	0.275		after 1 trick
	0.28846	0.42307	0.28846		after 2 tricks
	0.29509	0.40984	0.29509		after 3 tricks
	0.2985	0.403	0.2985		after 4 tricks
	0.3	0.4	0.3		after 5 tricks
3	0.11	0.39	0.39	0.11	after 0 tricks
	0.14706	0.35294	0.35294	0.14706	after 1 trick
	0.16463	0.33537	0.33537	0.16463	after 2 tricks
	0.17391	0.32609	0.32609	0.17391	after 3 tricks
	0.17857	0.32143	0.32143	0.17857	after 4 tricks
	0.18	0.32	0.32	0.18	after 5 tricks
4	0.04783	0.24870	0.40696	0.24870	after 0 tricks
	0.07377	0.24590	0.36025	0.24590	after 1 trick
	0.08835	0.24295	0.33742	0.24295	after 2 tricks
	0.09638	0.24096	0.32531	0.24096	after 3 tricks
	0.1	0.24	0.32	0.24	after 4 tricks
5	0.01957	0.14130	0.33913	0.33913	after 0 tricks
	0.03409	0.15340	0.31250	0.31250	after 1 trick
	0.04332	0.15884	0.29783	0.29783	after 2 tricks
	0.04839	0.16129	0.29033	0.29033	after 3 tricks
	0.05	0.162	0.288	0.288	after 4 tricks
6	0.00745	0.07267	0.24224	0.35528	after 0 tricks
	0.01430	0.08582	0.23600	0.32777	after 1 trick
	0.01897	0.09272	0.23181	0.31296	after 2 tricks
	0.02127	0.09574	0.22979	0.30638	after 3 tricks
7	0.00261	0.03391	0.15261	0.31087	after 0 tricks
	0.00535	0.04286	0.15714	0.29465	after 1 trick
	0.00723	0.04769	0.15896	0.28611	after 2 tricks
	0.00787	0.04921	0.15945	0.28346	after 3 tricks

after East's play these probabilities are increased proportionally, so that the probabilities of the king of trumps being held by East, or being a singleton with West, are in the proportion $0.11 + 0.26$ to 0.13, hence as 37 to 13, even a posteriori. So the probability of the finesse succeeding is considerably larger than that of the king of trumps dropping, and consequently South should here finesse by playing the queen.

Next we take the case that dummy has the king of trumps (and some small trumps) when South has the ace and jack of trumps, and the opponents hold four trumps including the queen. Now South first plays the king of trumps, because there is a possibility that the queen will drop, or that he will learn the distribution of the trumps if East or West fails to follow suit. We assume that East and West both follow suit without the queen dropping. South now plays a small trump from the dummy. If East has any choice, he will not play the queen on it. We assume that East plays a small trump. Then the following possibilities remain: East originally held either the queen of trumps and two small trumps, or just two small trumps. The a priori probability that East originally held three of the four trumps is (again according to Table 6) equal to 0.2487; the probability that he has three trumps and that the queen is one of them is $3/4 \times 0.2487 = 0.1865$, since the queen can be any of the four trumps which East and West hold between them (equally likely cases) and three of these cases are favorable (making the queen one of the three trumps that East holds); therefore, with the knowledge that East has three of the four outstanding trumps, the probability that the queen of trumps is one of these is 3/4. The a priori probability that East originally held two of the four outstanding trumps is 0.40696; hence, the probability that he originally held two small trumps is $1/2 \times 0.40696 = 0.20348$. After East and West have followed suit in the first round of trumps, and after East has followed suit in the second round, without the queen having dropped, these probabilities are increased proportionally, and therefore the a posteriori probabilities, like the others, are in the ratio of 11 to 12; hence, the probability that West has the queen of trumps is the larger, and therefore South should play the ace rather than finesse.

We now modify the foregoing to the effect that the opponents hold five trumps, with the queen included. After East has played two small trumps, and West one small trump, the following possibilities remain: East originally held the queen of trumps and three small trumps (probability $4/5 \times 0.14130 = 0.11304$); East originally held the queen of trumps and two small trumps (probability $3/5 \times 0.33913 = 0.20348$); East originally held three small trumps (probability $2/5 \times 0.33913 = 0.13565$); East originally held two small trumps (probability $3/5 \times 0.33913 = 0.20348$).

These probabilities (which have been derived from Table 6) are again to be taken as a priori probabilities, and they are all increased proportionally a posteriori. Hence, the probability of a successful finesse for the queen and the probability that the queen will drop to the ace of trumps in the second round, are in the ratio of 0.11304 + 0.20348 = 0.31652 to 0.13565, hence as 7 to 3, so that the finesse offers a much greater chance of success. Computations with ordinary fractions will show that the ratio 7:3 is not approximate, but exact. The same is true of the ratio 11:12 which we last found.

Chapter IX:
EVALUATION OF CONTINGENCIES AND MEAN VALUES

I. MATHEMATICAL EXPECTATION AND ITS APPLICATIONS

152. Mathematical expectation. One might ask: What is the value of having a known probability of winning a given sum of money? As an example we take the case of a raffle with 100 tickets, for a single prize of $100. The 100 tickets jointly have a value of $100, because together they provide a certainty of receiving $100; this implies that each ticket has a value of $1. Someone who has 37 tickets should be credited with the ownership of $37. On the other hand, what is provided by these 37 tickets is only a probability 37/100 of obtaining $100. This shows that the value of a certain number of tickets can be identified with the amount that is obtained when the sum that can be won with them is multiplied by the probability of winning this sum. In more general cases, to obtain the value of having a probability of acquiring some amount, we equate this to the product of the probability by the amount concerned. Hence, if $72 is offered for a throw of 6 made with two dice, an opportunity to throw has a value of $5/36 \times \$72 = \$10$, since the probability of having a total equal to 6 is 5/36 (from throws 1–5, 2–4, and 3–3).

The product of the probability and the amount to be won is called the mathematical expectation of that amount. A fair game should provide each player with a mathematical expectation equal to his stake.

153. Examples of mathematical expectation. Suppose that John throws a single die, and will receive from Peter a number of dollars equal to the number of pips for his throw. How large is his mathematical expectation in this case? In other words, how large should John's stake be if the game is to be fair?

John has a probability 1/6 of winning $1, a probability 1/6 of winning $2, and so on, so that his mathematical expectation (in dollars) is:

$$\frac{1}{6} \times 1 + \frac{1}{6} \times 2 + \frac{1}{6} \times 3 + \frac{1}{6} \times 4 + \frac{1}{6} \times 5 + \frac{1}{6} \times 6$$
$$= \frac{1}{6} \times 21 = 3\frac{1}{2}.$$

We can obtain this number without computation by the following reasoning. Suppose that Charles will give John a number of dollars which corresponds to the indication of the bottom of the die. Obviously, this involves the same mathematical expectation as applies to the agreement with Peter. Both agreements, taken together, offer a certainty that John will receive $7, since the sum of the upper and lower spot numbers is equal to 7. Hence, the value of each agreement is one-half of $7.

Next we take the case that John and Peter make the above agreement, with the difference that two dice are to be thrown. The total thrown can then be 2, 3, 4, 5, 6, 7, 8, 9, 10, 11, or 12. The probabilities of these are, in the same order, 1/36, 2/36, 3/36, 4/36, 5/36, 6/36, 5/36, 4/36, 3/36, 2/36, 1/36. Hence, John's mathematical expectation (in dollars) is:

$$\frac{1}{36} \times 2 + \frac{2}{36} \times 3 + \frac{3}{36} \times 4 + \frac{4}{36} \times 5 + \frac{5}{36} \times 6 + \frac{6}{36} \times 7$$
$$+ \frac{5}{36} \times 8 + \frac{4}{36} \times 9 + \frac{3}{36} \times 10 + \frac{2}{36} \times 11 + \frac{1}{36} \times 12 = 7.$$

As before, this result can be found immediately by having John and Charles agree that John will receive from Charles a number of dollars equal to the total of the hidden faces. This agreement has the same value as the one with Peter, and the joint effect of both agreements provides a certainty of receiving $14. Hence, the mathematical expectation corresponding to each agreement separately is one-half of $14. We might also observe that each of the two dice, separately, provides a mathematical expectation of $3.50.

154. More complicated example. We again take the case that John throws a single die and obtains from Peter a number of dollars corresponding to the spot number he throws, the difference being that John is allowed to throw once again if he thinks his throw is too low. Here, John will throw again if his throw is lower than $3\frac{1}{2}$, his mathematical expectation for one throw; that is, if he throws 1, 2, or 3, which have a joint probability of 1/2. John's mathematical expectation for the single throw or pair of throws is thus:

$$\frac{1}{2} \times \frac{7}{2} + \frac{1}{6} \times 4 + \frac{1}{6} \times 5 + \frac{1}{6} \times 6 = \frac{17}{4} = 4.25,$$

since he has a probability 1/2 of having a mathematical expectation 7/2 (from a second throw), equivalent to a probability 1/2 of having an actual amount 7/2, and additionally he has a probability 1/6 of obtaining a 4 on the first throw, and similarly for higher values.

We now take the case that John is allowed to make at most three throws. After the first throw, he will throw again if his throw was lower than 4.25, his mathematical expectation with two throws; that is, if he throws 1, 2, 3, or 4, the probability of which is 2/3. John's mathematical expectation with at most three throws is thus:

$$\frac{2}{3} \times \frac{17}{4} + \frac{1}{6} \times 5 + \frac{1}{6} \times 6 = \frac{14}{3} = 4\frac{2}{3}.$$

If John is allowed to throw four times, he throws again if his first throw is 4, but not if he has 5. John's mathematical expectation with at most four throws is thus:

$$\frac{2}{3} \times \frac{14}{3} + \frac{1}{6} \times 5 + \frac{1}{6} \times 6 = \frac{89}{18} = 4\frac{17}{18}.$$

Hence, when 5 throws are allowed, John will again be wise to stop if he has 5 with the first throw. Consequently, with at most five throws, John's mathematical expectation is:

$$\frac{2}{3} \times \frac{89}{18} + \frac{1}{6} \times 5 + \frac{1}{6} \times 6 = \frac{277}{54} = 5\frac{7}{54} = 6 - \frac{47}{54}.$$

It follows that when John is allowed six throws (and also, of course, when still more throws are allowed), he should continue to throw if he has 5 with the first throw. From this we find that John's mathematical expectation with at most six throws is:

$$\frac{5}{6}\left(6 - \frac{47}{54}\right) + \frac{1}{6} \times 6 = 6 - \frac{5}{6} \times \frac{47}{54} = 6 - \frac{235}{324}.$$

Similarly, we find John's mathematical expectation with at most seven throws to be $6 - \left(\frac{5}{6}\right)^2 \times \frac{47}{54}$, and so on. As his number of allowable throws is increased, John's mathematical expectation moves towards a value of 6 and eventually gets very close to 6. This result could be expected without any computation.

155. Modification of the example of §154. We now modify the example of §154 to the effect that John throws two dice, and is allowed to throw again a certain number of times if he thinks his throw is too low. Later throws have to be made using both dice. If John is allowed to throw twice at the most, he makes a second throw if he throws less than 7 on the first throw. If he throws 7 on the first throw, it is immaterial whether or not he throws again, because his mathematical expectation would be 7 if he were to do so. From the probabilities of the throws 2, 3, 4, . . . , 12 (see §153), the mathematical expectation in the case of at most two throws (since John has a probability 7/12 of throwing 7 or less on the first throw) is found to be:

$$\frac{7}{12} \times 7 + \frac{5}{36} \times 8 + \frac{4}{36} \times 9 + \frac{3}{36} \times 10 + \frac{2}{36} \times 11 + \frac{1}{36} \times 12 = \frac{287}{36} = 7\frac{35}{36}.$$

Hence, if John is allowed to throw three times, he should throw again if he throws 7 the first time, but not if he has 8. So if he is allowed to make at most three throws, his mathematical expectation is:

$$\frac{7}{12} \times \frac{287}{36} + \frac{5}{36} \times 8 + \frac{4}{36} \times 9 + \frac{3}{36} \times 10 + \frac{2}{36} \times 11 + \frac{1}{36} \times 12 = \frac{3689}{432} = 8\frac{233}{432}.$$

From this it follows that if he is allowed to throw four times, John should go on throwing if he throws 8 the first time, but not if he has 9.

156. Petersburg paradox. About 1730, the concept of mathematical expectation gave rise to a curious paradox. This is associated with the name of a famous family of mathematicians, the Bernoullis, who did most of their work at Basel. To distinguish among various mathematicians who were members of this family, it has even been necessary to attach Roman numerals to some of their names, similar to the customary practice for monarchs. The mathematical brilliance began with Jakob Bernoulli (1654–1705) and his younger brother Johann (1667–1748); the latter taught mathematics for ten years at the University of Groningen. These two are certainly the most eminent members of the learned family. They were followed by a younger generation, Nikolaus I (1687–1759), son of a brother of Jakob and Johann, and Nikolaus II (1695–1726) and Daniel (1700–1782), sons of Johann. A problem posed by Nikolaus I was modified by Daniel, in a way which produced the paradox in question. At that time Daniel was attached to the Academy newly founded by Peter the Great in St. Petersburg, and so the contradiction resulting from the game which Daniel considered became known as the Petersburg paradox. The game in question is the following:

John tosses a coin until it falls heads. If this occurs the first time, he pays \$1 to Peter. If it occurs the second time, he pays \$2 to Peter. If the third toss is the first to produce heads, John pays \$4, and so on; each time the amount to be paid is doubled. Now the question is: What payment should Peter make to John, in fair compensation?

The probability that heads will occur the first time is 1/2. Peter then receives \$1, so that he has a mathematical expectation of \$0.50 from the result of the first toss. The probability that heads will occur on the second toss is 1/4; Peter then receives \$2, and his mathematical expectation for the second toss is $1/4 \times \$2 = \$0.50$. Heads on the third toss has a probability of 1/8, and this gives a mathematical expectation of $1/8 \times \$4 = \$0.50$. Each of the later possibilities, infinite in number, provides a mathematical expectation of \$0.50, so that Peter's total mathematical expectation is infinitely large. Hence, his stake should also be infinitely large. Even if Peter stakes 100 million dollars on this game, the advantage is on his side. Yet it is hard to find anyone who will risk \$20, say, on these conditions.

Many different resolutions of this paradox have been offered, and the most natural of these is certainly the following. However rich John may be, he is not able to meet all the consequences which may result from the obligations he has taken upon himself. If heads did not appear until the sixty-fifth throw, he would have to give Peter more than \$18,000,000,000,000,000,000 (as many dollars as the number of grains the Indian monarch had to give to the inventor of the game of chess), a promise that no one will think worth \$0.50. Even if John is a billionaire, if he throws tails successively more often than 30 times he will no longer be able to meet his obligations, and Peter will have to content himself with a billion dollars. On the assumption that Peter can never win more than this figure, his mathematical expectation is approximately \$16.

The fact that Peter will still not want to risk \$16 on this game, because he rates his chances at a lower figure, is something that will be noticed in this game, and similarly in other games in which there is an extremely small probability of winning a very large sum of money, especially if this is so large that it would be difficult to know what to do with the greater part of it. Everyone will prefer a guaranteed possession of a million to a probability of 1/1000 of obtaining a thousand times as much.

The last remark links up, to some extent, with the solution that Daniel Bernoulli himself provided for the paradox. He sought to find a solution in terms of the relative value of money. The advantage of an increase in capital depends not only on the size of the increase but also on the capital that one already possesses; the larger this is, the smaller is the advantage. Daniel made this the basis of a complete theory, no doubt correct in conception, but worthless in practice because it cannot be used to provide a basis for stakes. This viewpoint brings out a disadvantage implicit in any game, even with equal chances and equal stakes, because when you have assets of $200, for example, the disadvantage of losing $100 is larger than the advantage which comes from a gain of $100.

The impracticality of using the theory of the relative value of money as a basis for determining stakes in a game is nicely illustrated in Bertrand's book (see §132) by the following imaginary conversation between two of Daniel Bernoulli's students.

"If I win," says Peter, who is poor, and invites John to play a game, "I can just pay for my dinner with your stake of $1.50."

"A meal for a meal," replies John; "so if you lose you should give me $10, for that is what I pay for a dinner."

"If I were to lose $10," Peter exclaims in a fright, "I would have nothing to eat today. You could lose $5000 before that would happen to you. Stake your $5000 against my $10, and the advantage is still on your side, as our teacher would confirm!"

Of course, they failed to reach an agreement.

II. FURTHER APPLICATION OF MATHEMATICAL EXPECTATION

157. Application of mathematical expectation to the theory of probability. Mathematical expectation can often provide assistance in the computation of probabilities for situations in which more direct computation encounters difficulties. As an example we take the following case:

John has $5 and Peter has $7. They play with equal probabilities of winning, having $1 at stake each time, until one of the players has lost all that he had. How large is the probability that John will be the winner?

Since the game is fair, John's capital of $5 is equal to his mathematical expectation. This derives from a probability of having $12 at the finish, and hence is equal to the product of this probability and the $12. From this it follows that John's probability of winning the game is 5/12. Peter's probability is 7/12, which shows that the winning probabilities for the players are in the same proportion as their assets. If Peter has a capital which is many times larger than John's, then Peter's victory is practically certain; on the other hand, what John can win is substantially more than what Peter can.

If John has not too small a capital, and the right to determine the amount to be played for (the stake) in each separate game, he may then fall victim to a misconception often found among gamblers, that he could make his chances more favorable by following a certain system, for instance the system whereby the stake is doubled after a loss. "Each winning round," John argues, "taken together with the immediately preceding losing rounds, gives a gain which is equal to the original stake, so I will therefore win slowly but surely." However, John here forgets that he cannot survive a long sequence of losses, because he can never stake more than the capital which he has left. If the game continues not until John or Peter has lost everything, but until John wants to stop, then no system John intends to pursue can make any difference to his mathematical expectation, which must be the product of his chance of success and the amount he seeks to win; this product is and remains equal to the capital he has to lose (so in the course of the game his mathematical expectation is obviously equal to the capital he has left). However, John can increase one factor of the product at the expense of the other. If he stops when he has lost everything or has won a predetermined amount, his probability of success is the greater, the lower he sets the amount concerned. By contenting himself with a small gain, he increases his probability of finishing as a winner; by aiming at a higher gain, he diminishes his chance of a favorable result. The manner in which he may intend to vary the amount of his stakes makes no difference to this. If John continues to play against Peter until either he or his opponent has lost everything, then with any system his probability of winning is equal to the quotient which is obtained when his capital is divided by the capital possessed jointly by them both.

In the foregoing it has been assumed that John cannot play on credit, and consequently cannot just run up a note of his losses on paper, to any arbitrary amount. A casino rightly takes a poor view of this procedure. If Peter knows that John can produce a larger sum than he has with him at the time (and that John is honest), he can allow John to write notes for his losses up to the amount that John can provide; when this sum is reached, the game is finished. However, this means that the result of the game is the same as if John brought the sum in question with him. To allow John to write notes for his unlimited losses would mean financial suicide for Peter, and by requesting this John would be guilty of an attempt to defraud, for (as in the Petersburg paradox) he would be assuming obligations when he knew he was unable to meet them.

158. Law of large numbers. If a die is to be thrown a large number of times, say 6000, everyone will be convinced a priori that there will be a throw of 1 about 1000 times, of 2 about 1000 times, and so on. It is true that we would not expect exactly 1000 occurrences of 1, and we would consider it perfectly possible that 1 would be thrown 950 or 1050 times, say. However, if there were 50 occurrences of 1, or 2000 such throws, we would no longer believe that the die was a fair one. Our intuition seems to tell us that the proportion of the number of 1's to the total number of throws will lie within certain limits on either side of 1/6, limits which will be narrower as the number of throws is increased. Something similar applies when a coin is tossed; here we would expect that the proportion of the number of heads to the total number of throws will lie between 0.49 and 0.51, provided a sufficiently large number of tosses is made. More generally, this applies with any often repeated event which has a variety of outcomes determined by chance, such as the spin of a roulette wheel, which each time produces some one of the numbers 0, 1, 2, . . . , 36.

It is difficult to say how this feeling originates. No doubt there is some basis of experience, even for someone who has never kept a record of throws of a die, or of numbers that appear in roulette, or of winning numbers in a lottery, or things of that sort.

The fact that the proportion of favorable cases in a number of experiments is approximately equal to the probability of a favorable result for each experiment separately, when the number of experiments is large, is called the law of large numbers. This law has been stated fairly explicitly by Cardano; but a mathematical explanation of this remarkable phenomenon was given for the first time by Jakob Bernoulli. He proved that the probability of finding (for example) that the number of heads divided by the total number of throws has a value between 0.49 and 0.51 becomes larger as the number of throws increases, and that this probability lies very close to 1 provided that the number of throws is sufficiently large. The same is true for the probability that the proportion in question will lie between 0.499 and 0.501, with the difference that a larger number of throws is needed to obtain a given probability close to 1 that the experimental proportion will lie between the limits in question.

159. Probable error. We would now like to have an estimate of the magnitude of the deviation to be expected between the result of a sequence of experiments (throws with a die, say) and a result exactly in accordance with the law of large numbers, that is, of the deviation between experimental and theoretical results. To be more concrete, we take a box with four white balls and six black balls. A ball is drawn from this box, and replaced, repeatedly, the balls being mixed before each new draw. After 1000 drawings, white would have appeared 400 times if the law of large numbers could guarantee an absolutely exact result, instead of an approximate one. A computation, which we cannot undertake to discuss here, shows that the probability that the number of white balls drawn will lie between 390 and 410 is the same as the probability that this number will lie outside this interval. The difference between each of these limits and the theoretical number of white balls drawn, here 10, is called the probable error.

The result can be expressed in another way by saying that the probability of finding that the number of white balls drawn can be divided by the number of drawings (1000) to give a value between 0.39 and 0.41, is the same as the probability for a quotient lying outside this interval. The difference of 0.01 which separates each of these limits from the probability 0.4 of obtaining white at one drawing is called the relative probable error. The difference 10 mentioned above can then be described as the absolute probable error.

If we take a number of drawings four times (or 100 times, or 400 times) as large, the absolute probable deviation becomes twice (or 10 times, or 20 times) as large, and so the relative probable deviation becomes twice (or 10 times or 20 times) as small; here it is assumed that the number of drawings is not itself too small. Hence, if we make 100,000 drawings, the absolute probable deviation is 100, and the relative probable deviation is 0.001. The probability that the number of white balls lies between 39,900 and 40,100 is equal to the probability that the number lies outside these limits. This means that the proportion of the number of white balls drawn to the number of drawings is then equally likely to lie between 0.399 and 0.401 as to have some value outside these limits.

The fact that the absolute probable deviation becomes twice (or 10 times or 20 times) as large, when the number of drawings becomes 2^2 times (or 10^2 times or 20^2 times) as large, also holds in other cases in which the probability of white is different, hence when the numbers of white and black balls in the box have other values.

160. Remarks on the law of large numbers. Without any full awareness of all the details, and without knowledge of any more or less exact formulation of the law of large numbers, everyone feels intuitively that it must be correct. Thus, the gambler who turned to Galileo for aid (see §132) apparently thought it self-evident that there must be a reason why 10 and 11 are thrown more frequently than 9 and 12 with three dice, and that this could not just be a freak of nature. De Méré's views about another dice game, which seemed to him to provide a contradiction of the laws of arithmetic, were also based on a similar conviction.

Yet, on the other hand, it frequently happens that a gambler or gamester holds an opinion that is entirely contrary to the law of large numbers. This happens when his own personality comes into play and when his vanity unconsciously represses a correct intuition of the law of large numbers. Thus, many a bridge player is heard to complain that he usually gets bad cards, and a person of this type is convinced that this will also happen to him in the future; he happens to be, so he says, unlucky at games, and this fate continues to weigh heavily upon him. In reality, however, his constant loss at the game arises because he is no match for his opponents and because he does not make the most of the chances that the game offers sometimes to one player, sometimes to another. However, it is much more flattering to his vanity to believe that he gets bad cards than to conclude that he plays badly. In bridge there is the additional factor that he does not appreciate how miserable an average hand of 13 cards looks, nor does he realize, additionally, that he can expect at times to get a still worse one; his lack of skill makes him not sufficiently aware of what can be achieved, in the defense, with such an average or slightly worse than average hand. If there were players who usually got bad cards, then there would also have to be players who generally got good cards. This is another clear proof that the curse of continually having bad hands must be imaginary. Of course, it can happen that on some particular evening a player may be dealt a run of bad cards.

161. Further relevance of the law of large numbers. The law of large numbers has a much wider scope than its application to games, as discussed in the foregoing. It is also fundamental to the mathematical theory of statistics, because it enables a distinction to be made between fortuitous events and systematic effects.

However, the law has much more extensive applications. It would not be too much to say that it governs the whole of nature. A tendency towards average expectations, which can be observed everywhere, is, after all, nothing but a consequence of this law. To take a very simple example, we can be sure that two adjoining paving-stones will be moistened approximately equally by a shower of rain; it is true that one stone will be hit by some tens of drops more than the other one will be, but relatively, that is to say, in proportion to the total number of drops, this difference becomes the smaller as the total number of drops becomes larger. Each drop formed in upper layers of the atmosphere has as much chance of hitting one stone or the other. Hence, it is conceivable that one of the stones will not be hit by even one drop of rain. The fact that this does not occur is to be explained by the extremely small probability of such an event.

If a gas is confined in a closed container where the temperature is the same everywhere, then each cubic millimeter of the container will contain practically the same fraction of gas molecules. Here the huge number of molecules secures that deviations from the mean are so small that we find them imperceptible. If we connect two vessels each containing a different gas, then the two gases will mix and be distributed equally over the combined volume. All this is based on the law of large numbers. In view of this law, the theory of probability, which originated from games of chance, has been developed to a level which puts it in a position to play a part in a number of other sciences in which the great number and the unpredictability of the relevant factors would make mathematics powerless if it could not leave the domain of absolute certainty, to enter that of probability.

III. AVERAGE VALUES

162. Averages. If some number can take different values, depending on chance, this, as we have seen, produces a corresponding mathematical expectation. For example, if the number has the probability

1/2 of taking the value 3, the probability 1/3 of taking the value 5, and the probability 1/6 of taking the value 11, this gives a mathematical expectation of $\frac{1}{2} \times 3 + \frac{1}{3} \times 5 + \frac{1}{6} \times 11 = 5$. This mathematical expectation is also called the average (mean) value of the number. This name finds its explanation in the law of large numbers. If the number turns up 6000 times, it will assume the value 3 approximately 3000 times, the value 5 approximately 2000 times, and the value 11 approximately 1000 times. The average of these 6000 numbers, that is, the sum of these numbers divided by 6000, is:

$$\frac{3000 \times 3 + 2000 \times 5 + 1000 \times 11}{6000} = \frac{1}{2} \times 3 + \frac{1}{3} \times 5 + \frac{1}{6} \times 11 = 5.$$

A simpler example is the following. John pays \$24 to Peter if the latter throws a 6 with 1 die. This represents a mathematical expectation of 1/6 × \$24 = \$4. If the game is played 6000 times, Peter will have thrown a 6 about 1000 times, so that then he receives approximately 1000 × \$24 in all. Hence, the average receipt per throw is $\frac{1000 \times \$24}{6000} = \frac{1}{6} \times \$24 = \$4$, which is nothing but his mathematical expectation for one throw.

An average number need not be an integer, of course. Thus, in throwing one die, the spot number that appears is 7/2 = 3.5, on the average. In throwing two dice the total of the spot numbers is 7 on the average; with three dice it is 10.5.

It requires a little more labor to give an answer to the following question:

Some throws are made with a die, and the spot numbers are noted down. This is continued until a previously obtained spot number appears again. What is the average number of throws needed for this?

A new appearance of a spot number that has appeared earlier will occur on the second throw at the earliest and on the seventh throw at the latest. The probabilities that it will happen on the second, third, fourth, fifth, sixth, or seventh throw are, in the same order:

$$\frac{1}{6}, \quad \frac{5}{6} \times \frac{2}{6} = \frac{5}{18}, \quad \frac{5}{6} \times \frac{4}{6} \times \frac{3}{6} = \frac{5}{18}, \quad \frac{5}{6} \times \frac{4}{6} \times \frac{3}{6} \times \frac{4}{6} = \frac{5}{27},$$

$$\frac{5}{6} \times \frac{4}{6} \times \frac{3}{6} \times \frac{2}{6} \times \frac{5}{6} = \frac{25}{324}, \quad \frac{5}{6} \times \frac{4}{6} \times \frac{3}{6} \times \frac{2}{6} \times \frac{1}{6} = \frac{5}{324}.$$

Hence, the average number of throws (that is, the mathematical expectation of somebody who gets a number of coins equal to the required number of throws) is:

$$\frac{1}{6} \times 2 + \frac{5}{18} \times 3 + \frac{5}{18} \times 4 + \frac{5}{27} \times 5 + \frac{25}{324} \times 6 + \frac{5}{324} \times 7$$
$$= \frac{1223}{324} = 3\frac{251}{324} = 3.7747.$$

163. Other examples of averages. We now require the average number of throws necessary to throw a 6 with a die. This question differs from the cases discussed in §162 in that the number of throws can become arbitrarily large, as in the Petersburg paradox (see §156). However, the average number of throws is not infinitely large.

The required average number is equal to the mathematical expectation of John, who gets a number of coins equal to the number of throws needed to produce a 6 for the first time; for example, if this occurs on the ninth throw, John gets nine coins. To avoid all arguments about the number of throws, a coin is paid to John at each throw, and this is terminated as soon as a 6 is thrown. When the first throw has been made, John receives a single coin if this throw is a 6, and he then has no expectation of anything more; if a spot number other than a 6 has been thrown, the ensuing situation is the same as applied before the start of the game, so that John then has one coin received already, and additionally he has a mathematical expectation the same as he had before the game began. The probability of a 6 on the first throw is 1/6, that of a different result is 5/6. Hence, John's initial mathematical expectation is the certainty of receiving an amount of one unit, increased by a 5/6 probability of receiving a sum equal to his mathematical expectation at the beginning; in other words, his mathematical expectation is equal to one unit, plus five-sixths of his mathematical expectation. As a consequence, one unit is equal to one sixth of his mathematical expectation, so that John has a mathematical expectation of six units. Hence, the average number of throws required to obtain a 6 is equal to 6.

This result can, of course, be transferred to all sorts of other cases. In throwing three dice the probability of three sixes is equal to 1/216. Hence, on the average, the three dice have to be thrown 216 times to produce three sixes. In roulette, the probability of zero appearing is equal to 1/37 (because one of the numbers 1, 2, 3, . . . , 36 may appear instead). Hence on the average 37 spins are required before zero appears. In view of the law of large numbers, we could expect no other result.

We now take a slightly more complicated case, which can be reduced to the foregoing. We make some throws with a die, and continue until all spot numbers have appeared, hence until each face of the die has been uppermost at least once. We require the average number of throws necessary to achieve this.

First we make one throw; its result is immaterial. If the first throw was a 4, say, then for the time being we continue to throw until a spot number different from 4 appears; as the probability that this will occur on the next throw is 5/6, the average number of throws needed to obtain a spot number different from 4 is equal to 6/5. If this spot number (differing from 4) is a 2, say, then we continue to throw until a spot number different from 2 and from 4 appears; the probability of this occurring on the immediately following throw is 4/6, so that the average number of throws required for this is 6/4. Likewise, the average number of throws necessary to obtain a spot number that differs from the three spot numbers already thrown is equal to 6/3. Continuing in this way, we find a value of

$$1 + \frac{6}{5} + \frac{6}{4} + \frac{6}{3} + \frac{6}{2} + \frac{6}{1} = 14.7$$

for the required number of throws.

164. Incorrect conclusion from the law of large numbers. An intuitive feeling for the law of large numbers has led to a misconception which is particularly common among gamblers. If such a gambler makes a large number of tosses with a coin, he will toss heads about the same number of times as he tosses tails; this is confirmed by experiment. Hence, if he has found heads twice and tails eight times in ten tosses, then, he argues, this result has to be corrected by the subsequent tosses in order to get heads as many times as tails. So in the following tosses heads will appear more often than tails; in other words, on the eleventh toss the probability for heads will be larger than that for tails.

Reasoning thus, he overlooks the fact that the numbers of heads and of tails will also become approximately equal if, in the subsequent tosses heads and tails appear equally often. Here "approximately equal" is to be taken in the sense that the proportion of the number of heads to the number of tails is close to 1, not in the sense that the difference of these numbers is small. If heads has appeared twice, and tails eight times, and if after that the player tosses heads 1000 times and tails 1000 times, the number of heads and tails obtained (1002 and 1008) are approximately equal in the above sense, and this is also true if the first ten throws are followed by 980 heads and 1020 tails, although then the excess of tails has even been increased by 40.

No less a man than d'Alembert, for all his sharp mind, seems to have understood nothing of the theory of probability, for when seeking a solution of the Petersburg paradox (see §156) he proceeded on the assumption that subsequent throws have, as it were, to balance up the preceding throws, and not merely to make deviations tend to vanish for proportions of the total number of throws. According to d'Alembert, the probability that the tenth throw will be the first to produce heads is even much smaller than $\frac{1}{2^{10}}$, since in the course of nine throws of tails the probability of tails is practically exhausted, and the probability that a head will appear becomes much larger than 1/2. He based this argument on the doubtless correct observation that a regular succession of heads and tails is much more improbable than an irregular one, but here he overlooked the fact that this is due only to the larger number of irregular successions; a given regular succession is no more improbable than a given irregular one. D'Alembert even went so far as to derive formulae for the probability of heads after a certain number of tails has been tossed. As already noted at the end of §132, similar views arise from common misconceptions of the notion of probability.

This misconception occurs frequently among regular visitors to casinos. It is not unwelcome to the management of these establishments, because a gambler will begin to play and continue to play all the more readily if he thinks he has found a system for increasing his chances. This explains why those in charge of the casino at Monte Carlo, who must certainly know very well that such a system does not exist, make it easier for their guests to perform their computations (futile though these are), by regularly publishing the numbers that have successively appeared at a given roulette table. Many a player studies these misleading records for entire nights, to find out which numbers offer the best chances for the following days. Often the results of the calculations are then combined with some method of determining the size of the successive stakes (see §157), all in vain however, as many a player has learned to his cost.

The enormous advantage which the bank derives from the seemingly small departure from fair odds which is produced by the presence of the zero, is also a consequence of the law of large numbers. This disadvantage to the player—which is considerably increased by tips given to croupiers by winners (something which is not repaid when they later become losers) ensures that the bank will beat the majority of the players with (one can safely say, absolute) certainty. No matter what system they invent, those who continue to play will be ruined sooner or later.

Chapter X:
SOME GAMES OF ENCIRCLEMENT

I. GAME OF WOLF AND SHEEP

165. Rules of the game of wolf and sheep. This game is played by two players on the 50 black squares of a checkerboard [Continental style, 10 × 10]. For ease of description, these squares have been numbered as indicated in Figure 8; this numbering is also used for the game of checkers. One player moves five white pieces, the sheep; the other player moves one black piece, the wolf. The board is set up with the sheep placed on the squares 46, 47, 48, 49, 50. The wolf is

	1		2		3		4		5
6		7		8		9		10	
	11		12		13		14		15
16		17		18		19		20	
	21		22		23		24		25
26		27		28		29		30	
	31		32		33		34		35
36		37		38		39		40	
	41		42		43		44		45
46		47		48		49		50	

Fig. 81

placed on any one of the remaining squares, with freedom of choice; this counts as the wolf's first move. Then there is a move for the sheep; this is a move of one of the sheep diagonally forward onto a black square that has a vertex in common with the original square—for example, from square 46 to square 41, which will be indicated here as 46–41. Now it is the wolf's move, which is a move to an adjacent square (adjacent by a vertex)—for example, from 28 to 22 or to 23, or to 32 or to 33; such a move will here be indicated by a single number, that of the new square. In this way the notation gives an immediate indication of whether a wolf's move or a sheep's move is in question. Then the sheep make another move. All moves must be made to free squares, and sheep are not allowed to move backwards, which prevents them from moving to squares with higher numbers. The wolf is free from this latter restriction.

The sheep win when they succeed in encircling the wolf (either in the middle of the board or on the border), in the sense of creating a situation in which the wolf has no move. If they do not succeed in doing this, then the wolf must have broken through the line of the sheep, so that the sheep will eventually reach a position in which they cannot move any more (because they cannot move backwards); the wolf, who can continue to move, then becomes the winner.

166. Correct methods for playing wolf and sheep. Provided they play carefully, the sheep win easily with no need for much theoretical knowledge of the best way to play. They have only to stick to the almost self-evident rule of advancing in close formation as much as possible, always beginning at the side where there is a sheep in contact with either the left-hand or the right-hand border of the board; otherwise a gap will arise in the line of sheep, or between their line and the border, through which the wolf can pass. Hence, if the position of the wolf does not hamper this, they will make successive moves: 46–41, 47–42, 48–43, 49–44, 50–45, 45–40, 44–39, etc.

If the wolf presses against the line of sheep, to make this strategy impossible for them, then they begin to move at the other side, starting at the border and taking care that no gap arises internally in their line, but only between their line and the border, and they should strive to close this gap in good time. If the sheep are to move in the position 40, 41, 42, 43, 44, 39 (where the last number indicates the wolf's square), then they cannot play 44–39; instead they play 41–37, intending to continue with 44–39 after the wolf's next move. In this way, no more difficult positions arise than the ones in *Figure 82*, where it is the wolf's move and where the sheep can win:

A position of this type undergoes no essential change when it is shifted up or down by one row, and is then reflected (to produce an interchange of left and right), or when it is shifted up by two rows without being reflected, and so on. But when the positions have a substantial shift, four rows, for example—the game becomes easier for the sheep, because then they will soon be able to make use of the upper edge of the board in their task of encircling the wolf. The left-hand diagram in Figure 82 can arise from the initial position thus:

38, 46–41, 42, 50–44, 38, 47–42, 43, 41–37,

and the right-hand diagram thus:

28, 46–41, 33, 47–42, 39, 48–43, 44, 41–37,

the difference being that now the positions are shifted down by 1 row, and reflected.

In the left-hand diagram, if the wolf tries to break through at the left by 32, then there is only one correct reply, namely: 43–38, 27, 37–31, 32, 42–37; if the next moves are then 28, 37–32, 33, 31–27, the left-hand diagram has arisen again (shifted up by 1 row), and the wolf has achieved nothing. If, in the left-hand diagram, the wolf tries

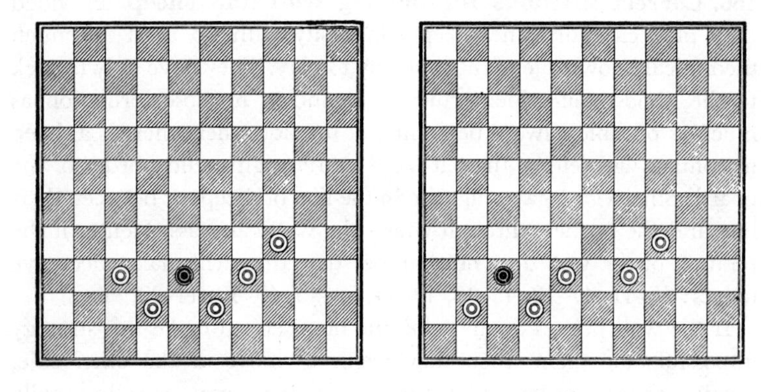

Fig. 82

to break through at the right by 33, then three of the six possible replies are correct, namely: 43–38, 42–38, and 34–29, the first two of which are more obvious than the third. After 33, the game may proceed as follows:

43–38, 28!, 38–32 (37–31 or 37–32 is equally good; 38–33? and 34–29? would be followed by 32; 39–33? would be followed by 23; and 34–30? would be followed by 33), 23! (to avoid forcing the move 42–38), 42–38!, 29, 38–33, 24, 34–30, 29, 39–34.

Or: 42–38, 29 (28 would be followed by 38–32), 38–33!, 24, 34–30, 29, 34–39, 23, 43–38, 28, 38–32.

Or: 34–29, 28, 37–32!, 22, 42–37, 28, 43–38, 23, 38–33, 19, 39–34.

Or: 34–29, 38, 37–32!, 33, 43–38, 28, 42–37 (38–33 is also correct), 23, 38–33, 19, 39–34.

The reply 34–29 to 33 requires very careful play on the part of the sheep. A strong move has been signalized by an exclamation point (although it may not lead to a win when it is a move by the wolf), and a weak sheep's move, which leads to a loss, receives a question mark.

Several variants of these lines of play are possible, but these chiefly amount to changing the order of moves. We advise the reader to play through some of these variants, so that he particularly may assure himself that the sheep's moves signalized by a question mark do indeed lead to a loss when the wolf plays correctly; this requires the subsequent moves to be played out in full.

In the right-hand diagram of Figure 82, the continuation: 32, 42–37, 28, 41–36 (37–32 is also correct), 33, 36–31, 29, 38–33, 24, 34–30, 29, 39–34 is a very obvious one. An alternative is 32, 41–37, 27, 37–31, 32, 42–37, 28, 37–32, 33, 31–27, after which the left-hand diagram has arisen, shifted up by one row. With the first line of play, the sheep make things much easier for themselves.

167. Some wolf and sheep problems. *Figure 83* shows two simple positions which can arise only when the sheep have played badly (or,

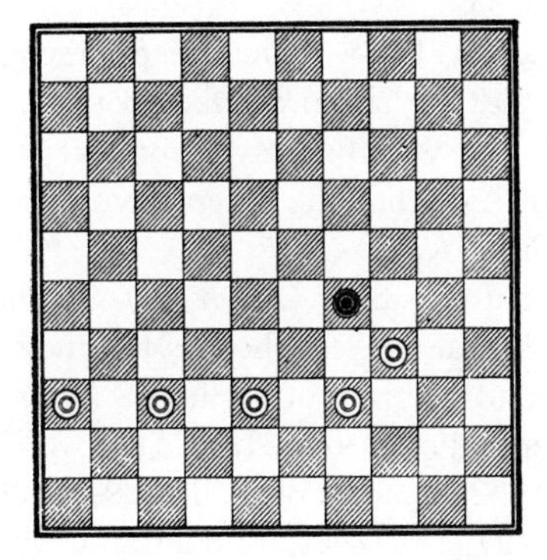

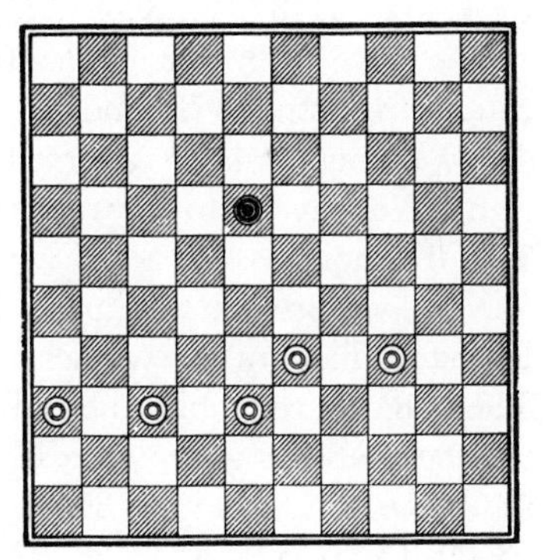

Fig. 83

at any rate, recklessly), because they have made moves at the wrong side of the board (at the side where the sheep are not in direct contact with the border). In both diagrams the player who has the move will win; if it is the sheep who are to move they have been able to indulge in their reckless play because the wolf was sufficiently far away.

If it is the sheep's move, in the left-hand diagram, they can win by only one of their seven possible moves, namely by 38–33. In the same diagram, if the wolf is to move, his only winning move is 24.

In the right-hand diagram, if the sheep are to move, two of the eight possible moves will lead to a win; 33–29 and 37–32. If it is the wolf's move, his only winning move is 23. We leave the reader to work out the further course of the play for the various continuations.

Two slightly more difficult problems are presented by the diagrams of Figure 82. The sheep are to move and win. With correct play, these situations cannot arise in the game, but this is irrelevant for the purposes of the problem.

In the left-hand diagram of Figure 82, three out of the five possible sheep's moves lead to a win. This becomes evident from the following sequences of moves: 37–32, 33, 43–38, 28 (or 29), 38–33, 22, 42–37 (or 24, 34–30); 39–33, 32, 43–38, 27 (28 would be followed by 37–32), 37–31, 32, 42–37, 28, 37–32; 34–29, 32 or 33, 43–38, 29 (27 would be followed by 37–31), 37–31, 23, 42–37, 19, 39–34, 23, 37–32 or 38–33.

In the right-hand diagram of Figure 82, the move 34–29 leads to a win thus: 34–39, 31 or 32, 42–37, 28, 37–32 (another possible move is 37–31, which serves as a reply to 27 as well), 23 (22 would be followed by 41–37), 41–37, 19, 39–34, 23, 38–33.

The move 38–32 does not lead to a win. The wolf then keeps moving to and fro between 37 and 31, so that the sheep are forced to play 42–37 sooner or later, after which the wolf breaks through at the right. We leave it to the reader to show that the sheep have a loss with the move 34–30, a win with the move 39–33.

Next, we present two more difficult positions (*Figure 84*). In the left-hand diagram the winner will be the player who has the move. The diagram resembles the right-hand diagram of Figure 82 (where the sheep always win), yet it is essentially different. If it is the wolf's move, he wins thus: 27, 36–31, 22, 32–28, 18, 28–23, 13, 29–24, 9, 33–29, 13, 37–32, 18, 32–28, 22, 31–27, 17, 27–21, 22. Here the wolf first forces the sheep to go to the right to keep him from breaking through there, and then he breaks through at the left. The course of the game includes several variants according to the different ways in which the sheep defend themselves against the attempted breakthrough at the right.

In the left-hand diagram, if the sheep are to move, they have two winning moves, 29–24 and 32–27. The game may proceed as follows: 29–24, 27, 36–31, 22, 32–28, 18, 37–32, 13 or 23, 33–29, 9 or 19,

28–23, 14, 24–20, 19, 29–24, 13, 32–28, 18, 28–22, 12, 31–27, where the sheep have made nearly all their moves at the right, which is the weak side. Alternatively, the game might run thus: 32–27, 26, 36–31, 21, 37–32, 17, 32–28, 22, 31–26, 18, 28–23, 13, 29–24, 9 or 19, 33–29, 14, 24–20, 19, 29–24, 13, 27–22, with several variants. The following sequence of moves shows that the move 29–23 leads to a loss for the sheep: 29–23, 27, 36–31, 22, 32–28, 18, 33–29, 13, 23–19, 9, 29–24, 13, 37–32, 18, 28–23, 22, 32–28, 27; because of the move 29–23 to the left, the other sheep have to make too many moves to the right, to prevent a breakthrough there, and after this the wolf can easily break through at the left.

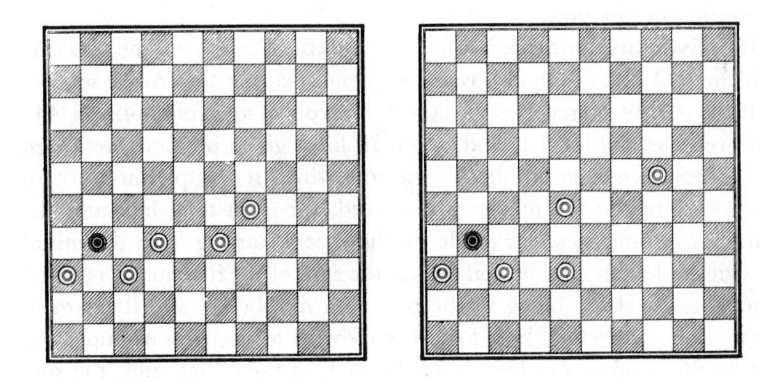

Fig. 84

The right-hand diagram has the peculiarity that the player who has the move will be the loser, something which practically never occurs in actual play of wolf and sheep. If the wolf begins, the sheep win, for example, as follows: 27, 36–31, 22, 38–33, 18, 28–23, 13, 37–32, 19, 33--29, 13, 32–28.

If it is the sheep who are to move, their only possible moves will weaken their position, because squares 36 and 37 are blocked. The moves 37–32, and 38–33 are fatal directly. 38–32 is followed by: 27, 28–22, 21, 36–31, 17, 32–28, 12, 22–18, 8, 28–23 (37–22 would be followed by 13), 12, 37–32, 17.

After the move 28–22, the game proceeds similarly except for an interchange of two moves made by the sheep; and it is of no avail to advance the sheep situated on 24, which has already been advanced quite far enough.

A somewhat simpler position still, in which the player with the move will be the loser (because some of the sheep in the rear are blocked), is 29, 37, 38, 39, 42, 32 (the wolf being on 32). If it is the sheep's move, they have to move either 29 or 39 (to avoid an immediate defeat), after which the wolf breaks through easily. If it is the wolf's move, the sheep can win easily, after 28 by 37–32, and after 27 by 37–31, but this involves them in slightly more careful play.

For anyone who wants to find the solution of the various problems without using the hints we have given, the most agreeable procedure is certainly to play against an opponent, starting from the given position, and giving the sheep sometimes to one player, and sometimes to the other. This brings out the different possibilities of the position in question.

168. Even and odd positions. We number the rows of the checkerboard 0, 1, 2, . . . , 9, in upward sequence, so that the row of squares 46–47–48–49–50 receives the label 0, the row of squares 41–42–43–44–45 receives the label 1, and so on. If, for a given position, we assign to every sheep the number of the row which it occupies and obtain the sum of the 5 numbers so involved, this sum then indicates the number of moves so far made by the sheep, starting from the initial position, for the initial position has the sum of the five numbers equal to 0, every sheep being in the row with number 0, and the sum in question is increased by 1 for every move of a sheep. As examples we take the diagrams of Figure 82. In each of these diagrams, the row numbers for sheep are 1, 1, 2, 2, 3, so that the sheep have already made $1 + 1 + 2 + 2 + 3 = 9$ moves.

If we increase the sum of the five row numbers for the sheep by adding the number of the row in which the wolf is, then we obtain what we shall call the position number. Every move of a sheep increases this number by 1, whereas every move of the wolf gives either an increase or a decrease by 1. We call the position an even one or an odd one according as the position number is even or odd. With every move, the position number changes either from even to odd, or from odd to even.

If the wolf has started on an even row (a row with an even number), there is an even position before the first move by the sheep. In every even position the move is then with the sheep, and in every odd position the move is with the wolf. If the wolf has started on an odd row, the opposite is true. So if there is information whether the wolf started on an even row or on an odd row, it becomes unnecessary even to ask whose move it is; this is immediately obvious from the position being odd or even.

If the wolf has started on an even row, the outcome will be different from what would happen if he started on an odd row. It is true that in both cases the same position can arise, but not with the same side to move. This means that only the very simplest positions will be equivalent in the two cases (regardless of whose move it is). If the wolf has started on an even row, the position in the left-hand diagram of Figure 82 can arise with the wolf to move. If the wolf has started on an odd square, that position is impossible, since then the wolf would have made the last move; however, he would not have gone to 38 in that case, but would have won easily by going to 27 or 29 instead.

The sheep have a much easier game when the wolf starts on an odd row, in which case they have to move when the position is odd. If they play carefully, no positions arise that are more difficult than: 31, 33, 34, 37, 38, 28, and 31, 32, 34, 38, 39, 29, with the sheep to move. In the first position the gap is easily closed by 37–32, 23, 33–29, and in the second position by 38–33, 23, 34–30. So the wolf is well-advised to start on an even row, and particularly on the row identified by the number 2, for example on 38, in order to have the best opportunity to take advantage of a mistake by the sheep.

169. Final remark on wolf and sheep. In a position in the game of wolf and sheep, the prospect of victory or defeat for the sheep will sometimes depend on whose move it is. However, wolf and sheep is not a game of timing. By a game of timing we mean one in which a player can use his choice of moves to arrange whether or not he will have the move when some particular position arises. In wolf and sheep this is impossible. The evenness or oddness of the position determines whose move it is; the way in which the position arises makes no difference to this.

In practical play it is an advantage to have the next move in a given situation; if this is one of inherent difficulty, there is the simple reason that the choice of move gives one a head start. The sheep can close their gap more quickly; the wolf can reach the gap in the sheep's line more quickly. Problems can be constructed in which it is a disadvantage to have to make a move. A case like this arises when sheep best suited to close the gap are unable to move because of the position of the wolf, and moves by the other sheep will only widen the gap or make it more difficult to close. If it is the wolf's move, it can be because he has to provide freedom of movement to one of the sheep that can make a good move. We have seen examples of this, (24, 28, 36, 37, 38, 31 and 29, 37, 38, 39, 42, 32). There are positions of some difficulty (mostly to be found in problems) where the sheep have to consider carefully whether they should move mainly to the left or to the right. Preferably, the sheep should line up in such a manner that they are evenly spread to left and right, not occupying too many different rows. Depending on the maneuvers of the wolf, the sheep are then able still to go safely to the left or to the right. The wolf should aim at preventing an even formation of this type, by forcing the sheep to go to one side, to allow him to return and break through at the other side.

II. GAME OF DWARFS OR "CATCH THE GIANT!"

170. Rules of the game. The game that we shall call "catch the giant!" or the game of dwarfs is played by two players on a board with eight numbered circles of a form as shown in *Figure 85.* One player has

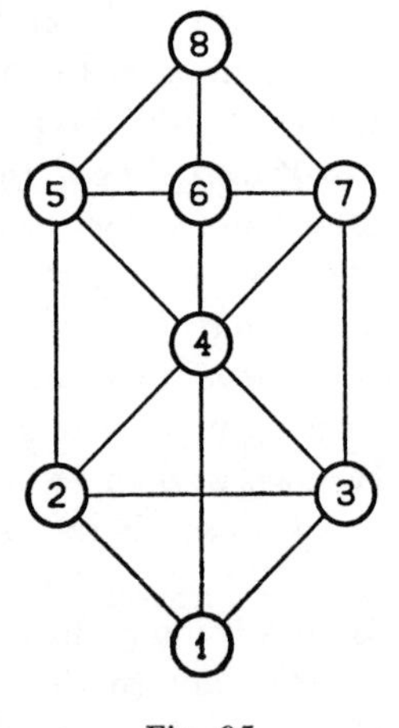

Fig. 85

three pieces, which we shall call the dwarfs, the other has one piece, the giant. The dwarfs are placed on circles 1, 2, and 3. Then the giant is placed on any one of the circles 5, 6, 7, or 8, with a free choice. After this first move of the giant, it is the dwarfs' move. Their move consists of moving one dwarf to an adjacent free circle, that is, to a circle which is connected to it by a single line-segment. The intersection of the lines 1–4 and 2–3 does not provide a circle, and so it is not possible to place a piece upon this intersection. For example, a dwarf is allowed to move from 1 to 4 (indicated by 1–4), but not from 1 to 6, since this path goes through circle 4. A dwarf is not allowed to move directly backwards, nor to move backwards diagonally, and the same condition of course applies to any subsequent move by a dwarf; however, a dwarf is allowed to move transversely, for example from 2 to 3, or from 6 to 5, and back. After the dwarfs make a first move (which may be 1–4, 2–4, 2–5, 3–4, or 3–7, so long as the move is not made to a circle which is occupied by the giant), the giant has a move. He moves to an adjacent circle according to the same rules, with the difference that the giant has the additional possibility of a backward move, so he can (for example) make a move from 4 to any one of the circles 1, 2, 3, 5, 6, or 7, (provided that the circle is not occupied by a dwarf). Then it is the dwarfs' move again, and so on.

The dwarfs have won when they have caught the giant, that is, when they have him encircled on circle 8; the dwarfs will then be occupying circles 5, 6, and 7. If they do not succeed in encircling the giant, they lose. This does not require the giant to break through. If a position arises in which the giant and one of the dwarfs both continue to make forward-and-backward moves (repetition of moves), this is regarded as a win for the giant. A repetition of moves is considered to have occurred when the same position has arisen three times in succession.

If the giant manages to break through the line of dwarfs, this gives him a double win.

171. Comparison with wolf and sheep. The game of dwarfs reminds one of wolf and sheep to some extent, because encirclement is the object of both games. The new game has, however, a quite different nature, because in wolf and sheep the wolf has to break through, to win, whereas a win by the giant will come from a repetition of moves, in a vast majority of cases. The dwarfs have to play very badly indeed if they are to give the giant an opportunity to break through, and consequently a double loss practically never occurs.

So the task of the dwarfs is not just to prevent the giant from breaking through, but to take care that they prevent a breakthrough without producing a repetition of moves. If the dwarfs make a mistake that gives the giant an opportunity to break through, they suffer immediate and evident damage from the reply by the giant, and they will not readily make such a mistake a second time. However, if they make a mistake that leads to a repetition of moves, the damage becomes evident only after the delay of a couple of moves, and then it is very difficult for the dwarfs to see what their actual mistake was; as a result, they are likely to make the same mistake or a similar mistake again, on another occasion. Besides, the moves that do the mischief are usually very natural ones. This is why the game of dwarfs with its 8 circles is a much more difficult game than the game of wolf and sheep with its 50 squares. Considering the simplicity of the game (in the matter of its design) and the very limited number of circles of the board, it is indeed surprising how difficult a game it is to play. This is what makes the game of dwarfs such an attractive game.

What we have said shows that the game of dwarfs has much of the nature of a game of timing. In this type of game you have to take care not to be the next to move when certain positions arise; hence, these are the positions that you should attempt to set up for your opponent, while, of course, you should try to prevent him from leaving them to you, in his turn. The possibility that a player can change the "opposition" is related to the fact that, in several cases, a piece can be transferred from one circle to another, not only directly (in one move), but also by a detour (in two moves); thus, you can go from circle 2 directly to circle 5, or reach circle 5 after the two moves 2–4 and 4–5. In wolf and sheep this was not possible. There you had only to watch the gaps in the line of the sheep, whereas in the game of dwarfs the only thing that matters is to retain or obtain the opposition, except when the dwarfs must hasten to close a gap.

172. Remarks on correct lines of play. Before going further into the correct modes of play, we shall make here some rather self-evident remarks. To avoid a breakthrough by the giant (bringing a double loss), the dwarfs should prevent the giant from occupying the central circle 4; once the giant has reached 4, he can proceed to one of the circles 1, 2, or 3, and then he can no longer be driven back. This is why one of the rules of the game is that the giant is not allowed to open on 4; for then the dwarfs could only play 2–5 or 3–7 and thus lose immediately, which would remove all interest from the game. To prevent the giant from reaching 4 later, the dwarfs have to occupy this circle themselves as long as the giant is not on 8; if the giant withdraws to 8, the dwarfs can evacuate 4 temporarily, but they should then occupy it again on the next move.

Further, the dwarfs have to make sure that the giant cannot reach circle 5 when circles 2 and 4 are both unoccupied (and likewise that he cannot reach 7 when circles 3 and 4 are unoccupied, which obviously amounts to the same thing); otherwise the giant will break through on his next move by going to 2 or 4. This shows that the dwarfs suffer a double loss when they occupy one of the groups of circles 1, 2, 5; 1, 2, 6; 1, 3, 6; or 1, 3, 7.

These rules are not nearly sufficient guidance for the dwarfs. They serve only to prevent the giant from making a breakthrough, but the real difficulty for the dwarfs lies in having to avoid repetition of moves, which means they have to obtain the opposition.

173. Correct way of playing. When we wish to investigate the various positions (locations jointly for the dwarfs and the giant), assuming that it is the giant to move, we consider the game in reverse sequence, and start from the final position, to find whether the dwarfs can win if they play correctly. We thus begin with advanced positions for the dwarfs, so few moves away from the final position that the whole of the subsequent course of the game can be worked out mentally; these are the positions in which the dwarfs already occupy two of the three circles 5, 6, 7.

We shall first consider a position 3, 4, 5 for the dwarfs (who do not have the next move), and take various positions in turn, for the giant, to determine whether the dwarfs can win, and if so, in how many moves. Then we can use this to find the corresponding results for the position 2, 4, 5 of the dwarfs, since this position can be changed into 3, 4, 5 in one move. From this, we can find the results for the position 1, 4, 5, which is in its turn one move away from 2, 4, 5. Continuing in this way, we can consider still less advanced positions of the dwarfs (when it is not their move), taking various possible positions for the giant, to determine which side wins (with correct play), and how many moves the dwarfs require, if they are the winners. A position will be called a winning position if the giant is to move, but the dwarfs can win.

Knowing the winning positions, we can easily consider some given position with the dwarfs to move, and determine whether they can win, and if so, in how many moves. To do this, we need only check whether the dwarfs can create a winning position in one move. When we have investigated all positions with the giant to move, then this will make us also able to decide whether the dwarfs can win from the initial position, and if so, by which first move and in how many moves.

It will turn out that the dwarfs can win, no matter which of the circles 5, 6, 7, 8, the giant chooses for his first move. In order to obtain an actual win, the dwarfs have to know their various winning positions (with the giant to move). All they then need do is to ensure that at each move they produce a winning position. Here, as in serious science, we can say that greater insight gives less need to use memory as a substitute. Someone who can look ahead only for a couple of moves has to memorize a few winning positions to have adequate starting points for his combinations, whereas a chess master is able to look sufficiently far ahead to find the correct move without more ado.

Among the positions which give the dwarfs a win when it is not their move, there is not a single one in which there is a dwarf on circle 6, unless in the final position, where the dwarfs have won already. If the dwarfs win, then 4–6 is always their last move. This shows that the dwarfs should not occupy circle 6 before the end of the game. Knowing this makes it easier to remember the winning positions and find correct moves.

174. Winning positions. If the dwarfs are on 5 and 7 and on one of the circles 1, 2, 3, while the giant is on 8, this clearly gives a winning position (that is, one where the dwarfs win when it is the giant's move). As well as this position, it turns out that there are seven other winning positions, provided we make no distinctions between positions that arise from one another by reflection (with interchange of left and right, and hence of 2 and 3, and of 5 and 7).

The seven winning positions are depicted in *Figure 86* in the order in which they are found, that is, starting with the position that is closest to the final position. The positions of the dwarfs have been indicated by double circles, the position of the giant by a number which also indicates the number of moves in which the dwarfs can win. In determining this number it has been assumed that the giant plays in such a way that he offers resistance as long as he can, and so requires the largest possible number of moves before he is encircled on circle 8. In the rare event that the dwarfs have a choice between two winning continuations, the number of moves indicated is the one for the continuation in which this number is least. For the sake of convenience the position numbers of the circles have been omitted in diagrams I–VII, but to aid the reader these numbers have been given again in the small diagram at the left.

In what follows we shall indicate a giant's move by a single number, namely the number of the circle to which the giant moves, whereas we shall indicate a move of one of the dwarfs by two numbers (giving the initial and final circles).

We start with a substantially advanced position 3, 4, 5 for the dwarfs (which of course is equivalent to 2, 4, 7); this position is indicated in diagram I. If the giant is on 7 and plays 8, the dwarfs win by 4–7 in three moves; if the giant plays 6, the dwarfs win by 3–7 in two moves. If the giant is on 6 or 8 and plays 7, the dwarfs have to reply with 5–6, since otherwise the giant breaks through. The giant next plays 8, after which the dwarfs play 6–7, in the hope that the

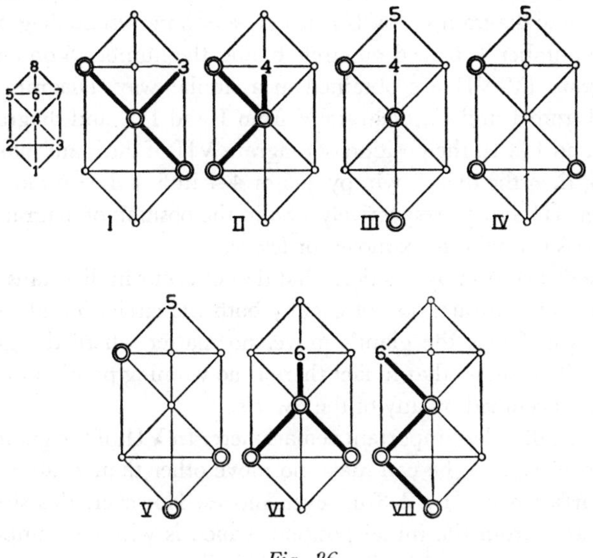

Fig. 86

giant will reply with 5, to let them win by 3–2; however, the giant will reply with 6, after which (for example) the continuation will be 3–2, 5, 7–6, 8, 6–7, 5. This gives a repetition of moves, and the giant wins. So in diagram I a number has been placed on circle 7 only, and this number is the largest number of moves that the dwarfs can require for a win.

Next we consider diagram II, where the dwarfs are on 2, 4, 5, which puts them a little further away from the final position; again we suppose it is the giant's move. If the giant plays correctly, the dwarfs can win only if he is on 6. By playing 7 here, the giant offers most resistance. According to diagram I (which arises from diagram

II in one move) the dwarfs then win by 2–3 in 1 + 3 = 4 moves. So the number 4 has been placed on circle 6 in diagram II. If the giant is on 7 or 8, and plays 6, the dwarfs can only play 2–3 (to keep the giant from breaking through), whereupon the giant replies with 7 and wins through a repetition of moves.

In the position 1, 4, 5 of the dwarfs, shown in diagram III, the dwarfs (who do not have the move) can only win when the giant is on 6 or on 8. In the first case, the giant has no better move than 7, after which the dwarfs can play 1–3, to win in 1 + 3 = 4 moves. If the giant is on 8, then 6 is his best move, after which the dwarfs can win, according to diagram II, by 1–2 in 1 + 4 = 5 moves. So diagram III has the number 4, placed on circle 6, and the number 5 on circle 8.

Diagram IV–VII are obtained in a similar way. Diagram IV is derived from I and II, diagram V from I and III, and diagram VI from I and IV. In the position of diagram VII, if the giant makes the move 6, then the dwarfs win by 2–5 or 4–7 in 5 + 1 = 6 moves (see diagrams III and V, respectively). So in the position of diagram VII, the dwarfs can win in six moves or fewer.

If the dwarfs occupy positions that do not occur in diagrams I–VII (so long as the dwarfs do not occupy both the circles 5 and 7), they cannot win if it is the giant's move, no matter where the giant is situated. This shows that in fact there is no winning position in which circle 6 is occupied by any of the dwarfs.

We can add a less important remark here. In VII, if the giant (who is to move) is on 3, he can make no move other than 7, after which the dwarfs can win by 2–3 in seven moves. However, this situation cannot arise from the initial position, which is why no number has been placed on circle 3 in diagram VII.

175. Positions where the dwarfs are to move. From the diagrams I–VII of Figure 86, we can consider the same positions of the dwarfs but with the dwarfs to move, and determine for which positions of the giant the dwarfs can win (see diagrams A–G *Figure 87*). These positions of the giant have been indicated by box symbols; within the box we have indicated how many moves the dwarfs will need for a win (when the giant puts up as persistent an opposition as possible), adding the first move of the dwarfs, in parentheses. For example, if the giant is on circle 6 in diagram C, the dwarfs create position II (Fig. 86) by playing 1–2, and win in 1 + 4 = 5 moves (or in four moves if the giant, by going to 8, offers less resistance).

In cases A–F, the dwarfs have only one good move, provided that in case F with the giant on 8 we make no distinction between the moves 4–5 and 4–7, which are mirror images of each other. In case G, however, the dwarfs have two winning moves, which have been indicated; if the giant is on 6, the dwarfs have their quickest win (in 1 + 4 = 5 moves) by 2–5, through which diagram III arises, but also another win by 1–3 in 1 + 6 = 7 moves (see diagram VI).

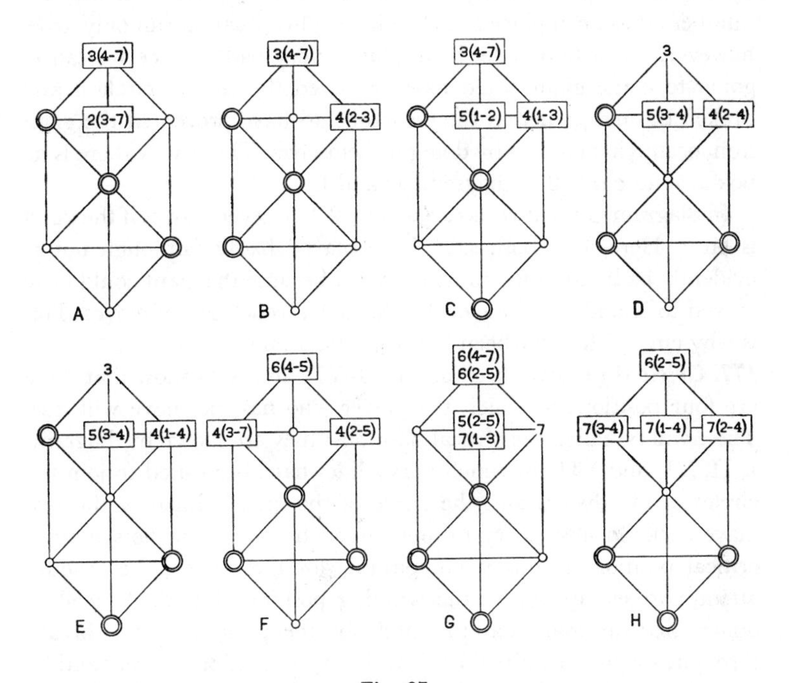

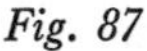
Fig. 87

In diagram H the dwarfs are in the initial position. If the giant occupies circle 8, the dwarfs win by 2–5 (or 3–7, which amounts to the same thing), through which diagram V arises, giving a win in 1 + 5 = 6 moves (assuming the best possible defense by the giant). If the giant occupies 6, the dwarfs create position VI by playing 1–4, and win in at most 1 + 6 = 7 moves. If the giant opens on circle 5, the dwarfs create position VII by playing 3–4 and again can win in at most seven moves; if the giant replies with 6, the dwarfs win by 2–5 in six moves, and by 1–3 in eight moves (in all). This completes the proof that the dwarfs can win.

III. FURTHER CONSIDERATIONS ON THE GAME OF DWARFS.

176. Remarks on diagrams D, E, and G. The remarks we shall now make have no importance when play proceeds correctly. If the dwarfs are in positions D and E (Fig. 87) they can win in three moves by 3–7 (if it is their move) when the giant is on 8, which is why the number 3 has been placed in this circle. This situation can only arise, however, when both sides have played very badly, since instead of going to 8 the giant could have occupied the central circle 4 and broken through, whereas the dwarfs could have prevented the giant from occupying circle 4 by doing it themselves. This is why there is no box around circle 8 in diagrams D and E.

In diagram G the dwarfs can win by 2–3 in seven moves if the giant is on 7. Here, too, both sides have played badly (although not so evidently badly as in the previous case), because the giant could have moved to 5, and could have won through a repetition of moves. This is why circle 7 has not been boxed in diagram G.

177. Critical positions. Diagrams I–VII and A–G show that there are four positions in which the player who has the move will lose. These will be called the critical positions; they are shown by diagrams I, II, VI, and VII (in which heavy lines have been used to join the circles of the dwarfs and the circle of the giant). Each of the two players should always try to move in such a way that he sets up a critical position. If this is not possible for the dwarfs, they should attempt to set up one of the winning positions III–V. As well as doing this, the giant has to watch for the possibility of a break-through; the opportunity for this will present itself only rarely and be immediately obvious when it occurs, so that the giant need only keep on the lookout for the possibility of creating a critical situation. The giant, too, should fix these positions firmly in his memory.

If the dwarfs have occupied circle 5 and the giant is on 8, it will be immediately obvious that the dwarfs can win by occupying circle 7. In the remaining cases, knowledge of the four critical positions is usually sufficient to enable the dwarfs to find the right move, even if they cannot themselves set up one of the critical positions. In this event they have only to move in such a way that the giant cannot break through and cannot reply with a critical position. This shows the dwarfs that when the giant begins on 8, they should reply with 2–5 (or 3–7), since otherwise the giant can set up one of the critical positions VI and VII; if the giant then replies with 6, the dwarfs can next see that they should play 3–4, because after 1–4 the giant could reply with the critical position I (by the move 7). In the position 1, 2, 4, 8 (that is, the dwarfs on 1, 2, 4 and the giant on 8) the dwarfs (who have to move) can see that the giant will reply to 1–3 with the critical position VI, and to 2–3 with the critical position VII, and that after 4–5 or 4–6 the giant will break through with 7; the two remaining possibilities for the dwarfs (2–5 and 4–7) are correct moves.

In all cases, knowledge of the critical positions will lead the dwarfs to a correct move if they also know that they should not occupy circle 6 before the final position. In this way the dwarfs can quickly see that in the position 2, 3, 4, 8, they have to reply with 4–5 (or 4–7) because the giant would reply to 2–5 with the critical position I.

The preceding considerations do not always serve to show the dwarfs the move that leads to victory in the quickest way, but this is naturally of lesser importance. When the position is 1, 2, 4, 6 they reveal the move 1–3, but not the move 2–5, which leads to success more quickly.

178. More about the correct way of playing. Diagrams A–H show that there are various positions where the dwarfs have to make a far from obvious move in order to keep the advantage. This is what makes the game so difficult for the dwarfs against an experienced giant, however simple it may seem to a casual onlooker. The dwarfs have a tendency to make their advance as much as possible in close formation, to prevent the giant from breaking through, instead of guarding against the danger that they will lose the opposition, and be unable to prevent a repetition of moves occurring thereafter.

For the giant, the game is hopeless against experienced dwarfs, but against less expert dwarfs his game is not so easy, either. The giant is inclined to move towards a gap in the line of the dwarfs, even when it can be easily seen that he will arrive too late, and that they can close their gap in time. By playing in this way, the giant virtually forces the dwarfs to make the correct moves, instead of disregarding the break-through (something which will in any event succeed only when there is very bad play by the dwarfs), to aim at capturing the opposition. The giant often can seize the opposition by making a retreat to circle 8, which makes it harder for the dwarfs to see the move which will let them retain the opposition.

179. Trap moves by the giant. The best prospect for the giant, who must lose if his adversaries play correctly, is to demand a long run of moves (including some improbable-looking moves) from the dwarfs, in the hope of taking advantage of a wrong move by them should this occur, to obtain the opposition (by creating one of the critical positions I, II, VI, VII). In the following discussions such trap moves by the giant are signalized by an exclamation point, likewise the correct replies that the dwarfs should make. The mistakes into which the dwarfs may be provoked by the giant's ingenuity are signalized by question marks.

How the dwarfs can avoid the traps, and win:

5, 3–4!, 6, 1–3, 8!, 4–5!, 6!, 3–4!, 7, 2–3, 8, 4–7, 6, 3–4, 8, 4–6;
5, 3–4!, 6, 2–5, 7, 1–3, 8, 4–7, 6, 3–4, 8, 4–6;
5, 3–4!, 8!, 2–5!, 6!, 1–2!, 7, 2–3, 8, 4–7, 6, 3–4, 8, 4–6;
5, 3–4!, 8!, 4–7!, 6!, 2–4!, 5, 1–2, 8, 4–5, 6, 2–4, 8, 4–6;
6, 1–4!, 8!, 4–5!, 6!, 3–4!, 7, 2–3, 8, 4–7, 6, 3–4, 8, 4–6;
8, 2–5!, 6!, 3–4!, 7, 1–3, 8, 4–7, 6, 3–4, 8, 4–6.

How the dwarfs can fall into a trap and lose:

5, 1–4?, 6!, 2–5, 7, etc.;
5, 3–4!, 6, 1–3, 8!, 2–5?, 7, etc.;
5, 3–4!, 8!, 1–3?, 6!, 2–5, 7, etc.;
5, 3–4!, 8!, 2–5!, 6!, 1–3?, 7, etc.;
5, 3–4!, 8!, 4–7!, 6!, 1–4?, 5, etc.;
6, 2–4?, 7!, 1–2, 6, etc.;
6, 1–4!, 8!, 2–5?, 7, etc.;
8, 1–4?, 6!, 2–5, 7, etc.;
8, 2–4?, 7!, 1–2, 6, etc.;
8, 2–5!, 6!, 1–4?, 7, etc.

By going through the moves, it will be appreciated how difficult it is for the dwarfs to give the correct reply to all the traps which the giant can set. The results we have given can provide us with the following examples.

With a position 1, 4, 5, 8 the giant (when it is his move) will not move toward the gap in the line of the dwarfs by playing 7 (he would arrive too late, anyway, and lose the opposition), but instead will make a reply of 6. The purpose of this is to induce a move 1–3 (to close the gap), which will let him obtain the opposition by the move 7. By playing 1–2 (which also leaves them time to close the gap), the dwarfs can retain the opposition.

This same situation will arise, for example, when the giant opens on 5, and the dwarfs have already twice succeeded in finding a correct, but not so obvious reply. It is quite likely that the dwarfs will go on to give a wrong reply 1–4; however, if they give the correct reply 3–4, the move 8 can set them another trap.

By opening on 5, the giant perhaps makes things hardest for the dwarfs, although the other openings, too, give him ample opportunities for setting traps. If the giant can manage to create one of the critical positions 1, 2, 4, 5; 2, 3, 4, 6; 2, 4, 5, 6; 3, 4, 5, 7, he then has a game which he can win.

If he is playing this game a number of times in succession, the giant (if he is an expert) will naturally be well-advised to open on a different circle every time, in order to vary his play as much as possible. It will be sufficiently clear from our discussion that the dwarfs will then need to know the game thoroughly to beat an expert giant.

On the other hand, the dwarfs cannot bring much variation into their play without losing their hopes of winning. In the situation 1, 2, 4, 8 the dwarfs have two correct replies (2–5 and 4–7), from which they will not always make the same choice, to avoid giving away the secrets of their strategy. The same is true for 1, 2, 4, 6, where 2–5 and 1–3 are the correct replies; the dwarfs will preferably choose the latter move, because this increases the number of moves needed to finish the game, and then there is less chance that the opponent will remember the appropriate sequences of moves.

180. Comparison of the game of dwarfs with chess. The game of "catch the giant" shares one characteristic with the game of chess, though in other respects there is no comparison between the two games. The common feature is the one which has been referred to as the "opposition," and this constitutes a particular aspect of chess which forms one of the many facets of this splendid game.

Especially in endgames which involve only pawns (apart from the Kings), the whole essence of the game is often the preservation or recapture of the opposition.

This is a matter of suitable maneuver of the King; for the King can affect the opposition in view of the fact that he can go to an adjacent square in one move, but can also take two moves, with an identical result.

In such endgames, even if they are of simple construction in the sense of having only few pieces on the board, it is often difficult to find the correct way of playing by analysis (let alone by unaided mental effort).

There are various situations where a loss is inevitable when you lose the opposition and your opponent makes no errors. Loss of the opposition often arises from a bad move of the King at a stage where this has consequences impossible to foresee (or extremely difficult to foresee). The bad position can also be the result of considerations which should have been taken into account when there were still other pieces on the board.

However, there are also cases where the King has lost the opposition but can maneuver in such a way that he regains it, for example by returning to his starting point in three moves, describing what is called a triangle as he does so. If the opponent's King cannot do the like, because he is confined so that he can only move to and fro, this produces the same position as the original one but with the other side to move. We had intended to devote a later chapter to the consideration of chess as a game of timing (or opposition game) but we have had to forgo this for lack of space.

Although the game of dwarfs does not allow the opposition to be captured compulsively in the same way as this can occur in chess, but only by tempting an opponent to make incorrect moves, it still seems to me that the game of dwarfs ought to arouse interest in chess circles.

This was also the opinion of our Dutch compatriot, the international chess-champion Dr. Max Euwe, to whom I showed the game. He could not possibly have seen the game before, for it had just been invented. I expected him to discover the correct replies which the dwarfs should make to the various possible openings by the giant, and indeed he did this in his head (without touching the pieces) in about 15 to 20 minutes, a feat that only few will be able to repeat.

On the other hand, the fact that Dr. Euwe required more than a quarter of an hour will certainly show that the game of dwarfs is far from childishly simple, even if it may look otherwise to an onlooker who watches others while they play it. Dr. Euwe told me that he thought the game of dwarfs could be extremely instructive for beginners in chess, and very suitable for illustrating the concept of the "opposition."

IV. MODIFIED GAME OF DWARFS

181. Rules of the game. We modify the game of §170 to the effect that a new circle is introduced at the intersection of the lines 2–3 and 1–4. The board then becomes entirely symmetrical, as is evident from

Figure 88, where the numbering of the circles has been adapted to the inclusion of the new circle.

If the game is played a number of times in succession, it is proper that the two players should take alternate turns to be the giant. With the game of §170 this would require the board to be turned every time; the modification now introduced makes this unnecessary, because of the symmetry.

For the rest, the newly added circle does not involve any modification in the rules of the game. A player is allowed to move only to an adjacent circle, so a single move cannot move a piece from 2 to 4, or from 1 to 5; he is allowed, though, to move (for example) from 2 to 3,

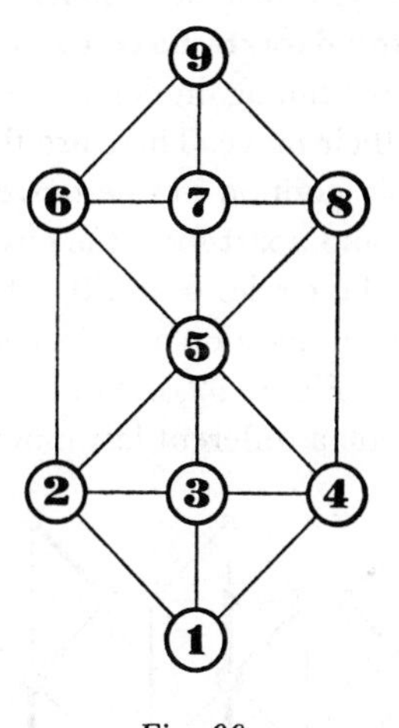

Fig. 88

or from 3 to 2. Here, too, a dwarf cannot go directly backwards or diagonally backwards. The dwarfs are initially on the circles 1, 2, and 4. Then the giant occupies any one of the six remaining circles. The dwarfs win when the giant is encircled on 9. If there is a continuous repetition of moves, the giant has won (and he has a double win if he breaks through). It is not necessary to forbid the giant to open on the central circle 5, because the dwarfs can prevent a breakthrough by the reply 1–3.

182. Winning positions of the modified game. The game of §181 is investigated in entirely the same way as that of §170, namely by beginning with the most advanced positions of the dwarfs, and then considering less advanced positions several times over. Apart from the positions in which the dwarfs occupy circles 6 and 8, there are seven winning positions (in the sense that the dwarfs win when it is the giant's move: positions which correspond by reflection are again considered to be identical), and these are shown in *Figure 89.*

The small diagram at the left gives a recapitulation of the scheme of numbering of the circles. In diagrams I–VII there is a number which indicates the circle for the giant, and also the number of moves in which the dwarfs can win (when it is not their move). All other positions for the dwarfs (assuming that they have not occupied both the circles 6 and 8), will make them always lose if it is the giant's move, no matter where he is. The positions I–III and V–VII are of the critical type; here the side which has to make a move will be the loser.

183. Case in which the dwarfs have to move. From the winning positions of Figure 89 we can again derive the positions in which the dwarfs can win if it is their move. These are the positions that can be changed into a winning position in one move. As well as the initial position of the dwarfs (and apart from the obvious cases in which the dwarfs have occupied the circles 6 and 8), the only positions of the dwarfs which deserve consideration are those which also occur in one of the diagrams I–VII of Figure 89, because otherwise the giant could have won if he had made a different last move.

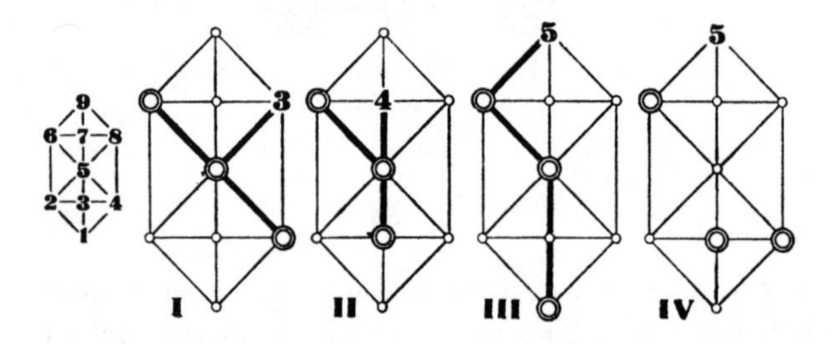

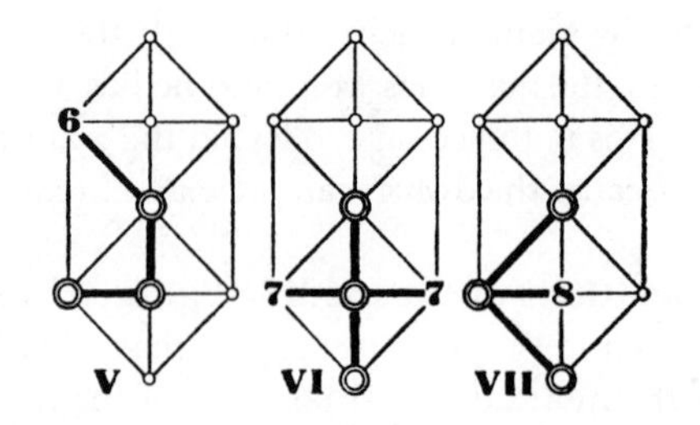

Fig. 89

This provides the cases shown in *Figure 90*, in which the giant is assumed to be on one of the circles where a box has been inserted. Inside the box we have indicated the number of moves in which the dwarfs can win, and, in parentheses, the first move the dwarfs have to make to achieve this.

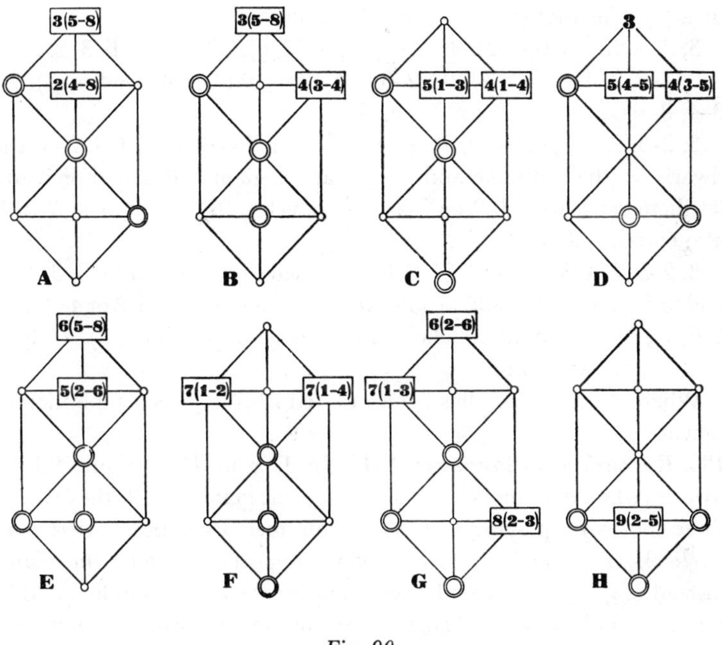

Fig. 90

184. Dwarfs puzzle. In diagram H of Figure 90 we observe that the dwarfs can win only if the giant opens on circle 3; they then win in nine moves at the most. This gives the solution to the following puzzle, which we shall call the dwarfs puzzle:

In the game of §181, where should the giant open in order to allow the dwarfs to win?

From diagrams A–H we see further that the dwarfs always have only one correct move, provided that in the initial position H we make no distinction between the moves 2–5 and 4–5, which are mirror images each of the other. This means that when they are able to win, the dwarfs can exert no influence on the number of moves which they will require. However, if the giant does not put up a strong enough defense, the number of moves can become smaller than the number which is indicated in the diagrams.

If the giant opens on 3, the best sequence of moves is 3, 2–5, 2, 4–3, 6, 1–2!, 9!, 5–8!, 7!, 2–5!, 6, 3–2, 9, 5–6, 7, 2–5, 9, 5–7. The giant's moves indicated by an exclamation point will not lead to a win, it is true, but the move they require from the dwarfs is less obvious. If the dwarfs make another move, they lose, thus:

3, 2–5, 2, 4–3, 6, 1–2!, 9!, 5–8!, 7!, 3–5?, 6, 8–7, 9, 2–3, 6, 3–2, . . . ;

3, 2–5, 2, 4–3, 6, 1–2!, 9!, 2–6? (3–4? would be followed by 7), 7!, 3–4, 8, 6–7, . . . ;

3, 2–5, 2, 4–3, 6, 3–2?, 7! (9? would be followed by 2–6!), 2–3 (the dwarfs see their mistake and hope that the giant will play 6 or 8 and let them reply with 1–2 or 1–4, respectively), 9!, 1–2, 8, 3–4, 7 (9 is also correct), 2–6, 8, . . . ;

3, 2–5, 2, 4–3, 6, 3–2?, 7!, 2–6 (1–3 would be followed by 8, 3–4, 7, and so on, and 1–4 would be followed by 9! and then 2–6, 8 or 4–3, 8 or 5–6, 7), 9! (8? would be followed by 1–4), 1–3 (tempting 8), 7!, 3–4, 8, 6–7, . . . , and so on.

Where "and so on" has been inserted, this implies a repetition of moves.

185. Remark on diagrams A–H. In diagram D of Figure 90 the dwarfs (who are to move) will also win if the giant is on 9; they win in three moves by playing 4–8; so the number 3 has been inserted in circle 9. However, this case is of no importance, since the giant, instead of going to 9, could have occupied the central circle 5 and in this way could have broken through the line of dwarfs. Hence, this case can arise only if there is very bad play by the giant; the dwarfs, too, must have played very badly, because they could easily have prevented the giant's breakthrough. This is why we have not put a box on circle 9 in diagram D.

This indeed is also why we made no mention in §183 of positions in which the dwarfs are to move and can win only because the giant's last move was wrong. As an example of this we take the case in which the dwarfs are on 2, 3, 4, and the giant on 6 or 8 (or the giant on 9, in which case he has missed a chance to occupy circle 5 and to break through); the dwarfs can then win by 4–5 or 2–5, respectively (or by 2–6 if the giant is on 9). Another example (again needing bad play by the giant) has the dwarfs on 2, 4, 5 and the giant on 6 or 8; in this position the dwarfs win by playing 4–8 or 2–6, respectively.

186. Other opening moves of the giant. In the game of §181 the main difficulty for the giant lies in choosing the circle which he should occupy at the beginning of the game. But even if he chooses a good starting point, that is, one of the circles 5, 6, 7, 8, 9, he still has to be on constant watch to give correct replies to trap moves made by the dwarfs, as appears from the following games:

5, 1–3, 7! (6? would be followed by 4–5!), 3–5 (2–5 would be followed by 6), 9! (6? would be followed by 4–8!), 5–6 (2–6 would of course be followed by 8), 7! (8? would be followed by 2–5!), 2–5, 8, . . . ;

6, 2–5 (tempting a move 2?), 7! (9? would be followed by 4–8!), 4–3, 9! (6? would be followed by 1–2!), 3–2 (1–2 would be followed by 6 or 8), 8, 1–4, 7, 2–6, 8, . . . ;

7, 2–5, 6! (8? would be followed by 1–3 and 9? would be followed by 4–8!), 1–2, 7, . . . and so on; 9, 1–3, 7! (6? would be followed by 4–5!), 3–5, 9! (6? would be followed by 4–8), 5–6, 7!, . . . ;

9, 2–5 (after 2–3, any reply by the giant would be correct), 6, 1–2, 7, ...

V. THE SOLDIERS' GAME

187. Rules of the game. The games of §§170 and 181 are simplifications of the so-called soldiers' game, which was published in a French military magazine in 1886, when it attracted much attention because its simplicity of design is coupled with an extraordinary variety of possibilities which make it a substantially more difficult game than the games of dwarfs of §§170 and 181.

The soldiers' game is played on a board with eleven circles, as shown in *Figure 91*. Three white pieces are placed on circles 1, 2, and 4, after which a black piece is allowed a choice of one of the eight remaining circles. White wins if he manages to encircle Black on 11; otherwise White loses. For the rest, the rules are the same as for the games of §§170 and 181.

The increase in the number of circles makes it an extremely difficult matter to arrive at a comprehensive view of the game without making a preliminary analysis. This will establish that White can win, no matter on what circle Black chooses to open the game.

However, this can indeed become an extraordinarily difficult task for White if Black makes several retreats to the circle 11. White then need not be concerned to prevent a breakthrough by Black, and if this is all he considers, White has a choice of several moves; there are some positions where only one of these moves will allow him to retain the opposition.

Because this strategy forces White to make a larger number of moves, there is a fair chance that he will lose the opposition and there will be a repetition of moves, if Black plays correctly, all the more so since in many cases the moves that will allow White to retain the opposition are far from obvious ones.

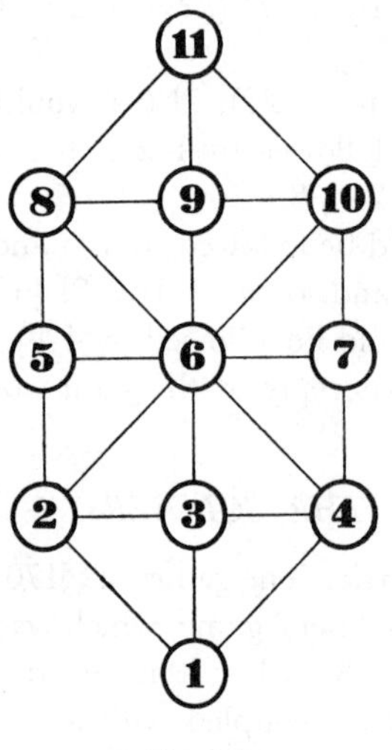

Fig. 91

188. Winning positions. To be sure of following correct lines of play, White must be aware of a number of winning positions (positions in which White will win, or more accurately can win, if it is not his move), since these are the positions that White should aim at producing.

These positions, which are found by the same principle as before (starting out with the final situation) are shown in *Figure 92* (where winning positions that White does not need are omitted); Black's circle has been indicated by a number which also serves to give the number of moves in which White can win, while the double circles indicate the positions of the white pieces.

From the initial position, White can attain one of the winning positions of diagram 17, unless Black opens on circle 6. If Black begins with 6, White can secure one of the two winning positions which are indicated in diagram 18, and this is true also when Black

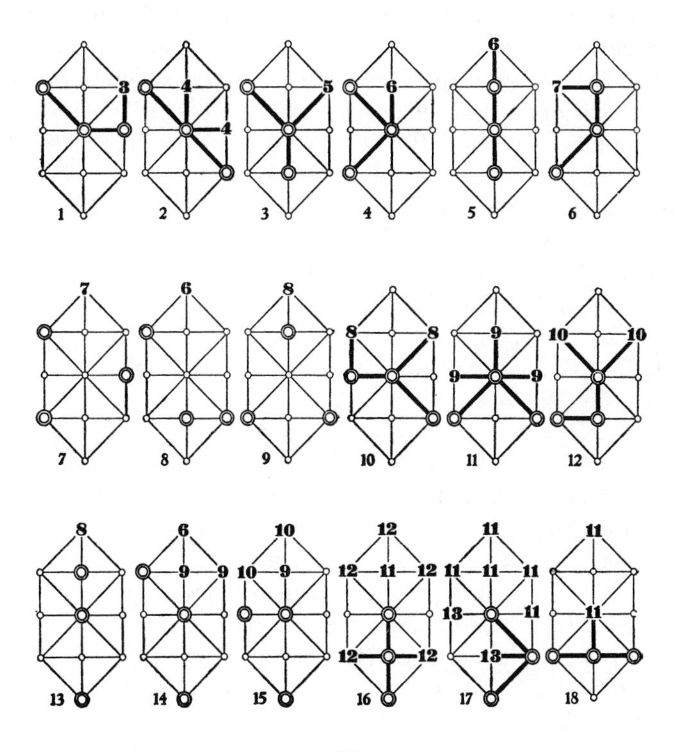

Fig. 92

opens with 11. This makes it clear that White can indeed win against every opening move by Black, in at most 14 moves (14 moves if Black opens on 3, and 12 moves otherwise).

Among the winning positions there are some of the critical type, in which the player who has the move will be the loser. These positions have again been indicated by using heavy lines to connect the circles for the white pieces and the circle for the black piece.

189. Course of the game. To play a successful game, all that White has to do is to keep securing one of the winning positions of Figure 92. The best course for Black is to make his moves in the way which keeps making White need the largest possible number of moves for a win, since this gives White a larger number of opportunities to make a mistake, and likewise involves him in a need to make several transverse moves (for instance, from 2 to 3 or from 8 to 9), something which he may overlook in pressing onwards to his goal.

We shall now set out the sequences of moves for the six possible opening moves of Black, on the assumption that White and Black will always choose their moves with the respective aims of curtailing and prolonging the game as much as possible. If White has only one correct move, and this a very obvious one, we have signalized this by attaching an exclamation point. Each move by White is followed by parentheses which enclose a reference number for the associated winning positions (Fig. 92).

8, 4–6 (17), 11, 2–5! (15), 9, 5–8! (14), 11, 8–9! (13), 8, 1–2 (6), 11, 2–3! (5), 8, 9–10! (3), 9, 3–2 (2), 8, 2–5 (1), 11, 6–8, 9, 5–6, 11, 6–9 (12 moves);

6, 1–3 (18), 8 or 10, 4–6! (12), 9, 3–4! (11), 11, 6–9! (9), 8, 4–6 (6), and so on (12 moves);

6, 1–3 (18), 8 or 10, 4–6! (12), 9, 3–4! (11), 8 or 10, 4–7 (10), 11, 6–8! (7), 9, 7–6! (4), 10, 2–3 (3), and so on (12 moves);

5, 4–6 (17), 8, 2–5 (15), 9, and so on (12 moves);

5, 4–6 (17), 8, 1–3 (12), and so on (12 moves);

9, 4–6 (17), 8 or 11, and so on (12 moves);

9, 4–6 (17), 10, 1–3 (12), and so on (12 moves);

11, 4–6 (17), 8 or 10, and so on (12 moves);

11, 1–3 (18), 8, and so on (12 moves);

3, 2–6 (17), 2, 4–3 (16), 5, 3–2! (17), and so on (14 moves).

The phrase "and so on" indicates that an equivalent position has been discussed earlier in the list.

Black can prolong the game the most by beginning on circle 3. Yet this is not a way for him to make things more difficult for White, because it virtually forces White's first two moves. Black can also play in a way which allows White to win in a smaller number of moves, which require White to set up the winning position 8 (which has not yet appeared in our sequences of moves), as in the following example: 6, 1–3 (18), 8 or 10, 4–6 (12), 11, 6–10! (8), 8, 2–6 (3), and so on (9 moves).

The previous results will show us how easy it can be for White to lose the opposition. When Black retreats repeatedly to circle 11, the game becomes extremely difficult for White, and something which is beyond the mental analysis even of a great chess master.

190. Other winning positions. Apart from the winning positions shown in Figure 92, there are other winning positions for White (when it is not his move), as indicated in *Figure 93.* By making use of these further positions (which in several cases increase the required number of moves), White can make it still more difficult for Black to discover the correct way of playing.

The last diagram shows that White could reply with 2–3 (or 4–3) to each of Black's opening moves (except 3); but this would not break any new ground. Furthermore, in the position 3, 6, 9, 8 (with Black on 8) White can reply with 3–2, which does not get White anywhere forward, either, because this is followed by 11, 2–3, 8. So there should be a stipulation that White becomes the loser when the same position has occurred three times, say. White has to play still more carefully if there is a stipulation that he becomes the loser when the same position has arisen twice over.

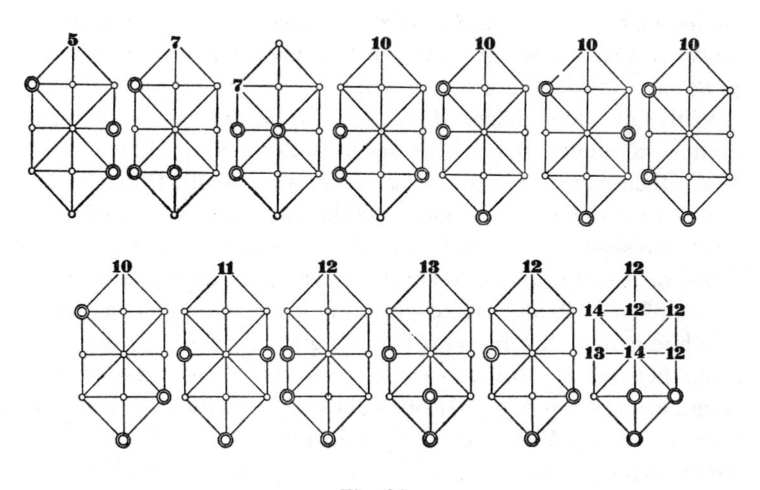

Fig. 93

191. Modified soldiers' game. We modify the soldiers' game of §187 by requiring White to move every one of his three pieces, in any order he likes, whenever it is his turn to play. White's goal is to encircle Black on circle 11, which requires the White pieces to be on the circles 8, 9, 10. Even in his last play, when Black is on 11, White has to move every one of his three pieces in order to win; if White reaches the final position by moving one or two of his pieces only, this makes him the loser. Initially the white pieces are placed on circles 1, 2, and 4, and the black piece on any other circle, after which it is White's move. Now the question is: What initial positions of Black's piece will allow White a win, and how will White obtain this?

Black begins by occupying any circle he cares to choose. However, Black should be forbidden to begin on 3 or on 6, because failing this prohibition he can break through immediately after White's first move. After Black has opened on one of the circles 5, 7, 8, 9, 10, 11, the game is not always an easy one for White (who can always win as we shall see) in spite of the fact that the end will come after six moves at the most. This makes the game a very interesting one indeed.

Black makes it hardest for White by choosing 8, 9, or 10, as his starting position. White then has only the following way to win, in 6 moves: 8, 9, or 10, 2–3–6, 11, 2–5–6 (of course, 4–6–7 is also correct), 10, 5–6–7, 9, 5–6–8, 10, 6–7–9, 11, 8–9–10.

White's move has been indicated each time by giving the three circles which are occupied by the white pieces at the finish of their move; it is easy to see in what order the pieces have to be moved, to put them in their indicated positions.

With any other defense by Black, White can win in a smaller number of moves. If Black replies to 9, 2–3–6 with either 8 or 10, White then plays 5–6–7 and wins in 5 moves (in all). If Black replies to 8 or 10, 2–3–6 with the move 9, White then plays 5–6–8 and wins in four moves (or in three moves if Black continues with 11). With 8, 2–3–6 or 3–4–6, if Black makes the move 5, the continuation is then 2–6–10, 8, 5–6–9, and White wins in four moves (in total, as before).

White wins in five moves if Black begins on 11. White then has a choice between 2–3–6 and 2–5–6 (or between 3–4–6 and 4–6–7, which comes to the same thing). If Black then replies with 10, which is the best he can do, White wins by 5–6–7 in five moves; if Black replies with 9, White wins by 5–6–8 in four moves; and if Black replies with 8, White wins by 5–6–9 in three moves.

White wins in three moves if Black opens on 5, by playing 2–3–6 followed by 5–6–9.

This shows that an opening move 2–3–6 by White is correct for every opening move of Black (when Black is not allowed 3 or 6), provided that a reflected move 3–4–6 is substituted when Black opens on 7.

Chapter XI:
SLIDING-MOVEMENT PUZZLES

I. GAME OF FIVE

192. Rules of the game. There are five numbered cubes in an open box. The box has room for just six such cubes, so there is a vacant space which allows the cubes to be moved around. They may not be taken out of the box in the course of the puzzle, when the problem is to make sliding movements of the cubes in such a way as to change a given initial position (for example, the one in the bottom diagram of *Figure 94*) into a given final position (the one in the upper diagram).

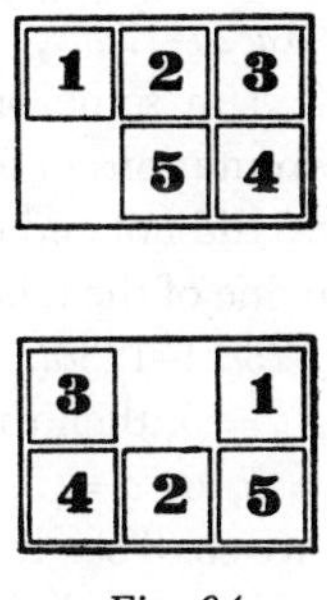

Fig. 94

We shall specify a position by listing the numbers in the first row, taken from left to right, followed by the numbers in the second row, taken from right to left. We shall call this the symbol of the position. For our two diagrams, the symbols are 12345 and 31524. The location of the vacant space is immaterial, and this can be chosen arbitrarily so long as the order of the numbers is not altered. It will be evident, also, that the important consideration is the cyclic order of the numbers, which assumes the last number to be followed by the first one over again, for cyclic movement of the cubes can change the position from 31524, say, to 15243, 52431, 24315, and 43152.

Readers will find later explanations easier to follow, if they cut out five small cardboard squares and number them. A container suitable for these squares can be easily imagined.

193. Some general advice. Without moving a cube from above to below, or in the reverse direction (hence, without changing the symbol of the position), we can place two cubes against the left-hand side of the box (if they are not there already). This then allows us to give a cyclic permutation to the three remaining cubes. This produces a cyclic permutation of the middle three numbers in the position symbol.

By alternately making cyclic permutations of all five numbers and of the middle three numbers, we can transform one of our two positions into the other, thus; 31524–15243–12453–24531–23451–12345.

It is always possible to bring any desired numbers into the first three places (the first three numbers of the symbol). To prove this, we may safely assume 12345 to be the desired final position. We can locate the 1 in the first place (that is, the upper left-hand corner) by cyclically permuting all five numbers. Bringing the 2 into the second place (that is, to the right of the 1) is the most difficult when the 2 is the last number in the symbol. The transposition of 2 takes place as follows: 1*abc*2–*abc*21–*ac*2*b*1–1*ac*2*b*–12*abc*; here the letters *a*, *b*, *c* represent the numbers 3, 4, 5 in some order. Here, as before, the alternate steps are cyclic permutations of all five numbers and of the middle three numbers. If the 3 did not end up in the third place, it can be transposed to this place in one of the following ways: 12*a*3*b*–2*a*3*b*1–23*ba*1–123*ba* or 12*cd*3–2*cd*31–23*cd*1–123*cd*.

If $a = 5$, $b = 4$, or $c = 4$, $d = 5$, then the desired final position has been reached. If $a = 4$, $b = 5$, or $c = 5$, $d = 4$, it is impossible to reach that final position, as we shall see in §195.

194. Moving a single cube. The symbol of a position for the cubes is some permutation (see §126) of the numbers 1, 2, 3, 4, 5. Every time a lower number is preceded by a higher number (whether immediately or not), we call this an inversion, so that, for example, the permutation 31524 of §192 involves four inversions. According as the number of inversions is even or odd, we speak of an even or odd permutation. Two even permutations are said to be of the same parity, or similar, and the same is said of two odd permutations.

When a cube is moved, the parity of the permutation does not change. This is immediately obvious for cases where the cube is moved horizontally, or along the third vertical row (on the extreme right), because when this is done the symbol of the permutation is unchanged.

If we move a cube along the middle vertical row, it passes over two other numbers in the symbol (those of the extreme right-hand cubes);

so it happens twice over that an inversion either arises or disappears, and hence the permutation remains of the same parity. For instance, if we move the number 2 up, in the lower diagram of Figure 94, then 2 passes over the number 1, which makes an inversion arise, and it also passes over the number 5, which makes an inversion disappear; this is evident in the permutation, which changes from 31524 to 32154.

If we move a cube along the first vertical row (on the extreme left), it passes over four other numbers, so that now, too, the permutation remains of the same parity. For instance, if we move the number 1 down, in the upper diagram of Figure 94 (changing the permutation 12345 into 23451), four inversions then arise.

195. Condition for solvability. The conservation of the parity of the permutation shows immediately that we cannot get from one (initial) position to another (final) position if the corresponding permutations are dissimilar. If they are similar, and if we have brought the desired numbers in the first three places (which can always be done, according to §193), then the two other numbers will automatically be in the correct order; if they were reversed, a permutation would have arisen in which there was either one inversion more or one inversion less, than in the desired final position, and this would be dissimilar to the permutations of the initial and the final positions. This shows that we can move from one position to the other if the corresponding permutations are similar, and that otherwise we cannot make the necessary transformation.

We see also that it is not possible to reach the final position if we have brought the desired numbers to the first three places only to find that the two other numbers are then reversed. This then implies that with a given initial position there are just as many positions obtainable as unobtainable. Out of the total of $5! = 5 \times 4 \times 3 \times 2 \times 1 = 120$ permutations (see §126), there are 60 which can be achieved and 60 which are impossible. The permutations that can be reached from the position 12345 are: 12345, 12453, 12534, 13254, 13425, 13542, 14235, 14352, 14523, 15243, 15324, 15432, and those that arise from these by cyclic permutation.

II. EXTENSIONS OF THE GAME OF FIVE

196. Some results summarized. The game of five can be generalized by making the two rows of cubes longer, again in such a way that there is one vacant space. As an example we take a box measuring 2 by 5 which contains $2 \times 5 - 1 = 9$ numbered cubes (*Figure 95*). As before, a position will be denoted by a permutation of the numbers, giving the numbers of the top row from left to right, followed by the numbers of the bottom row from right to left, with no account taken of the vacant space. These results follow also for arbitrary double-row lengths:

(*a*) When a cube is moved, the parity of the permutation does not change;

(*b*) it is always possible to reach a given final position from a given starting position, provided that we can disregard the last two numbers of the permutation for the final position;

(*c*) after this last operation, the last two numbers will be correct or reversed, according as the permutations of the starting and final positions are similar or dissimilar;

(*d*) the positions that can be reached from a given starting position amount to half the total number of positions.

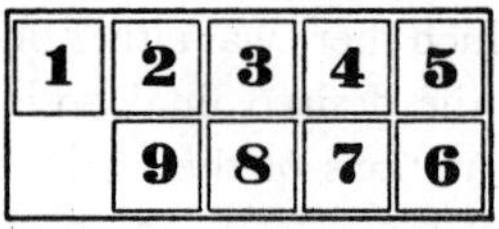

1	2	3	4	5
	9	8	7	6

Fig. 95

197. Proof of the assertions of §196. The correctness of (*a*) again follows from the fact that a number, when moved, passes over an even number of other numbers.

To prove (*b*), we assume that 123456789 is the desired final position. If the cyclic order has the 1 immediately following the 2, it takes little trouble to change this situation; so we can assume that the 1 does not follow the 2. We move the numbers cyclically to bring the 1 into the upper left-hand corner, with another number below (which then is not the 2). After that we move the remaining numbers cyclically to make the 2 immediately follow the 1. Then we move all numbers cyclically to put the 1 in the lower left-hand corner, and the 2 in the upper left-hand corner, after which we cyclically move the remaining numbers in such a way that the 3 comes next after the 2. Placing the 4 after the 3 gives trouble only when the 4 is below the 3, as in

$$\begin{matrix} 2\,3\,a\,b\,c \\ 1\,4\;\;e\,d \end{matrix};$$

here a, b, c, d, e are the numbers 5, 6, 7, 8, 9 in some order. We can then leave the numbers 1, 2, c, d undisturbed, and move the numbers 3, a, b, e, 4, as we found for the game of five, in such a way that eventually the 3 comes back to the same place but with the 4 after it. Next, we place the numbers as follows:

$$\begin{matrix} 3\,4fgh \\ 2\,1\;\;ji \end{matrix},$$

where f, g, h, i, j are the numbers 5, 6, 7, 8, 9 in some order. Without moving the numbers 1, 2, 3, 4, we can (as in the game of five) move the remaining numbers in such a way that 5, 6, 7 will be in the place of f, g, h, respectively, after which we can cyclically move all the numbers in such a way that the following position results:

$$\begin{matrix} 1\,2\,3\,4\,5 \\ \;\;l\,k\,7\,6 \end{matrix}.$$

Even when both rows consist of a larger number of cubes, we can use this method to place all the numbers as for the given final position, provided we disregard the last two numbers.

Justification of (c) and (d) of §196 then follows as in §195.

III. FATAL FIFTEEN

198. Further extension of the game of five. This consists in taking not two rows of cubes, but an arbitrary number. An example of this is the well-known "boss-puzzle" (in French, *diablotin* or *jeu du taquin*) of anonymous origin, which became very popular about 1880. A square open box contains 15 numbered cubes, in a space which has room for $4 \times 4 = 16$ cubes. Here, too, the results (a), (b), (c), and (d) of §196 hold. The permutation that specifies the position of the

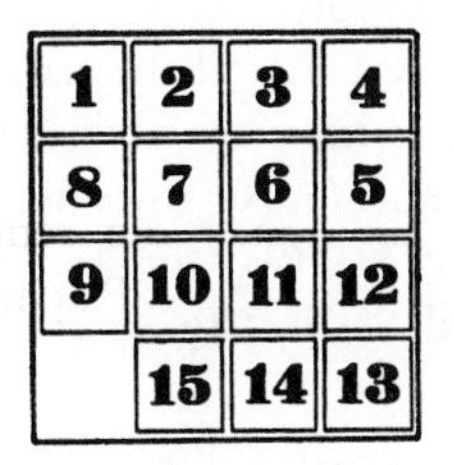

Fig. 96

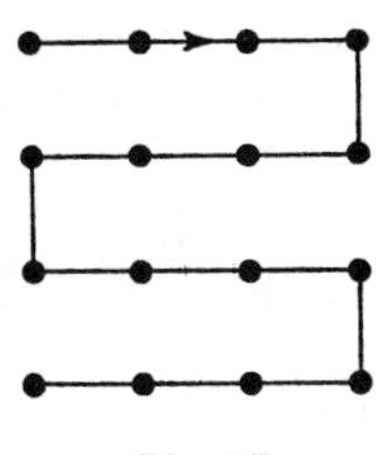

Fig. 97

numbers is now taken in such a way that (for example) the position shown in *Figure 96* is indicated by 1 2 3 4 5 6 7 8 9 10 11 12 13 14 15. This means that the numbers are written down in the order in which they occur on the zigzag line of *Figure 97*, where the dots represent the centers of the 16 spaces.

199. Proof of corresponding results. The only need is to prove the correctness of (*b*) of §196 for the extended game of §198, because the correctness of (*a*), (*c*), and (*d*) can be proved in exactly the same way as before. As an example we again take the boss-puzzle.

First we move the four numbers of the first (top) horizontal row to their positions. In each case the number next to be placed has first to be brought to the first or second row, accompanied by the vacant space, without removing any previously placed numbers from these two rows (this can be done without difficulty). The number concerned

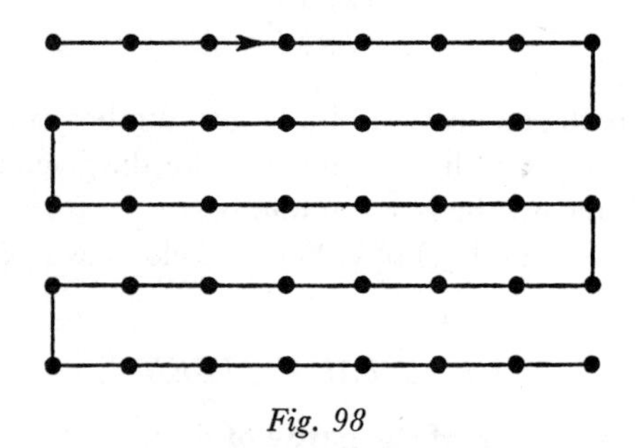

Fig. 98

is next moved into its position in the manner indicated in §197 (by moving only cubes of the first two rows) in such a way that earlier numbers resume their correct positions. When all four numbers of the first row have been placed as desired, they are not moved any more. We then use the same system to place the four numbers of the second row; each new number to be placed is first transferred to the second or third row, together with the empty space, and is then brought to its proper place by moving only cubes of the second and third rows. When the four desired numbers have been placed in the second row also, only cubes of the third and fourth rows are moved in what follows. In the manner indicated in §197, this leads to the desired placing of the numbers, provided the last two numbers are disregarded.

It is clear that all this will remain practically unchanged for a rectangular box which contains a larger number of cubes, for example five rows of eight cubes each, inclusive of the empty space, containing 39 numbered cubes. The permutation that indicates a given position of the numbers is then derived with regard to the zigzag line of *Figure 98.*

IV. FURTHER CONSIDERATIONS ON INVERSIONS

200. Property of inversions. To decide beforehand whether we can get from a given starting position to a given final position, with the box containing 39 cubes referred to in §199, we have to determine whether a permutation of 39 numbers is even or odd. This can be done more quickly than by a laborious count of inversions.

To do this, we first observe that the parity of a permutation changes when we interchange two adjacent numbers; for according as these numbers originally did or did not form an inversion, the result of the interchange is to make a single reduction or a single increase in the number of inversions. If we interchange two non-adjacent numbers, this also makes the permutation change its parity. For example, if there are three numbers between the numbers to be exchanged, then we can accomplish the interchange by $2 \times 3 + 1 = 7$ interchanges of pairs of numbers which in each case are adjacent. This is shown by the following example, in which the 3 and the 6 are finally interchanged: **3**152**6**4–1**3**52**6**4–15**3**2**6**4–152**36**4–152**63**4–15**6**2**3**4–**6**152**3**4.

This shows that the permutation changes parity an odd number of times, and therefore finally has a change of parity. This also holds when there is a different number of cubes between the cubes which are to be interchanged.

If we take a smaller number of cubes, say 15, and replace the numbers by letters (which have to be transposed to produce some given text) then all that we have said will still apply, so long as all the letters are different; the number of positions which allow the desired text to be obtained by legitimate transposition is then equal to half the number of possible positions. However, if there are two copies of the same letter, for instance two *e*'s, then the final position can be reached from every position (this also applies, of course, when there are several repeats of letters), since if we number the cubes, giving the two *e*'s the numbers 3 and 7, say, and then interchange the cubes which carry the two *e*'s, then the permutation will change from even to odd or from odd to even. This means that if the desired text cannot be achieved with *e*3 (the *e* with the number 3) preceding *e*7 (this will be deduced from the fact that all letters can then be placed correctly, except for two letters different from *e*), then we can succeed if we make *e*3 occur later in the text than *e*7.

***201. Cyclic permutation.** In some permutation, 37851426 say, we can cyclically permute four numbers, for example 7251. By this we mean that 7 is replaced by 2, 2 by 5, 5 by 1, and 1 by 7; this cyclic permutation can be denoted by (7251). The cyclic interchange can be reduced to $4 - 1 = 3$ interchanges, each involving two numbers only, namely first 7 and 2, then 7 and 5, and then 7 and 1, thus: 3**7**8**51**4**2**6–3**2**8**51**4**7**6–3**2**8**71**4**5**6–3**2**8**17**4**5**6.

So after this cyclic interchange, the permutation has changed parity three times, and thus finally has a change of parity. If a cyclic permutation involves a different number of cubes, 5 say, then it can be reduced to $5 - 1 = 4$ transpositions of two numbers each time, which will not produce a change in parity. This shows that an even cycle (with cyclic permutation of an even number of numbers) will give a change of parity, whereas an odd cycle will not. This is in keeping with the fact that a cycle of one number leaves the permutation unchanged, and that a cycle of two numbers is nothing other than a transposition of these two numbers.

***202. Parity determination in terms of cyclic permutations.** The results of §201 provide a quick method of determining the parity of a permutation. As an example we take the following permutation of 15 numbers (derived from the boss-puzzle):

15 6 8 11 14 10 3 5 12 1 2 9 13 7 4.

We proceed from the so-called identity permutation:

1 2 3 4 5 6 7 8 9 10 11 12 13 14 15,

which contains not a single inversion, and we examine the cyclic permutations which are needed to transform this into the permutation first mentioned. These cyclic permutations are:

(1 15 4 11 2 6 10) (3 8 5 14 7) (9 12);

that is, starting with the identity permutation, we replace 1 by 15, 15 by 4, 4 by 11, 11 by 2, 2 by 6, 6 by 10, and 10 by 1 (first cycle); after that we replace 3 by 8, 8 by 5, and so on. Hence we obtain two odd cycles (which cause no change of parity) as well as one even cycle, and this shows that the given permutation is of odd parity. This means that if the cubes have a position corresponding to the permutation in question, it is not possible to obtain the position of the cubes shown in Figure 96.

If we have two arbitrary positions of the cubes and want to know whether we can get from one position to the other, we do not in fact have to decide whether each of the corresponding permutations is even or odd. We need only find a set of one or more cyclic interchanges which will convert one of the permutations into the other, and count how many of the cycles are even; if the number of even cycles is even (which may also mean that there are no even cycles at all), then one position can be transformed into the other one; otherwise it cannot. For example, if the initial and the final positions are represented by the following permutations:

6 9 2 15 13 7 4 14 12 1 8 5 10 3 11,

13 11 14 4 7 8 1 6 15 12 2 9 5 10 3,

then the cycles that turn the first permutation into the second are

(6 13 7 8 2 14) (9 11 3 10 5) (15 4 1 12).

There are two even cycles, so that the desired transformation is possible by moving the cubes. To find the cycles which establish that both permutations have even parity would need at least twice as much work; we leave this to the reader.

V. LEAST NUMBER OF MOVES

203. Determination of the least number of moves. In §§192–199 we gave no thought to the question of how we could transform the initial position into the final position (when feasible) in the least possible number of moves, assuming that each movement of a single cube counts as one of these moves. To determine how to do this, we use symbols which take note also of the vacant space, indicated by 0, so that (for example) we have a symbol 301524 for the lower diagram in Figure 94.

To find the least number of moves, we construct a tree which has the initial position for its base. The number of branches at the base is equal to the number of opening moves. Every later position produces a number of branches equal to the number of allowable continuations; of course, we ignore a move that merely cancels the previous move; also, a branch is left out if it would lead to a position that has already been shown to require a smaller number of moves; this can result in a branch having no continuation. The tree is first continued to take note of two moves, then three moves, then four moves, and so on, until the final position appears after some number of moves. It is possible that some number of moves produces the final position more than once. This then means that there are several ways to reach the final position in the least number of moves.

Since the tree has many branches, there is a useful economy if we construct two trees, one starting from the initial position, and one from the final position. We then work on both trees alternately, taking note of one more move on each tree every time. We keep on in this way until both trees produce some common position. This is less tedious than constructing one big tree. This procedure of constructing two smaller trees cannot be applied when we want to know which positions can be reached from the initial position in a given number of moves, or for what position (or positions) the greatest number of moves is required.

With the boss puzzle, we cannot use this way of determining a least number of moves for arbitrary initial and final positions, because of the large number of attainable positions; this number amounts to one-half of 16!, that is, more than 10^{10}. In the most unfavorable case, a complete tree would have to contain all these positions; splitting it up into two smaller (but still extravagantly large) trees would be of no avail, either. So we shall here restrict ourselves to the simplest form of sliding puzzle, that is, to the game of five, with 360 positions obtainable from a given initial position (in not too large a number of moves), which makes this game feasible to discuss. Yet even for this simple case the complete analysis is more laborious than one might at first think.

204. First example. For the game of five we require the least number of moves which can transform the position in *Figure 99*

3	2	4
5		1

Fig. 99

(symbol 324105) into the first position of Figure 94 (symbol 123450). For this purpose we construct two trees, one starting from 123450 and one from 324105. After seven moves in the first tree and six moves in the second, the trees provide displays of a common symbol, showing how we can use 13 moves (but not fewer) to pass from one position to the other.

The trees in question are:

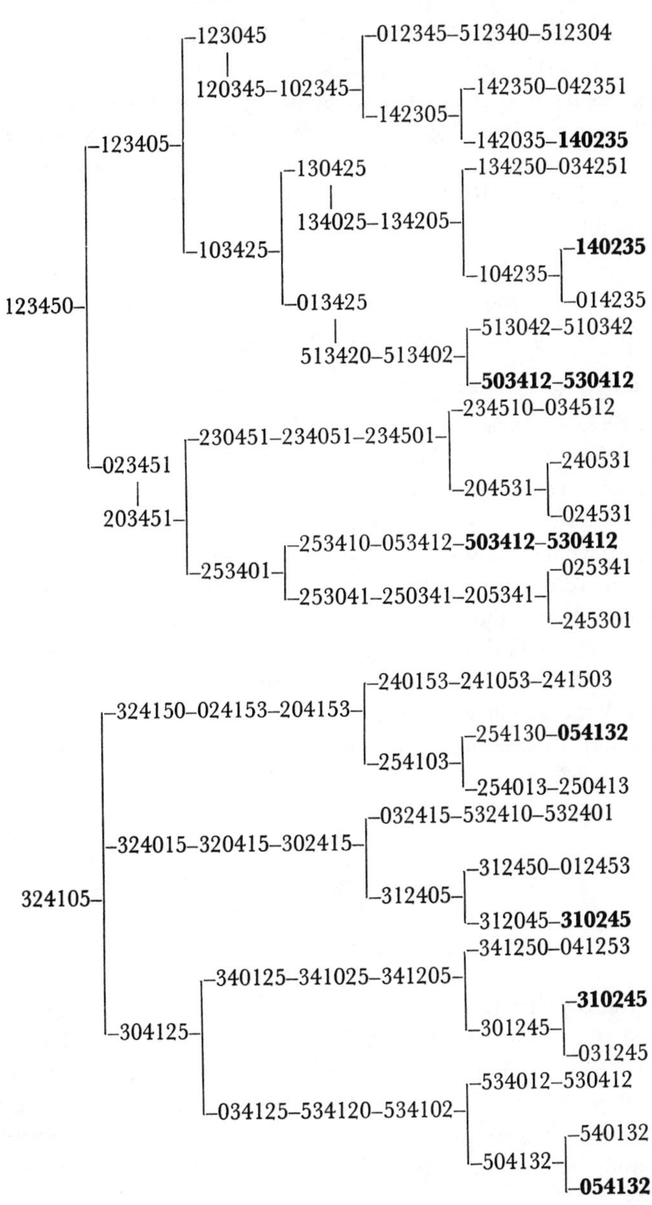

Positions that appear twice in the same tree after the same number of moves have been printed in bold type. In the first tree (seventh move), two branches have dropped out because they lead to the positions 053412 and 513402, which appeared earlier. When the second tree is continued up to and including the sixth move, it involves 38 different positions. Only one of these positions, 530412, occurs in the first tree; at the seventh move, and in two places. This means that we can get from 324105 to 123450 in 13 moves (but not fewer), in two ways, as follows:

324105—304125—034125—534120—534102—534012—530412—

—503412 ⟨ 053412—253410—253401—203451—023451 ⟩ 123450
—503412 ⟨ 513402—513420—013425—103425—123405 ⟩ 123450

***205. Some more examples.** We require the smallest number of moves which can transform the second position of Figure 94 into the first position. The trees that start from the two positions have to be continued here for nine moves before we obtain a common position, 045312, for the first time. This position occurs only once in each of the trees, which shows that a transfer from 301524 to 123450 is possible in 18 moves, in only one way. We have to move horizontally and vertically in alternation, beginning horizontally and to the right. After that (if we limit our attention to the horizontal moves only) we have to make double moves to the left and to the right in alternation.

In the last example we require the least number of moves which can transform the first position of Figure 94 to the position of *Figure 100*, in

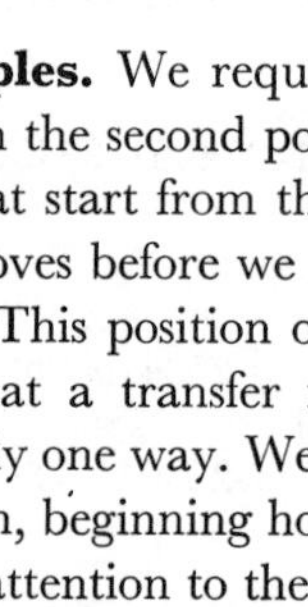

	5	4
1	2	3

Fig. 100

which the two horizontal rows of the first position have been interchanged. By continuing the tree starting from 123450 for 11 moves, and the tree starting from 054321 for 10 moves, we obtain four common positions: 120453, 150324, 350421, 423501. These occur only once in each of the trees. Hence, the smallest number of moves is 21, and there are four possibilities. We indicate these four ways as follows:

.2.1.1.2.1.1.2.1.1 .1.1.2.1.1.2.1.1.2
1.1.2.1.1.2.1.1.2. 2.1.1.2.1.1.2.1.1.

Here a dot means a vertical move, a 1 means a horizontal move, and a 2 means two successive horizontal moves. The double horizontal moves take place in the direction opposite to that of the single ones. The directions which the horizontal moves have to take follow from the fact that the first horizontal move has to be made to the left. If we arrange to have a constant alternation of two single horizontal moves and one double move, with a vertical move intervening at every change, we can easily demonstrate the four solutions without further assistance, beginning with .2 or .1 or 1. or 2. as desired.

Each solution can produce another solution if we first consider how the solution proceeds in a reversed direction (from the final position to the initial position) and then apply a corresponding procedure, beginning with the initial position. This corresponds to reversing the direction in which we read the notation of the solution.

We conclude by remarking that among all the 360 positions that can be reached from the initial position 123450, this final position 054321 is the one which needs the greatest number of moves.

VI. PUZZLES IN DECANTING LIQUIDS

206. Simple decanting puzzle with three jugs. The movement puzzles which we have just discussed have some resemblance to the following decanting puzzle, which is of some antiquity (it dates from the fifteenth or sixteenth century): We have three jugs, for 8, 5, and 3 liters (*Fig. 101*). The biggest jug is entirely filled with wine, the other

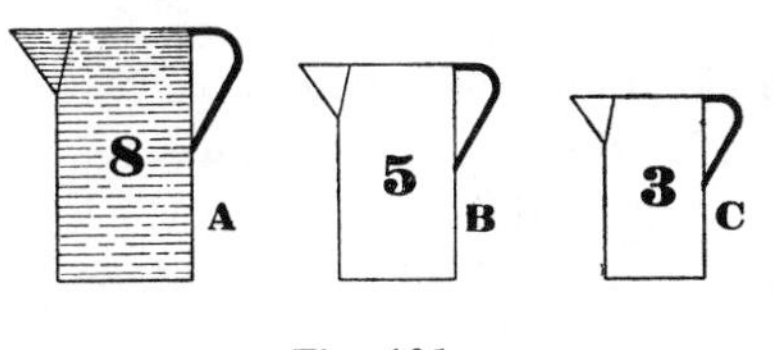

Fig. 101

two jugs are empty. By decanting, we want to divide the wine into two equal parts (leaving the smallest jug empty). The jugs are not calibrated, so that we can do nothing but pour wine from one jug to another until the first jug is entirely empty, or the second jug entirely full. We also assume that the decanting can be done accurately and hence that no wine is spilled. We require the least number of decantings in which our goal can be achieved.

We can begin by pouring wine into jug B, until B is full; the resulting situation will be indicated by 350, and we shall use a similar notation for other situations. We can continue in one way only, if we want to avoid retracing our steps. We also have to avoid a situation in which B and C are entirely full, because then we can do nothing but pour the contents of B and C entirely into A, to produce a situation that can be achieved more quickly in another way. Since we can also begin by pouring wine from A into C, we have only the following two possibilities:

800 [–350–323–620–602–152–143–440
    [–503–530–233–251–701–710–413–440

Hence, the quickest way is to start by pouring into the middle jug; the number of decantings is then seven. Also, the total amount of wine to be poured is then the least possible, 22 liters, as against 23 liters when we start by pouring into the smallest jug. The situations which require the largest number of decantings are 440 and 413.

207. Another decanting puzzle with three jugs. We replace the jugs of §206 by jugs for 12, 7 and 5 liters (*Fig. 102*) and again require

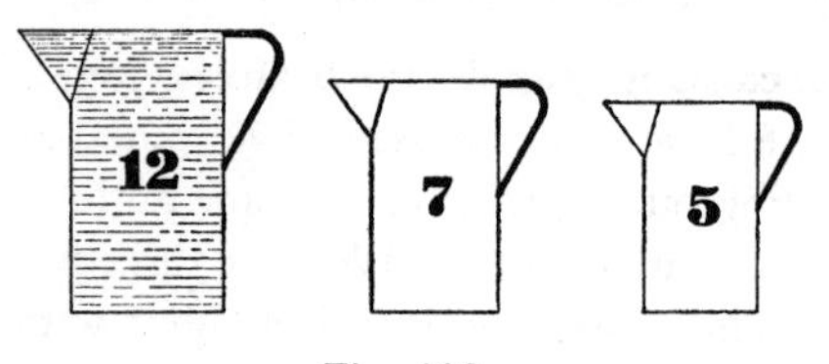

Fig. 102

the wine to be divided into two equal parts. Now, too, we can begin in two ways, and after the first decanting we can continue in one way only, if we are to avoid obtaining situations which we achieved earlier. In this way we obtain the following tree with only one branching:

12 00 [–570–525–10 20–10 02–372–345–840–804–174–165–660
      [–705–750–255–273–903–930–435–471–11 01–11 10–615–660

This shows that the division into two equal parts is achieved after 11 decantings, provided we start off by pouring into the middle jug; otherwise, 12 decantings are required, and the amount of wine to be decanted is then larger, too.

208. Remarks on the puzzles of §§206 and 207. If what we want is not to divide the wine equally between two jugs, but only to get half of the wine in one of the jugs, the goal of the puzzle of §206 can also be achieved in six decantings, by beginning with 350; this then also gives the least amount of wine to be decanted. With the puzzle of §207 in the modified versions we reach the goal most quickly by beginning with 570; half of the wine is then in the middle jug after 10 decantings.

With the modified version it is not necessary for the middle jug to be bigger than half the largest. If the jugs have capacities of 8, 3, and 2, then we can measure out 4 liters in three decantings as follows: 800–602–620–422. With jugs for 10, 4, and 3 liters we reach the goal in four decantings, thus: 1000–640–613–910–901–541.

Finally it may be noted that ineffective decanting can never spoil things to such an extent that the division into two equal parts can no longer be achieved, since we can always pour back all the wine into the biggest jug. Of course, we then have to decant needlessly often.

209. Changes of the three jugs. If we replace the jugs by others, it makes no sense to take fractional numbers, because in that event we can eliminate the fractions by assuming a smaller unit (1/4 liter, say). We can also avoid having all three numbers divisible by a number greater than 1; for example, we can reduce 24, 15, and 9 liters to the numbers 8, 5, and 3, which are three times as small, by taking 3 liters as the unit.

We require the wine (which initially fills the largest jug entirely) to be divided equally between two of the jugs. For that purpose, the largest jug should be even (that is, should contain an even number of liters) and the middle jug should have a capacity in excess of half the capacity of the biggest jug (it could be equal to half this capacity, but then there would be no fun in the puzzle, since the goal could then be reached by decanting only once).

From a situation in which one of the jugs contains at least as much wine as the capacity of the smallest measure, it is not possible in one decanting to create a situation in which each of the three jugs contains less wine than the capacity of the smallest jug. So a situation with the latter property cannot arise at all. Hence, if the wine is to be divided into two equal parts, the smallest jug must have a capacity less than half that of the biggest one.

210. Further remarks on the three-jug puzzle. Suppose that the capacities of the two smallest jugs (in liters) are both divisible by 5, say, when we can assume that this is not true for the largest jug. Of the three amounts of wine (in liters), two are then always divisible by 5, when we consider 0 to be a number which is divisible by 5. In the desired final position, however, it is only the empty jug which has an amount of wine which is divisible by 5. Hence, if the final position is to be attainable, the capacities of the two smallest jugs must be represented by mutually prime numbers, that is, numbers that have no common factor greater than 1. We illustrate this with the case of jugs with capacities of 12, 10, and 5 liters. The corresponding tree is:

```
                   ┌─0 10 2–10 02–10 20
          ┌─2 10 0─┤
          │        └─255
12 00─────┤
          │        ┌─075–570–525
          └─  705──┤
                   └─750
```

Any extension of this would only produce situations that have appeared before. It includes all the positions for which at least one of the jugs is full or empty and two of the three numbers of liters of wine are divisible by 5, and no positions of any other type.

If the capacity of the largest jug is equal to or greater than the sum of the capacities of the other two jugs (as in §§206 and 207), then the various possibilities will display branching only at the first decanting (into the middle or into the smallest jug).

If the capacity of the largest jug is less than the sum of the two other capacities, then the possibilities also branch at the second decanting, but not again after that. We illustrate this with the example of jugs for 12, 9, and 5 liters:

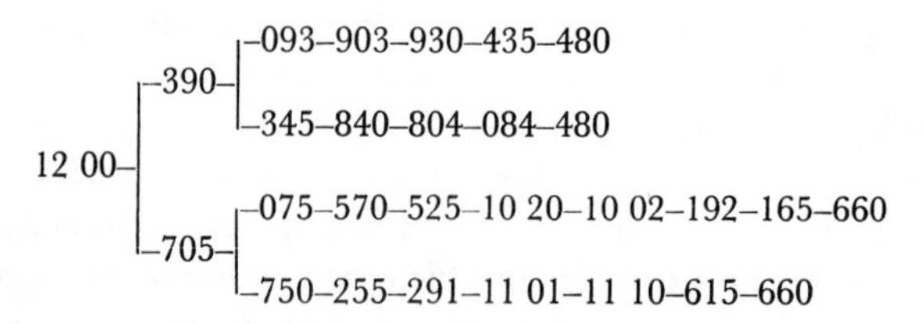

It is not possible to reach other positions. The positions concerned include all the positions in which at least one of the cans is empty or full.

211. Decanting puzzle with four jugs. We now take four jugs with capacities of 9, 5, 4, and 2 liters (*Fig. 103*). The biggest jug is full of wine. By decanting repeatedly, always with the complete filling or emptying of a jug, we have to divide the wine into three equal parts, in as few decantings as possible. If we pour from one jug into two others, or from two jugs into a third, then each of these counts as two decantings.

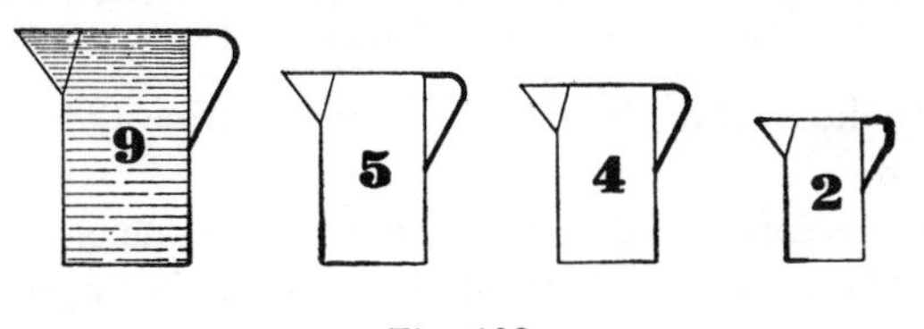

Fig. 103

This causes more branching in the tree, because there are more possibilities every time. By constructing the tree, we find that the goal can be achieved in six decantings, and in four ways:

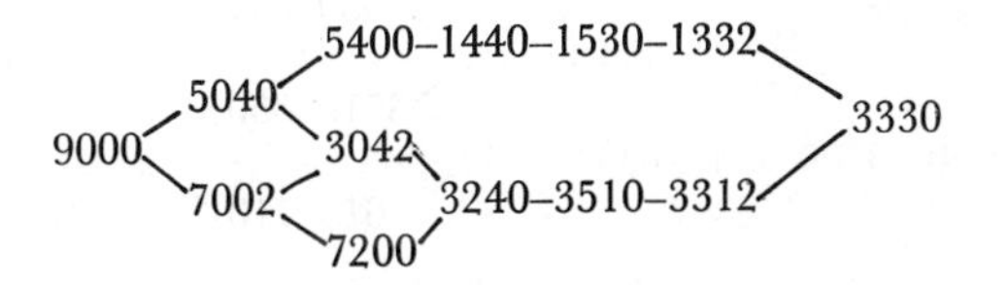

We can reach all situations in which at least one of the jugs is full or empty. The situation 5031 requires the largest number of decantings, namely, eight; by decanting eight times, this situation can be reached in seven ways, thus:

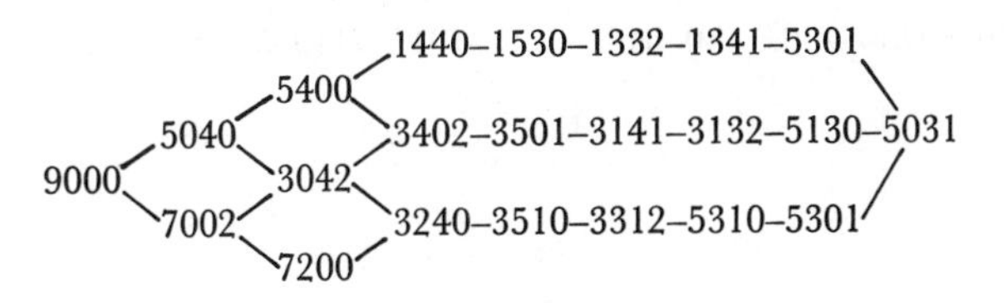

212. Another puzzle with four jugs. We take the same puzzle as that of §211, but with jugs for 9, 7, 4, and 2 litres (*Fig. 104*). In eight

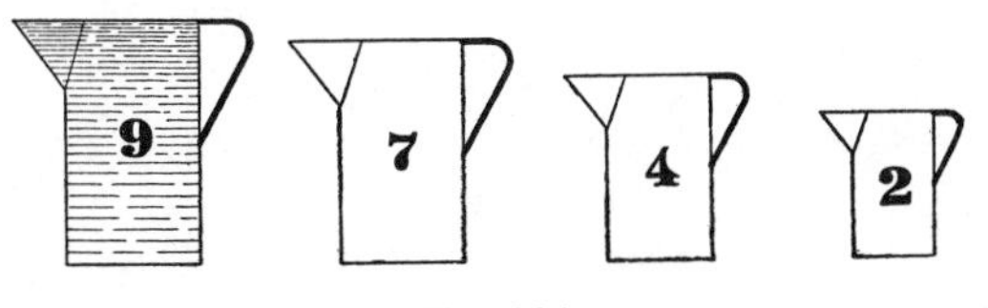

Fig. 104

decantings (and not sooner) the wine can be divided into three equal parts, and that in three ways, as follows:

```
                                            3510–3312
9000–5040–5400–1440–1710–1512–1530–1332–3330
                                       1701–1341
```

In each of these ways the same amount of wine (23 liters) has to be decanted.

All positions can be reached in which at least one of the jugs is full or empty. The situations 3132 and 5130 require the largest number of decantings (namely, nine). The situation 3132 can be reached in six ways in nine decantings—in three ways via the situation 3330, and in three ways via the situation 3141, thus:

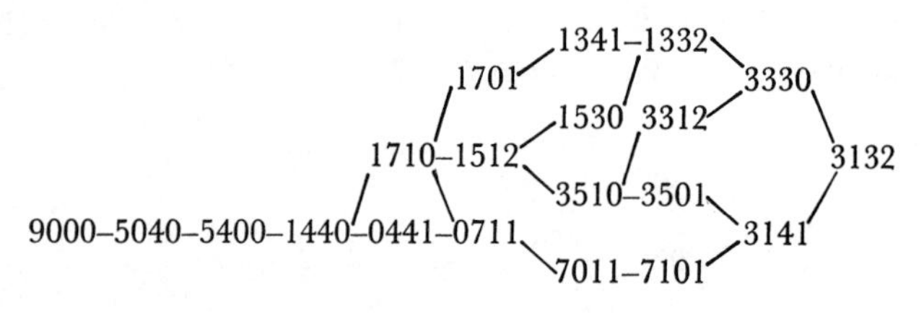

The situation 5130 can be reached after nine decantings in the following four ways:

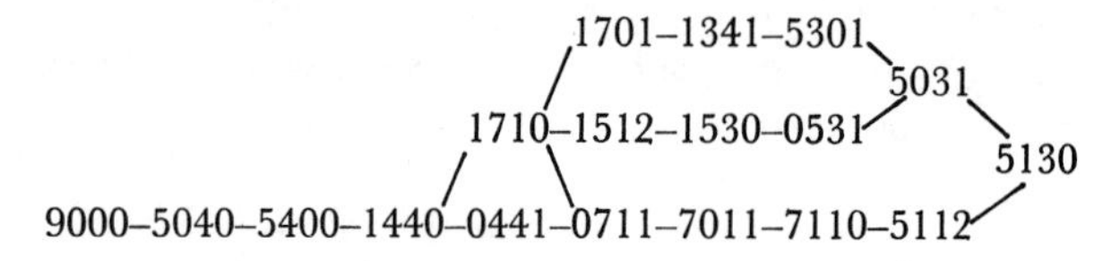

Chapter XII: SUBTRACTION GAMES

I. SUBTRACTION GAME WITH A SIMPLE OBSTACLE

213. Subtraction games in general. Two persons, John and Peter, play the following game: John begins by mentioning not too small a number. Peter has to diminish this number by subtracting some number, which we shall call his deduction. After that, John has to make some new deduction from the number Peter produced, after which it is Peter's turn again, and so on. The deductions available to Peter and John each time must satisfy certain conditions (rules of the game). The subtractions must not produce negative numbers (that is, numbers which are less than zero). A zero result calls for the use of an agreed rule for determining the winner and the loser.

The game may also be played as an addition game; in this event the total of successive additions is not allowed to exceed a certain number, 100 say. If we replace every cumulative total by the number that indicates its difference from 100, the addition game becomes a subtraction game. This shows that we may limit our consideration to subtraction games.

The subtraction game can also be considered as a match game to be played with one pile, which John and Peter have to reduce, by taking turns to remove one or more matches according to some given rule. The match games of §§99–101, therefore, can be played equally well as subtraction games, and, conversely, match-game interpretations are possible for the subtraction games next to be discussed.

In a subtraction game we must attempt to reply with certain numbers, which we shall call winning numbers. For example, if we are allowed to choose our deduction from among the numbers 1, 2, 3, 4, or from among the numbers 1, 2, 3, 4, 5, 6, 7, in cases where reaching 0 means a win, then the winning numbers are the multiples of 5 or of 8, respectively; if 0 means a loss, then the winning numbers are the multiples of 5 plus 1, or the multiples of 8 plus 1, respectively. If the deductions can be chosen from among 1, 2, 3, . . . , 9, where 0 means a win, the winning numbers are 10, 20, 30, and so on; even if Peter is not an expert at the game and is not sufficiently intelligent to find these numbers himself, he will immediately notice the regularity in John's replies.

214. Subtraction games with obstacles. In the example mentioned in §213, the correct way of playing for John, once he is in a winning position, is to choose that deduction which would have to be added to Peter's last deduction, to turn it into the number which is one unit greater than the largest allowable deduction. For instance, if 1, 2, 3, 4, are the allowable deductions, John chooses a deduction which could be added to Peter's last deduction, to give a sum of 5. We can make the game less transparent, and thus more interesting, by prohibiting this. Then the player will not be allowed to choose a deduction which can be added to his opponent's last deduction, to produce a sum which is one more than the largest allowable deduction. If this latter is equal to 4, say, and if the reply made to 48 has been 46 (a deduction of 2), then the reply 43 is not allowed, but only the replies 45, 44, and 42; the deduction 3 is then blocked (because of $2 + 3 = 5$), in the sense that it is made temporarily unavailable.

Here we also have to decide whether reaching 0 means a win or a loss, and also, what the result is when it is necessary to reply to "1 with the deduction 1 blocked." If 0 means a win, then this can be interpreted as a loss for a player who has no move left; the most natural arrangement, therefore, is to agree that a player loses if he has to reply to 1 when the deduction 1 is blocked. On similar lines, we should conclude that the same situation would count as a win, in cases where 0 implies a loss. As an alternative, however, we could have a provision that the rule for blocked deductions ceases to apply when it becomes necessary to reply to 1; the reply 0 would then become obligatory. Hence, in summary, we have the following possible arrangements for deciding the result:

0 wins	*A*: 1 is a winning reply when the deduction 1 is blocked; *B*: 1 is always a losing reply;
0 loses	*a*: 1 is a losing reply when the deduction 1 is blocked; *b*: 1 is always a winning reply.

215. Winning numbers when 0 wins. Let us consider the game of §214 with 0 winning. To be more concrete, we assume that the deduction is not allowed to be larger than 7, so that a deduction is blocked if it gives a sum of 8 when it is added to the previous deduction. The winning numbers are then found to be the multiples of $(7 + 2)$, hence of 9.

For example, if John has replied with a multiple of 9, say 36, then he can respond to every reply of Peter by replying with the next lower multiple of 9, namely 27, except when Peter chooses the deduction 1, to make a reply of 35. In that case, John too chooses the deduction 1 to give a reply of 34, after which Peter is not allowed to use the deduction $8 - 1 = 7$ which would let him reply with 27. John, however, can say 27 after any reply available to Peter. In the same way John can reach 18, and then 9 and 0. It is immaterial here which of the arrangements *A* and *B* of §214 is chosen because neither John nor Peter will arrive at 1.

This way of playing for John (when he is in a winning position) is not the only way in which he can win. He also wins when he applies the following rule: Reply with a multiple of 9, if possible. If this is not possible because Peter has mentioned a multiple of 9 minus 1 then John's reply is immaterial. Peter cannot then reply with a multiple of 9 because the required deduction is blocked. John's next reply can then be a multiple of 9, unless Peter has again mentioned a multiple of 9 minus 1; John's reply is then immaterial once again, and so on. For instance, if Peter has replied with 35 to John's 36 and if John now says 32, then Peter cannot follow it up with 27, and John's reply can be 27 or 18, unless Peter replied with 26, and so on. There is only one exception, namely when arrangement *B* of §214 has been made and Peter says 8; in that case John should not reply 1, because Peter will follow it up with 0.

With the first-mentioned strategy for John (use the deduction 1 for the reply to a multiple of 9 minus 1), John will be sure to arrive, after one interruption, at a new multiple of 9. However, John would do better to take all advantage possible of the fact that his reply is immaterial when Peter produces a multiple of 9 minus 1; when this occurs, he should vary his choice of deductions as much as possible. John then will not need to mention so many winning numbers, which will probably keep them hidden longer from Peter.

It is evident that the foregoing will also apply with an obvious modification when the largest allowable deduction is 5 or 9, say. Then the winning numbers are the multiples of 7 and of 11, respectively.

216. Winning numbers when 0 loses. In the game of §214 with 0 losing, we are forced to consider the rule about reaching 1 with the deduction 1 blocked, and it is essential to know which of the arrangements *a* and *b* of §214 has been chosen. Again we take 7 as the largest deduction. Then, with arrangement *a*, the winning numbers are the multiples of 9 minus 1 (8, 17, 26, and so on).

If we assume that John was able to reply with a winning number (or started with one), his jumps to the next smaller winning numbers (reducing by 9) will occur in the same way as in §215. If John wants to arrive at a new winning number as soon as possible, he uses the deduction 1 when he cannot reply with a winning number. In this way, John can reach the number 8 with rule *a*, and the number 10, with rule *b*. When John has reached 8 with rule *a*, he wins when Peter replies with 1, because then John has no reply left. If Peter replies to 8 otherwise than with 1, John replies with 1; after that, Peter has to say 0, so that he loses here, too.

If John has reached 10, with rule *b*, and if Peter then replies with 8 or a smaller number (hence 8, 7, 6, 5, 4, or 3), then John can reply with 1, after which Peter has to say 0 and lose. If Peter replies with 9 to 10 from John, John's reply is immaterial; Peter then has to reply with 0 or a number larger than 1 (since the deduction that leads to 1 is blocked; in the latter case, John replies with 1, after which Peter has to say 0, so that in either case Peter will be the loser.

II. SUBTRACTION GAME WITH A MORE COMPLICATED OBSTACLE

217. Rules of the game. Again we take the case in which the deductions have to be greater than 0 and not greater than a given number. Here we introduce the obstacle that the same deduction cannot be applied twice in succession; that is, a player is not allowed to use the deduction that has just been used by his opponent. We shall speak of a 5-subtraction game, to mean that the largest deduction is 5. This leaves scope for choosing the remaining rules of the game in any of four different ways: 0 wins, with one of the arrangements *A* and *B*, or 0 loses, with one of the arrangements *a* or *b* (see §214). Hence, we can speak of the 5-subtraction game *A*, the 5-subtraction game *B*, the 5-subtraction game *a*, and the 5-subtraction game *b*. To ensure that there is proper observance of the obstacle rule (for the blocked deductions), it is recommended that each player in turn would not only mention the result, but also his deduction; thus, in reply to 48, for example, he should not merely say "44," but "minus 4 makes 44;" the other player then immediately knows that he is not allowed to reply "minus 4 makes 40."

218. Even-subtraction game. The subtraction game of §217 is very simple when the largest deduction is even, say 8. If John is in a winning position, he can retain it by following the strategy indicated in §213, that is, by choosing at each turn a deduction which can be added to the deduction last used by Peter, to the sum of $8 + 1 = 9$. This is always possible, since 9 is odd and hence the two deductions cannot be equal.

In the case in which 0 wins, the winning numbers are the multiples of 9. John, who begins with a multiple of 9, can always reply with a multiple of 9, and can thus reach the number 0 eventually. By doing this, John avoids any application of rule *A* or *B* of §214. John can afford to make various departures from this strategy without giving up his victory; however, these departures are different in case *A* from what they are in case *B*. For an arbitrary even value of the largest deductions, it is difficult to give a complete account of all the allowable deviations. However, John need not know all these deviations to win (once he is in a winning position), or to make use of a mistake of Peter, if he is in a losing position. For this, John only has to follow this rule: Reply with a multiple of 9, if possible, and otherwise halve the distance to a multiple of 9, to prevent Peter from attaining a multiple of 9. There is no guarantee that this will enable John to attain a multiple of 9, because Peter has not necessarily made a mistake in mentioning a number not divisible by 9; however, if Peter's reply was wrong, this strategy will enable John to take proper advantage of Peter's mistake.

In the case where 0 loses, the winning numbers are the multiples of 9 plus 2, with rule *a* of §214, and the multiples of 9 plus 1, with rule *b*. In case *a*, if John has started with a multiple of 9 plus 2, he will finally reach the number 2 (by making pairs of deductions add to 9 each time). Peter then has to reply with 1 or 0, after which John has no reply left, and therefore wins. If John has started with a multiple of 9 plus 1, in case *b*, he finally attains 1, after which Peter has to reply with 0.

In the case where 0 loses, also, you can afford to make many different departures from the given strategy, once you are in a winning position. Once again, you have no need to know these to stay in a winning position or to get there (when this is possible), for you need only reply with a winning number, when you can, and otherwise, halve the distance to a winning number.

What we have said will still apply when the largest deduction is some other even number, say 6. In that case we have to replace "multiple of 9" by "multiple of 7."

219. Odd-subtraction game. For an arbitrary odd value of the largest deduction, I have not been able to find a general rule for the winning numbers. This does not, however, prevent us from finding the winning numbers when we take some specified odd number, 5 say, for the largest deduction. The reader may therefore expect that the odd-subtraction game (when the largest deduction is larger than 3) will be much more difficult than the even-subtraction game of §218, or the subtraction game with the obstacle of §§214–216.

This applies not only to the determination of the winning numbers, but also to the practical application of the rules which specify strategies by which John, who began, can win, and by which Peter can tempt John into a mistake, to benefit from a possible error. This makes the odd-subtraction game a much more interesting one. The 5-subtraction game seems to me to be the most attractive one, because it is difficult enough, and still not too complicated.

III. 3-, 5-, 7-, AND 9-SUBTRACTION GAMES

220. 3-subtraction game *A***.** We consider the 3-subtraction game with 0 winning, in which rule *A* of §214 applies (the player who cannot reply loses). The strategy which, if feasible, leads to a win, is shown in the accompanying diagram. The first column contains the numbers

1		1
2		1, 2
3	↑	3
4	│	—
5	│	1
6	↓	1, 2, 3
7		3
8		—
9		1
10		1, 2, 3
11		3
12		—

to which a reply has to be given. The second column gives the deductions that lead to a win; these we shall call the good deductions.

The diagram indicates, for instance, that when Peter has said 7, John can win only by replying "minus 3 makes 4." A horizontal dash indicates that no possible reply can lead to a win. So these dashes signalize the winning numbers; in this case these are the multiples of 4. The sequence to the right of the vertical line with the arrowheads recurs periodically.

Because only one deduction is blocked every time, a winning answer can always be given to a number which has two or more good deductions next to it. A winning answer can be given to a number with only one good deduction next to it only when that deduction is not blocked. Thus, the winning reply "minus 3 makes 4" (to 7) can be given only when 7 has not arisen from 10 by "minus 3 makes 7"; this makes 3 a good deduction for 10.

The foregoing also shows how good deductions can be found from good deductions for smaller numbers. We have 1 as a good deduction for 1. Alternatively to 2, 1 is also a good deduction for 2, since it blocks 1, the only good deduction for 1. The only good deduction for 3 is 3, since the deductions 1 and 2 do not block all good deductions for 2 and 1, respectively. The number 4 has no good deductions, since the deduction 1 does not block the good deduction for $4 - 1 = 3$, while the deduction 2 does not block all good deductions for $4 - 2 = 2$, and so on. When the good deductions have been found in this manner for the numbers up to and including 9, say, deductions for 10 are then found as follows: 1 is a good deduction for 10, since it is the only good deduction of 9 (for which it is blocked in view of "minus 1 makes 9"); 2 is a good deduction for 10, since 8 has no good deductions; 3 is a good deduction for 10, since 3 is the only good deduction for 7. Next, the good deductions for 11 are found as follows: 1 is not a good deduction, since 10 has a good deduction which is different from 1; 2 is not a good deduction for 11, since 9 has a good deduction different from 2; 3 is a good deduction for 11, since 8 has no good deductions. The good deductions are quickly found in this way. This is continued until periodicity has been established.

From the diagram it is immediately evident how John, who begins, has to play in order to win: he begins with a multiple of 4, and replies, if possible, with the next lower multiple of 4; if this is not possible (because Peter has just used the deduction 2), then John's reply is immaterial.

221. The other 3-subtraction games. In an entirely similar manner we find the table of good deductions for the 3-subtraction game when rule B of §214 is accepted, or when 0 means a loss and one of the rules a and b of §214 applies. Table 9 (which includes an abridged version of the diagram of §220, with the assumption that there is periodic repetition for the numbers to the right of the vertical lines with the arrowheads) gives the good deductions for the four cases A, B, a, and b:

TABLE 9

0 wins				0 loses			
A		*B*		*a*		*b*	
1	1	1	1	1	—	1	—
2	1, 2	2	↑ 2	2	—	2	1
3	↑ 3	3	3	3	1, 2	3	1, 2
4	—	4	2	4	2, 3	4	↑ 3
5	1	5	—	5	↑ 3	5	—
6	↓ 1, 2, 3	6	↓ 1, 2, 3	6	—	6	1
7	3	7	2	7	1	7	↓ 1, 2, 3
8	—	8	3	8	↓ 1, 2, 3	8	3

In the 3-subtraction game B, the winning numbers are the multiples of 5. John, who has begun with a multiple of 5, should use the following strategy in order to win: If possible, reply with a multiple of 5; if this is not possible, then halve the distance to a multiple of 5 (so a multiple of 5 plus 4 requires a reply with the deduction 2, and a multiple of 5 plus 1 requires a reply with the deduction 3). However, John can deviate from this and reply with a deduction 2 to a multiple of 5 plus 1.

In the 3-subtraction game a, the winning numbers are the multiples of 4 plus 2, and in the 3-subtraction game b they are the multiples of 4 plus 1. When John is in a winning position, he wins by applying the following strategy: If possible, reply with a winning number. If this is not possible, his reply is immaterial.

With rule A the winning numbers are different from what they are with rule B, and with rule a they are different from what they are with rule b. This shows that when John is in a winning position, he cannot arrange to play in such a way as to avoid the case where he or Peter will say 1 and by doing so will block the deduction 1 (that is, the case where 1 is given as the reply to 2).

222. 5-subtraction game. Table 10 shows the good deductions in the 5-subtraction games A, B, a, and b (these are found in the same way as in §220 for the 3-subtraction game A). The section on the right is a continuation of the section on the left. So we should mentally transfer the right-hand columns to a position under the corresponding left-hand columns.

TABLE 10

	0 wins		0 loses	
	A	*B*	*a*	*b*
1	1	1	—	—
2	*1*, 2	2	—	1
3	3	3	1, 2	*1*, 2
4	4	*2*, 4	2, 3	3
5	5	5	3, 4	4
6	3	3	4, 5	5
7	—	—	5	3
8	1, *4*	1	—	—
9	2, *3*	*1*, 2, *3*	1	1, *4*
10	3, 5	3, 5	*1*, 2	2, *3*

11	4	4	3	3, 5
12	5	5	4, *5*	4
13	—	—	5	5
14	1	1	3	—
15	*1*, 2, *4*	*1*, 2, *4*	—	1
16	3	3	1	*1*, 2, *4*
17	4, *5*	4, *5*	*1*, 2, *3*	3
18	5	5	3, 5	4, *5*
19	3	3	4	5
20	—	—	5	3
21	1	1	—	—
22	*1*, 2, *3*	*1*, 2, *3*	1	1
23	3, 5	3, 5	*1*, 2, *4*	*1*, 2, *3*
24	4	4	3	3, 5

If 0 means a win, the winning numbers are the multiples of 13, and the multiples of 13 plus 7, both in case *A* and in case *B*. For the numbers 10 and larger, the good deductions are the same for *A* as for *B*. John, who has begun with a winning number, makes things easiest for himself by keeping to the rule: If possible, reply with a winning number, and otherwise halve the distance to a winning number. This strategy makes it quite unnecessary to know whether the agreed rule is *A* or *B*. John can also afford to make deviations from this strategy, without giving up his winning position; these deviations (which consist of halving the distance to a winning number, in some cases in which John can also reply with a winning number, or of reducing that distance by a factor of 3) have been indicated in the tables by italicizing the good deductions. When the number which demands a reply becomes less than 10, it becomes necessary to consider whether one is playing according to *A* or to *B* before choosing a deduction. If 0 means a loss, then the winning numbers with rule *a* are the multiples of 13 plus 2, and the multiples of 13 plus 8; with *b* they are the multiples of 13 plus 1, and the multiples of 13 plus 8. Here, too, if John has begun with a winning number, he will win when he follows the above-mentioned strategy (with rule *a* he should count 1 as a winning number too, except as a reply to 2). This strategy, indeed, is a very obvious one, since halving the distance to a winning number is enough to prevent Peter from replying with a winning number. Hence the important point is only to know the winning numbers.

This also holds for Peter, who is in a losing position. Peter, too, is well-advised to halve the distance to a winning number, if possible, to make it hard for John to retain his winning position, and to have more hope of obtaining a winning position as a result of a mistake by John. If John makes a mistake which allows Peter to reply with a winning number, Peter should prefer this to halving the distance to a winning number, although that is sometimes correct, too; however, this is a little more difficult to judge.

It may finally be remarked that for every number the good deductions are the same, for rule *A*, as they are with rule *b* for a number increased by 1. This can be seen at once from the way of determining the good deductions (which are derived from those for 1 and for 2). This is true for every odd-subtraction game (and indeed also for any even-subtraction game).

223. 7-subtraction game. Table 11 shows the good deductions in the 7-subtraction games *A*, *B*, *a*, and *b*. When rule *A* applies, the

TABLE 11

	0 wins		0 loses	
	A	*B*	*a*	*b*
1	1	1	—	—
2	1, 2	2	—	1
3	3	3	1, 2	1, 2
4	4	2, 4	2, 3	3
5	5	5	3, 4	4
6	3, 6	3, 6	4, 5	5
7	↑ 7	↑ 7	5, 6	3, 6
8	4	—	6, 7	↑ 7
9	—	1	7	4
10	1, 5	1, 2, 5	—	—
11	2	3	1	1, 5
12	3, 4!	4	1, 2	2
13	2, 4	5	↑ 3	3, 4!
14	5, 7	3, 6, 7	4	2, 4
15	6	7	5	5, 7
16	7	4	3, 6, 7	6
17	—	—	7	7
18	1	1, 5	4	—
19	1!, 2	2	—	1
20	3	3, 4!	1, 5	1!, 2
21	4, 6!	2, 4	2	3
22	5	5, 7	3, 4!	4, 6!
23	3, 6, 7	6	2, 4	5
24	7	7	5, 7	3, 6, 7
25	—	—	6	7
26	1	1	7	—
27	1, 2, 5	1!, 2	—	1
28	3	3	1	1, 2, 5
29	4	4, 6!	1!, 2	3
30	5	5	3	4
31	↓ 3, 6, 7	↓ 3, 6, 7	4, 6!	5
32	7	7	5	↓ 3, 6, 7
33	4	—	3, 6, 7	7
34	—	1	7	4
35	1, 5	1, 2, 5	—	—
36	2	3	1	1, 5
37	3, 4!	4	↓ 1, 2, 5	2
38	2, 4	5	3	3, 4!

winning numbers are the multiples of 25, the multiples of 25 plus 9, and the multiples of 25 plus 17. With rule *B*, the winning numbers are the multiples of 25, the multiples of 25 plus 8, and the multiples of 25 plus 17. With rule *a*, the winning numbers are the multiples of 25 plus 2, the multiples of 25 plus 10, and the multiples of 25 plus

19. With rule *b*, the winning numbers are the multiples of 25 plus 1, the multiples of 25 plus 10, and the multiples of 25 plus 18. Hence, with rule *A* the winning numbers are different from those with *B*, and those with *a* are different from those with *b*. This shows that when John is in a winning position, he cannot play in such a way that it is immaterial whether the rule is *A* or *B* (when 0 wins) or *a* or *b* (when 0 loses).

The repetition periods for the good deductions are the same for *A*, *B*, *a*, and *b*, but they have different starting points (which is also true for the 5- and 9-subtraction games). In order to win, John should aim at 9 with *A*, and with *B*, *a*, and *b*, he should aim at 17, 19, and 10, respectively, in exactly the same manner. If John, in a winning position, cannot reply with a winning number, it is usually, but not always, correct for him to halve the distance to a winning number. For this reason, the good deductions to which John should pay special attention have been signalized by an exclamation point in the tables. Thus, with rule *A*, John's correct reply to 37 with the subtrahend 3 blocked is not "minus 6 makes 31," but "minus 4 makes 33;" that is, he should use a deduction which is not one-half, but one-third, of the distance to the winning number 25. Because of this, Peter cannot halve the distance to 25 (which he would have been able to do if John had used the deduction 6). In replying to 21 (still with rule *A*), John should halve the distance to 9, and not that to 17; with 19, it is just the other way round. Hence, to retain his winning position against an expert opponent, John has to know the game thoroughly. He wins if he keeps to the following strategy: If he cannot attain a winning number, then he should reduce the distance to a winning number by a factor of 3 if he can do this by subtracting 4. If this is not possible, either, then he should halve the distance to a winning number, giving preference to the deductions 1 and 6. This strategy applies with rule *A* of §214, as well as with *B*, *a*, or *b*.

Peter can attempt to dislodge John from his winning position in the following way, where we assume that rule *A* is applied. John 25, Peter 21, John 15, Peter 12, John 8, Peter immaterial, John 0, where we can imagine all numbers to be increased by a multiple of 25. There are ingenious traps concealed in Peter's replies 21 and 12 since John would lose by replying with 19 and 6, respectively (Peter would then say 18 and 3, respectively). If John begins with the winning number 42, then Peter can lead John to 25 (without John making an error) by a route: John 42, Peter 38, John 36, Peter 35, John 30, Peter 29, 28, or

26, John 25; all the time, here, Peter has made things as difficult as possible for John. If John begins with the winning number 34, Peter can lead him to 25 by a route: John 34, Peter 32, 30, 29, or 28, John 25; then Peter has an opportunity to set the two earlier discussed traps. All this also hold with the rules *B*, *a*, and *b*, provided that we increase all numbers by 8, 10, and 1, in the respective cases.

224. 9-subtraction game. Table 12 gives the good deductions in the 9-subtraction games *A*, *B*, *a*, and *b*. This table includes only those good deductions that do not produce a winning number, since deductions that do this can be found immediately. If a number has no good deduction other than one which produces a winning number, we have left it out, to keep the table shorter.

TABLE 12

0 wins				0 loses											
A		*B*		*a*		*b*									
2	1	4	2	1	—	1	—	26	2, 5	29	4, 9	33	5	24	6!
6	3	6	3	2	—	3	1	27	8	31	5	34	—	27	2, 5
8	4	10	5	12	—	7	3	28	3	32	—	35	6!	28	8
10	↑ 5	11	—	14	1	9	4	29	9	34	1	38	2, 5	29	3
11	—	13	1	18	3	11	↑ 5	32	—	35	7	39	8	30	9
13	1	14	7	20	↑ 4, 9	12	—	34	1!	36	5!	40	3	33	—
14	7	15	5!	22	5	14	1	38	3, 8	39	9	41	9	35	1!
15	5!	16	8	23	—	15	7	40	↓ 4, 9	42	5	44	—	39	3, 8
18	9	18	9	25	1	16	5!	42	5	43	—	46	1!	41	↓ 4, 9
21	5	21	—	26	7	19	9	43	—	44	6!	50	↓ 3, 8	43	5
22	—	23	1, 6	27	5!	22	5	45	1	47	2, 5	52	4, 9	44	—
23	6!	27	↑ 3, 8	30	9	23	—	46	7	48	8	54	5	46	1
								47	5!	49	3	55	—	47	7
								50	9	50	9	57	1	48	5!
								53	5	53	—	58	7	51	9
								54	—	55	↓ 1!	59	5!	54	5

With *A*, the winning numbers are the multiples of 32, the multiples of 32 plus 11, and the multiples of 32 plus 22; with *B*, the winning numbers are the multiples of 32, the multiples of 32 plus 11, and the multiples of 32 plus 21; with *a* they are the multiples of 32 plus 2, the multiples of 32 plus 12, and the multiples of 32 plus 23; with *b*, they are the multiples of 32 plus 1, the multiples of 32 plus 12, and the multiples of 32 plus 23.

The good deductions which require special attention, have been signalized by exclamation points. From the table one can derive the correct strategy (for someone in a winning position). This consists of halving the distance to a winning number, or dividing this by 3 (if one cannot attain a winning number at once), giving preference to the deductions 1 and 5, and, choosing the deduction 6 when these others are of no avail.

IV. SUBTRACTION GAME WHERE THE OPENER LOSES

225. Modified subtraction game. In the subtraction games so far discussed, the winner can always be the player who chooses a number to open the game (assuming good play), provided he is unrestricted in his choice of the number. We now should like to discuss a game where the situation is precisely the reverse, that is, where the one who opens the game should be the loser, no matter what number he may select for his first choice. The deductions in this game consist of the numbers 0, 1, and so on, up to some limiting number, but the same deduction cannot be used twice in succession (and no negative numbers are allowed to be mentioned). The player who attains 0 may win or lose, according to what is initially agreed. Hence, the difference from the subtraction games discussed in §§217–224 lies in the fact that 0, too, can be chosen as a deduction. If 5, say, is the largest allowable deduction, we speak of the modified 5-subtraction game.

In this game there is no possibility that a player will find all replies barred at some stage earlier than when he reaches 0; even with 1, one or other of the replies 1 or 0 can be made, for the two deductions 1 and 0 cannot both be blocked.

When reaching 0 means a loss, the game proceeds very simply, and thus has little interest. The player who does not begin replies with a deduction 0 every time; by doing this he will win because he compels the other player to mention a smaller number every time, which means that finally he cannot avoid giving a reply of 0.[1]

When reaching 0 means winning, which we assume in what follows, John will begin, and then Peter will reply with the deduction 0 to every number which has no good deduction greater than 0. If there is one, of course, then he chooses a good deduction greater than 0. By doing this. Peter always remains in a winning position. However, this is not a sufficient rule for Peter to play the game successfully (nor for John to make use of any mistake made by Peter), for it is necessary to know which numbers have good deductions greater than 0, and

---

[[1] There seems to be an implicit assumption that the first player cannot use 0 for his *first* deduction —T.H.O'B.]

which deductions are then the good ones. These are determined in exactly the same way as for the previous subtraction games, by beginning with the smallest numbers.

226. Modified 2-, 3-, 4-, and 5-subtraction games. Table 13 shows the good deductions for the cases in which the largest allowable deduction is 2, 3, 4, or 5 (when 0 wins). The top row indicates the largest deduction, and the left-hand column shows the number to which a reply has to be given. The numbers that recur periodically have been given to the right of the double arrow.

TABLE 13

	2	3	4	5
1	1	1	1	1
2	1, 2	1, 2	1, 2	1, 2
3	↕ 0	↑ 3	3	3
4	0	0	4	4
5	0	↓ 0	0	5
6	0	3	3	3
7	0	0	0	0
8	0	0	4	4
9	0	3	3	3
10	0	0	0	5
11	0	0	0	0
12	0	3	3, 4	3, 4
13	0	0	↕ 0	↑ 0
14	0	0	0	0
15	0	3	0	5
16	0	0	0	0
17	0	0	0	↓ 0

In the modified 2- or 4-subtraction game, Peter, who has not opened, replies with the deduction 0 when the number is sufficiently large (larger than 2 or 12, respectively). In the modified 4-subtraction game, when a reply has to be given to a number greater than 2 and at most 12, then Peter chooses the deduction 0 if the number is divisible neither by 3 nor by 4, and otherwise chooses the divisor, 3, or 4, for his deduction; when replying to 12, the deductions 3 and 4 are both good.

With the modified 3- (or 5-) subtraction game, Peter replies with the deduction 0 when the number is not divisible by 3 (or 5), and otherwise with the deduction 3 (or 5), provided the number is larger than 2 (or 12). In the modified 5-subtraction game, if the number to which he has to reply is greater than 2 and at most 12, then Peter replies with the deduction 0 when none of the numbers 3, 4, 5 is a divisor of the number, otherwise he takes one of these numbers for his deduction, in cases where the number concerned is indeed a divisor.

227. Modified 6-, 7-, 8-, and 9-subtraction games. Table 14 presents the good deductions when the largest deduction is 6, 7, 8, or 9. In each of these four cases, we have shortened the table by omitting the numbers for which 0 is a good deduction (it is then the only one, of course).

TABLE 14

	6	7	8	9
1	1	1	1	1
2	1, 2	1, 2	1, 2	1, 2
3	3	3	3	3
4	4	4	4	4
5	5	5	5	5
6	3, 6	3, 6	3, 6	3, 6
7	0	7	7	7
8	4	4	4, 8	4, 8
9	0	0	0	9
10	5	5	5	5
12	4	4	0	0
14	0	7	7	7
15	5	5	5	5
16	4	4	0	0
18	0	0	0	9
20	4, 5	4, 5	5	5
21	↕ 0	↑ 7	7	7
25	0	0	5	5
27	0	↓ 0	0	9
28	0	7	7	7
30	0	0	5	5
35	0	7	5, 7	5, 7
36	0	0	↕ 0	↑ 9
42	0	7	0	0
44	0	0	0	↓ 0

These tables provide what we need for the direct derivation of results like those which we detailed in §226. For example, if the largest deduction is either 8 or 9, and if the number is larger than 35, then in the first case 0 is the good deduction, and in the second case 0 or 9, depending on whether the number is not, or is, divisible by 9. If the number is larger than 8 and at most 35, then 0 is the good deduction, when the number is divisible neither by 5 nor by 7, if 8 is the largest deduction, and also when the number is not divisible by 5, by 7, or by 9, if 9 is the largest deduction; in cases where a number is divisible by 7, say, then this divisor 7 is a good deduction. In both cases (largest deduction either 8 or 9) the good deductions are the same, except that the good deduction is 0 in the first case and 9 in the second case for numbers which are divisible by 9.

We find entirely similar results, but with different numbers, when the largest deduction is either 6 or 7. According to §226, the same holds when the largest deductions are 4 and 5, and when the largest deductions are 2 and 3.

***228. Modified subtraction game with larger deductions.** As appears from the findings in §§226 and 227, a number can have no good deductions different from 0, which are not included among its own divisors. As soon as two good deductions appear for some number which is divisible by both deductions (as 4 and 5 do for the number 20 when 6 or 7 is the largest deduction), then these deductions cease to be good deductions for larger numbers. In consequence of this destructive pairing of the good deductions, no good deduction, or only one, will finally remain, depending on whether the number of deductions different from 0 (the same as the largest deduction) is even or odd. The last qualification ("depending on . . .") is certainly true when the largest allowable deduction is not too large, but there is some doubt whether it will still hold for larger maximum deductions; we shall return to this presently.

The pairing of two good deductions occurs very early for an odd deduction whose double is also a deduction, for a deduction, divisible by 4 but not by 8, whose double is a deduction, and, in general, for a deduction that contains an odd number of factors 2, and for a deduction containing an even number of factors 2 if its double is also a deduction.

We have seen that the only numbers that matter are the numbers in which good deductions meet, that is, the numbers that contain more than one good deduction as a factor. These numbers can be easily found (as numbers that are divisible by both deductions), without writing down the good deductions for other numbers.

Table 15 presents the numbers with more than one good deduction, for largest deductions 10, 11, . . . , 21. Nothing has been filled in

TABLE 15

	10 or 11	12 or 13	14 or 15	16 or 17	18 or 19	20 or 21
2	1, 2	1, 2	1, 2	1, 2	1, 2	1, 2
6	3, 6	3, 6	3, 6	3, 6	3, 6	3, 6
8	4, 8	4, 8	4, 8	4, 8	4, 8	4, 8
10	5, 10	5, 10	5, 10	5, 10	5, 10	5, 10
14			7, 14	7, 14	7, 14	7, 14
18					9, 18	9, 18
36		9, 12	9, 12	9, 12		
48					12, 16	12, 16
60						15, 20
63	7, 9					
77		7, 11				
143			11, 13	11, 13	11, 13	11, 13
240				15, 16		
255					15, 17	
323						17, 19

when there is only one good deduction. The complete table can be read off from the abridged one without much trouble; the good deduction for a number is a divisor that has not yet dropped out and that does not exceed the maximum deduction; if such a divisor does not exist, 0 is the only good deduction.

The preceding results give the impression that when the largest deduction is even, the abridged table (of numbers with more than one good deduction) is the same as for a largest deduction given by the next following (odd) number, this largest deduction will eventually be the only good deduction (but only, of course, for numbers that are divisible by this largest deduction). However, this does not hold for still larger maximum deductions. A departure occurs for the first time when 26 and 27 are the largest deductions, as is shown by Table 16. If the largest deduction is 27, and if the number is sufficiently large, then the good deduction (for numbers divisible by it) is not 27, as one might think after the foregoing, but 25. However, when there is an odd largest deduction which is a prime number, the largest deduction does indeed become the last remaining good deduction.

It is not certain, either, that for larger maximum deductions there will be either no good deductions, or just one, left over, depending on whether the maximum allowable deduction is even or odd. This correspondence does indeed exist when the maximum deduction is

TABLE 16

	26	27
2	1, 2	1, 2
6	3, 6	3, 6
8	4, 8	4, 8
10	5, 10	5, 10
14	7, 14	7, 14
18	9, 18	9, 18
22	11, 22	11, 22
24	12, 24	12, 24
26	13, 26	13, 26
60	15, 20	15, 20
189		21, 27
272	16, 17	16, 17
399	19, 21	
437		19, 23
575	23, 25	

reasonably small, less than 50, say. However, it is not inconceivable that three good deductions could meet in the same number, for example 35, 55, and 77 in the number $5 \times 7 \times 11 = 385$. However, I have not been able to find a maximum deduction for which this type of aberration arises.

Chapter XIII:

PUZZLES WITH SOME MATHEMATICAL ASPECTS

I. SIMPLE PUZZLES WITH SQUARES

229. Puzzle with two square numbers of two or three digits. A square is a number that results from multiplying a number by itself, and so the two-digit square numbers are 16, 25, 36, 49, 64, and 81. It is not at all difficult to write two 2-digit squares, one below the other, in such a way that digits read downwards will again form two squares. This can be done in four ways:

16	36	64	81
64	64	49	16

It takes a little more trouble to write two 3-digit squares, one below the other, in such a way that reading downwards will give three squares each of two digits. The first puzzle could admittedly have been solved by writing two squares, one below the other, in all possible ways (although even there the labor could be reduced by attention to what we shall next discuss), but for two 3-digit squares, it becomes desirable to supplement trial by reasoning, in order to obtain some simplification.

In the first place we note that a square can end only in one of the digits 0, 1, 4, 5, 6, 9, as is immediately evident if successive squares are written down. A 2-digit square, however, cannot end in 0. Now the 3-digit squares are:

100	121	**144**	**169**	**196**	225	256	289	324	361	400
441	484	529	576	625	676	729	784	841	800	**961**

The squares printed in bold type are the only ones here which contain no digit other than 1, 4, 5, 6, 9, so that the lower of the two 3-digit squares must be chosen from these five numbers; otherwise we would not have three 2-digit squares when reading downwards. Since the uppermost of the two 3-digit squares cannot end in 3 or 8, we get the following four possibilities (when we pay attention only to the three 2-digit squares):

866	814	834	**841**
144	169	169	**196**

The last case is the only one where the uppermost 3-digit number is a square, so that this is the case which gives the solution.

230. Puzzle with three 3-digit squares. We require three 3-digit squares to be written one below the other in such a way that, when reading downwards, we again have three squares. Since there is now a larger number of possibilities, it is advantageous to take additional note of the following properties of squares (which also become evident if squares are written down):

If a square ends in 0 or in 5, its second-last digit must then be 0 or 2, respectively. If a square ends in 6, its second last digit is odd. If a square ends in 1, 4, or 9, its second-last digit is even.

In the present problem, the square number at the bottom, and also the one at the right, can contain no digits other than 0, 1, 4, 5, 6, 9 (those which can be last digits of squares), and each of these numbers must therefore be one of the numbers 100, 144, 169, 196, 400, 441, 900, 961. These two squares must have the same last digit (0, 1, 4, 6, or 9). If this is 1, say, then it is not possible for the lower square to be 441 and, at the same time, for the right-hand square to be 961. For an

. . 9
arrangement . X6 would require the digit indicated by the X to be
441

an odd one for the middle horizontal square, and an even one for the middle vertical square (because of the above-mentioned characteristics of the second-last digit of a square). So the bottom and right-hand squares are both 441, or are both 961. In the first case the middle square (horizontal or vertical) must begin with an even digit and end in 4; only 484 satisfies these conditions, so that the upper and the left-hand squares must end in 44, and so must both be 144. In the second case (the bottom and right-hand squares both equal to 961), the middle horizontal square (and also the middle vertical square) must begin with an even digit and end in 6, so that the middle squares must both be equal to 256 or both be equal to 676 (since they have their middle digit in common); in the first case the upper square and the left-hand square must both be 529 or both be 729, while in the second case the squares in question must both be 169. So the case in which there is a 1 in the lower right-hand corner leads to the four solutions:

144	529	729	169
484	256	256	676
441	961	961	961

A similar investigation can be undertaken for the cases in which the number in the lower right-hand corner is 0, 4, 6, or 9. Here we find nine further solutions:

441 841 144 441 841 121 121 361 961
400 400 400 484 484 289 256 676 676
100 100 400 144 144 196 169 169 169

Of course, we here must also investigate, for example, the case in which the lower number is 400, and the right-hand number is 900. The middle horizontal square and the middle vertical square must then both be 400, so that the upper square must end in 49, and the left-hand square in 44; however, this is impossible since the last-mentioned squares must have the same initial digit.

231. Puzzle of 230 with initial zeros. The thirteen solutions of the puzzle of 230 are all symmetric with respect to the main diagonal, which we take to be the diagonal from the top left-hand corner to the bottom right-hand corner; this means that the three horizontal squares and the three vertical squares are the same numbers. Indeed, this need not surprise us, since for a symmetric solution we have to make three squares from six digits, and for an asymmetric solution (that is, a solution which has no symmetry with regard to the main diagonal) we must make six squares from nine digits, a much more demanding task.

If we also allow two initial zeros to be prefixed to a 1-digit number, or one initial zero to a 2-digit number, to make these into 3-digit numbers, then there are also asymmetric solutions, and in fact three of these, if we make no distinction between two solutions that arise from each other by reflection in the main diagonal. These three asymmetric solutions are:

100 400 400
400 400 441
441 144 196

By admitting initial zeros we also obtain 29 additional symmetric solutions. These are:

049 004 004 009 009 100 400 900 100 400
400 004 064 016 036 016 016 016 036 036
900 441 441 961 961 064 064 064 064 064

900 001 001 100 400 900 100 400 900 001
036 004 064 001 001 001 081 081 081 009
064 144 144 016 016 016 016 016 016 196

001	100	400	900	100	400	900	001	001
049	004	004	004	064	064	064	016	036
196	049	049	049	049	049	049	169	169

If we also admit the square 0 (in the form 000) we obtain 45 further solutions, among which is the trivial solution in which all digits are 0; this makes the total number of solutions rise to 90. Among the 45 solutions which use zero values for one or more of the squares, there are 25 symmetric solutions, all more or less trivial, and 20 asymmetric solutions. The latter set of solutions is:

000	000	000	001	001	004	049	049	049	049
000	000	000	000	000	000	000	000	000	000
100	400	900	400	900	900	000	100	400	900

441	144	000	000	000	100	400	900	000	000
000	000	000	000	000	000	000	000	001	841
000	000	441	144	049	049	049	049	196	196

When we delete the 000, the last of these solutions becomes the solution of the last puzzle of §229.

II. PUZZLE WITH 4-DIGIT SQUARES

232. 4-digit squares. To determine the 4-digit squares, it is not necessary to make a separate computation of the squares of each of the 68 numbers 32, 33, . . . , 99. First of all, we can observe that the final digits of the squares repeat regularly at intervals of ten steps (0, 1, 4, 9, 6, 5, 6, 9, 4, 1) with repetition in inverted order after the halfway position; hence the final digits of the squares can be filled in without any trouble. The truncated square, that is, the number that arises when the last digit of the square is deleted, increases regularly with each passage to a new square, a few times by 6, first of all (if we begin with the square of 30), then a few times by 7, next a few times by 8, and so on. The number that must be added to a truncated square, to produce the next truncated square, has to be increased by 1 every time there is a passage from a square ending in 9 to a next following square.

The numbers formed by the last two digits of the squares repeat periodically, too, with a period of 50 numbers, showing repetition in inverted order after the halfway position. So after writing down the squares of the numbers 1, 2, . . . , 25, we can immediately fill in the last two digits of the next following squares. This produces still greater economy of effort, so that we can quickly construct the arrangement shown in Table 17.

TABLE 17

30	900	40	*1600*	50	2500	60	3600	70	*4900*	80	*6400*	90	8100
31	961	41	1681	51	2601	61	3721	71	*5041*	81	**6561**	91	8281
32	1024	42	1764	52	2704	62	3844	72	5184	82	6724	92	8464
33	1089	43	1849	53	2809	63	3969	73	5329	83	6889	93	8649
34	**1156**	44	1936	54	2916	64	*4096*	74	5476	84	7056	94	8836
35	1225	45	2025	55	3025	65	4225	75	5625	85	7225	95	9025
36	1296	46	2116	56	3136	66	4356	76	5776	86	7396	96	9216
37	1369	47	2209	57	3249	67	4489	77	5929	87	7569	97	*9409*
38	**1444**	48	2304	58	3364	68	4624	78	6084	88	7744	98	*9604*
39	1521	49	2401	59	3481	69	4761	79	6241	89	7921	99	9801

233. Puzzle of the four 4-digit squares. We now proceed to the problem: write down four squares, one below the other, in such a way that, reading downwards, we again have four squares. To reduce the number of solutions we add the further requirement that none of the four squares is to contain a digit 0.

The horizontal square at the bottom and the vertical square at the right must each contain no digits other than 1, 4, 5, 6, 9 (the final digits of squares), so each of these numbers can only be one of the three squares printed in boldface in Table 17 (1156, 1444, or 6561). Since these three squares concerned have different final digits, the bottom square must be the same as the right-hand one.

First we take 1156 for the bottom square and the right-hand square. In view of the properties of the second-last digit of a square, which were mentioned in §230, the third square from the top must begin with two even digits (and end in 25), so that it must be 4225; this must also be the third square from the left. The upper square (and also the left-hand square) must then end in 41, and must therefore be 6241 (again in the light of Table 17), so that the second square from the top would then have to begin with 2 and end in 21. However, there is no such square.

Next, we take 1444 for the bottom square and the right-hand square. The third square from the top must then begin with two even digits and end in 4, and so must be either 4624 or 8464; since these numbers do not have the same next-to-last digit, the third square from the top must be the same as the third square from the left. If we choose 4624, we find 6241 for the left-hand square, and then reach an impasse in finding the second square from the top. We similarly come to a dead end after a choice of 8464.

This means that we have to take 6561 for the bottom square and the right-hand square. The third square from the top must then have 2 as its second digit, must begin with an odd digit, and must end in 6, so it must be either 1296 or 9216; then the third square from the left must be the same as the third square from the top. If we choose 1296, we find that the top square can be 2116, 2916, or 9216. If we take 2116 for the top square, then the left-hand square cannot be equal to 2916, because otherwise we reach an impasse (even if we were to admit the digit 0) in finding the second square from the top and the second square from the left (which have the same second digit). Making the top square 2116 (with the same choice for the right-hand square) now leads to the solution:

2116
1225
1296
6561

Choosing 2916 or 9216 for the top square (and for the left-hand square) does not allow the choice of a square without a 0 for the second row from the top. If we take 9216 as the third square from the top, we again reach an impasse, which shows that the solution given is the only solution.

234. Puzzle of §233 with zeros. If we admit zeros in the puzzle of §233, but not an initial zero, then there are further solutions. To find these, we note that the square at the bottom (as well as the square at the right) can then also be one of the seven italicized numbers in Table 17. Now we have to take into account the possibility that the bottom square is 5041 and the right-hand square 6561. The third digit of the third square from the top is then odd. This is also the third digit of the square in the third column, where, however it is required to be even (because the last digit of that number is 4). So this case comes to a dead end.

If the bottom square is 1156 and the right-hand square is 4096, then the third square from the top must begin with two even digits and end in 29. Such a square, however, does not exist, so that this case, too, comes to a dead end. Continuing in this way we find the following twelve solutions which involve one or more zeros:

number b contains no 0, does not end in 1, and, if a is odd, does not end in 5, either. The final digit of b is not equal to any other digit of b (for if it were, the final digit common to c and f would also be a final digit either for d or e). If b contains the digit 1, then the six digits of a and b must all be different. The first digit of b is not 9 (for it it were, the initial digits of b, c, and f would be the same, and all would in fact be 9), so we now have proof that b does not contain two digits 9.

If b contains a single digit 9, then a is smaller than 111 (because c, d, and e consist of three digits), and so c, d, and e must begin with digits which occur in b. So if b contains a 9, the three digits of b must all be different. Similarly, b cannot contain an 8 and two equal digits less than 5, and cannot consist of a 7 and two digits 2. The first digit of b and another digit of b cannot both be 8 or more (for if they were, the first digits of e and f and the first digit of c or d would be at least 8, and either 8 or 9 would then have to occur three times); we now have proof that b does not contain two digits 8. The three digits of b are not all greater than 6 (for if they were, the first digits of c, d, e, and f would be greater than 6, and then one of the digits 7, 8, 9 would have to occur three times). If b contains the digit 1 (which cannot then appear in a because in this case the six digits of a and b have to be all different), then the initial digit of a is at least 2, and so the largest digit of b is at most 4; b cannot here contain the digit 2, because b must also contain a digit greater than 2 which differs from the initial digit of a (which would be greater than 2, if b contained both digits 1 and 2); this would be incompatible with the fact that c, d, and e are three-digit numbers. So if b contains the digit 1, it follows that 3 and 4 are the other digits of b, and 2 the initial digit of a.

236. Connection with remainders for divisions by 9. If the numbers a and b of §234 are multiplied together, their product $a \times b$ (or f) is the sum of the products of the numbers c, d, and e by 1, 10 and 100, respectively, and hence, when divided by 9, it must have the same remainder as $c + d + e$. It follows that the sum $a + b + c + d + e + f$, when divided by 9, must have the same remainder as $a + b + (2 \times a \times b)$. Since a number divided by 9 has the same remainder as the sum of its digits (see §23), and since we know that a, b, c, d, e, and f jointly involve two digits 0, two digits 1, and similarly for the rest, it follows that $a + b + c + d + e + f$ has the same remainder when divided by 9 as $(2 \times 0) + (2 \times 1) + (2 \times 2) + (2 \times 3) + (2 \times 4) + (2 \times 5) + (2 \times 6) + (2 \times 7) + (2 \times 8) + (2 \times 9) = 90$. It follows that $a + b + c + d + e + f$, and hence also

$a + b + (2 \times a \times b)$ must be divisible by 9. If a is a multiple of 3 plus 1, then $a + b + (2 \times a \times b)$ has the same remainder when divided by 3 as $1 + b + (2 \times b)$, which is to say, as $1 + (3 \times b)$, so that it must have the remainder 1; this is consequently also incompatible with the fact that $a + b + (2 \times a \times b)$ must be divisible by 9, and consequently also by 3. So a cannot be a multiple of 3 plus 1, and so is not a multiple of 9 plus 1, nor a multiple of 9 plus 4, nor a multiple of 9 plus 7. If a is a multiple of 9, then $a + b + (2 \times a \times b)$, when divided by 9, has the same remainder as b, so that b, too must be a multiple of 9. If a is a multiple of 9 plus 2, then $a + b + (2 \times a \times b)$, when divided by 9, has the same remainder as $2 + b + (2 \times 2 \times b)$, which is to say, the same remainder as $2 + (5 \times b)$; since $a + b + (2 \times a \times b)$ is divisible by 9, so is $2 + (5 \times b)$, which makes b a multiple of 9 plus 5. Continuing in this way, we find the following six cases (where 9m is put as an abbreviation for "multiple of 9," 9m + 2 for "multiple of 9 plus 2," and so on):

	A	*B*	*C*	*D*	*E*	*F*
a:	9m	9m + 2	9m + 3	9m + 5	9m + 6	9m + 8
b:	9m	9m + 5	9m + 6	9m + 2	9m + 3	9m + 8

By these arguments, the number of cases to be investigated is reduced by a factor of $13\frac{1}{2}$.

237. Combination of the results of §§235 and 236. If, for the moment, we disregard the digits of the number f of §235, the order of the digits of b becomes irrelevant temporarily, and we can meanwhile assume these digits to be in non-decreasing order. When account is taken of the results of §235 for the digits of b, the cases mentioned in §236 then lead to these possibilities for b:

A: 225 234 279 369 378 459 468 477 558 567;
B: 239 248 257 266 347 356 446 455 689;
C: 249 258 267 348 357 366 447 456;
D: 236 245 335 344 389 479 569 578 668 677;
E: 237 246 255 336 345 489 579 678;
F: 134 224 233 269 278 359 368 377 458 467 557 566.

If two of the digits of b are equal, this determines the order of the digits: for the other digit must then be used for the last digit.

If the digits of b are not 332, then (in view of our list of remaining possibilities) the largest digit of b must be at least 4, so that a must be less than 250. If $b = 332$, then a cannot contain the digit 3, and so a must be less than 300. If a is a multiple of 9 plus 5, which requires b to be a multiple of 9 plus 6 (case C), then the largest digit of b is at least 6, and a must be less than 166. If a is a multiple of 9 plus 2, or plus 6, when b must be a multiple of 9 plus 5, or plus 3 (case B, or case E), then the largest digit of b is at least 5, and a must be less than 200. If a is a multiple of 9 or a multiple of 9 plus 5, then a must be less than 250. When we connect this with the fact that a does not end in 0, 1, or 5 (see §235), this leaves the following values of a to be investigated:

A: 108 117 126 144 153 162 189 198 207 216 234 243;
B: 119 128 137 146 164 173 182;
C: 102 129 138 147 156;
D: 104 113 122 149 158 167 176 194 203 212 239 248;
E: 114 123 132 159 168 177 186;
F: 107 116 134 143 152 179 188 197 206 224 233 242 269 278 287 296.

For a number b in which a high digit occurs, we evidently can dispense with examining some of the numbers a in our lists. If we take further note of the fact that no digit can occur three times in a and b jointly, that a must be even when b contains a 5 and two equal digits (in which case the 5 must be the last digit of b), and that the six digits of a and b must be different when b contains a 1 (see §235), then our remaining task is the examination of only 171 cases which associate a number b (disregarding the order of its digits) and a number a. If we carry out the multiplications as far as the formation of the partial products c, d, and e (omitting the formation of the total product f) then in most cases we soon obtain three equal digits. If no three equal digits are to occur (still with disregard of f), only the following five cases remain:

A	D	D	D	F
108	122	158	176	**179**
369	578	245	245	**224**
972	976	790	880	**716**
648	854	632	704	**358**
324	610	316	352	**358**
				40096

The last multiplication is the only one in which the order of the digits of b is determined (because b has two equal digits). So the complete product f has also been calculated there; it turns out that this multiplication provides a solution. For the other multiplications, the partial products have been written without indentation, because their order is still undecided. In the first multiplication, the final digit of the total product must be 2, 8, or 4, and in the second-last multiplication it must be 0, 4, or 2, and this digit must then occur three times in all. In the third multiplication, 0 must be the final digit of the total product (and also of c), since otherwise this final digit would occur three times; but no matter how we choose the order of the two other digits of b, there will still always be a triple repetition of some digit. It turns out that the second multiplication does not lead to a solution either, and so the problem has only a single solution, the one which has been printed in bold type.

IV. PROBLEM ON REMAINDERS AND QUOTIENTS

238. Arithmetical puzzle. A number less than one million is known to be such that diminishing it by 3 makes it divisible by 7. From the number diminished by 3 we subtract the seventh part. We then obtain a number that also becomes divisible by 7 after 3 is subtracted from it. From this number we derive another in the same way, namely by subtracting a seventh part from the number diminished by 3. This time, too, a number results that is divisible by 7 after subtracting 3. This occurs four more times, so that, in all, it happens seven times over, that a number is divisible by 7 after subtracting 3.

A non-mathematician will not find it easy to determine this number, and this may be true for a mathematician of moderate accomplishment, also. Even so, no mathematical knowledge is needed to understand the solution which follows, although it may well be impossible to appreciate where the idea of the solution came from, without some basis of mathematics. The original number and the numbers derived from it (by successive subtractions of a 3, and of a seventh part of what remains) will be called the first number, the second number, and so on. We increase these numbers by $3 \times (7 - 1)$, hence by 18, and we shall refer to the numbers so obtained as the new first number, the new second number, and so on. The new first number is larger by 21 than the first number diminished by 3, which shows that it too is divisible by 7. Diminishing a multiple of 7 by the quotient of a division by 7 (hence by its seventh part) amounts to a multiplication by 1 − 1/7, that is, by 6/7. The result of multiplying the new first number by 6/7 is larger by 6/7 × 21 (that is, by 18) than the product of 6/7 and the first number diminished by 3, which shows that the result is equal to the new second number. The new second number, the new third number, and so on, are also divisible by 7, and each of these numbers must arise from the preceding number by multiplying the latter by 6/7. Since we can continue in this way up to a seventh new number (according to the terms of the problem), we can multiply the first number seven times by 6/7 without any appearance of fractions. So the new first number must be some multiple (possibly a unit multiple) of 7^7, that is, of 823,543, and the new seventh number is then the same multiple of 6^7, that is, of 279,936. Since the original number (the first number) is less than one million, this makes the new first number less than 1,000,018, so that the new first number must be 823,543, which shows that the required number is equal to 823,543 − 18 = **823,525**. The following working shows that this is correct:

823525	705876	605034	518598	444510	381006	326574
823522	705873	605031	518595	444507	381003	326571
117646	100839	86433	74085	63501	54429	46653
705876	605034	518598	444510	381006	326574	279918

239. Variants of the puzzle of §238. We modify the puzzle of §238 to the effect that the sixth number is formed in the same way as before but has the remainder 5, instead of 3, when divided by 7. We assume further that the required number lies between two million and three million.

After being increased by 18, the required number is a multiple of $7^6 = 117{,}649$, and thus can be written as $(117{,}649 \times v) - 18$. The new sixth number is $6^6 \times v = 46{,}656 \times v$, and so the sixth number is $46{,}656 \times v - 18$. Now the remainder of 46,656 for division by 7 is equal to 1*, so that the sixth number, when divided by 7, has the same remainder as $v - 18$, and hence the same remainder as $v + 3$. This remainder is 5 when v is a multiple of 7 plus 2, that is, when v can be written as $(7 \times w) + 2$. So the required number can be written as $(823{,}543 \times w) + 235{,}298 - 18$, that is, as $(823{,}543 \times w) + 235{,}280$.

* More generally, the relation $6 = 7 - 1$ implies that 6^2, 6^4, 6^6, and so on, have the remainder 1 when divided by 7, and that 6^3, 6^5, 6^7, and so on, have the remainder 6.

From the limits between which the required number has to lie, it turns out that we must have $w = 3$, which makes the required number equal to **2,705,909**. The following working shows that this is correct:

2705909	2319348	1988010	1704006	1460574	1251918	1073070
2705906	2319345	1988007	1704003	1460571	1251915	1073065
386558	331335	284001	243429	208653	178845	153295
2319348	1988010	1704006	1460574	1251918	1073070	

If we ignore the limits, then the smallest number that satisfies the requirements is given by $w = 0$; hence, this smallest number is then 235,280.

It is clear that we can make other choices for the number by which we divide each time, and for the remainder that appears each time. For instance, if we divide by 5 each time, and if 1 is the remainder, then the new numbers arise by adding $1 \times (5 - 1)$, that is 4. If the remainder 1 appears five times, the new first number is a multiple of 5^5, which we can write as $3125 \times v$, when the original number becomes $(3125 \times v) - 4$. The smallest number that satisfies the conditions is then 3121.

If a remainder 1 for division by 5 appears four times only, after which the fourth number has the remainder 3 when divided by 5, then the required number can be written as $(625 \times v) - 4$. The fourth number is then equal to $(256 \times v) - 4$ and therefore will have the remainder 3 when divided by 5, if v is a multiple of 5 plus 2, that is, if $v = (5 \times w) + 2$. The original (first) number is then equal to $(3125 \times w) + 1246$, so that 1246 is then the smallest number that satisfies the conditions.

Next we take the case in which the required number lies between 3000 and 4000, while the remainder 1 arises four times on division by 5, but not the fifth time (when we diminish each number both by the remainder and by the quotient). The number is then a multiple of 625 minus 4, but not a multiple of 3125 minus 4, and therefore must be **3746**.

One of the preceding problems once appeared in the following form. A number of nuts, less than 5000, is intended to be divided equally among five boys. However, in the division, one nut would be left over; this is intended for their dog. One of the boys goes to the nuts secretly, takes away his share, and gives the dog a nut that is left over from a division into five equal parts. After that, another boy does the same thing, and once again, it turns out that one nut is left over after a division into five equal parts; he gives this nut to the dog, and takes away what he considers to be his share. Each of the other boys does this, too, and no one suspects that anyone else has been there before him, because after every division into five equal parts, one nut is left over. What is the number of nuts that they must have had to divide? The preceding arguments show that it must be **3121**.

We modify the foregoing to the effect that only the last boy notices that someone else has been there before him, because his division of the nuts into equal parts produces no nut left over for the dog. The number of nuts is then **2496**.

***240. Mathematical discussion of the puzzle.** We now give a discussion of the puzzle of §§238 and 239, in terms of formulae, for readers who have some grounding in mathematics. An unknown (positive integral) number x_1 is known to have the remainder r when divided by a given number a ($a \geqslant 2$). The number x_2 which arises from x_1 by subtracting the quotient $\frac{x_1 - r}{a}$ and the remainder r is:

$$x_2 = x_1 - \frac{x_1 - r}{a} - r = \frac{a-1}{a} x_1 - \frac{a-1}{a} r.$$

We now determine a number p such that

$$x_2 + p = \frac{a-1}{a} (x_1 + p),$$

which gives $x_2 = \frac{a-1}{a} x_1 - \frac{p}{a}$. For this, we must have $\frac{p}{a} = \frac{a-1}{a} r$, so we must have $p = r(a - 1)$. Consequently,

$$x_2 + r(a-1) = \frac{a-1}{a} \{x_1 + r(a-1)\}.$$

It is further given that x_2 in its turn has a remainder r when divided by a. Hence, the number x_3 which is derived from x_2 in the same way as x_2 was derived from x_1 must satisfy the equation:

$$x_3 + r(a-1) = \frac{a-1}{a} \{x_2 + r(a-1)\} = \left(\frac{a-1}{a}\right)^2 \{x_1 + r(a-1)\}.$$

It is also given that x_3, too, has the remainder r when divided by a, as does the number x_4 derived from it, and so on, up to and including

x_n. From x_n we can go on to form a number x_{n+1} in exactly the same way. For this number, we have:

$$x_{n+1} + r(a-1) = \left(\frac{a-1}{a}\right)^n \{x_1 + r(a-1)\}.$$

Since $a - 1$ and a are relatively prime, we may assume relations

$$x_1 + r(a-1) = ca^n, \qquad x_{n+1} + r(a-1) = c(a-1)^n,$$

where c is a positive integer. Without further data it is not possible to determine c. If x_1 has to be as small as possible, then $c = 1$, and so $x_1 = a^n - r(a-1)$. If x_1 has to lie between given limits, then these lead to limits between which c must lie; from this we find a set of numbers that satisfy the problem, possibly only one number if the limits are sufficiently close to each other, or possibly no number at all.

If it is given that x_{n+1} has the remainder s when divided by a, this is written as $x_{n+1} \equiv s \pmod{a}$. We then have

$$s \equiv c(a-1)^n - r(a-1) \equiv (-1)^n c + r \pmod{a},$$

and this leads to $c \equiv (-1)^n(s - r) \pmod{a}$, which is only another way of saying that $c = ga + (-1)^n(s - r)$ for some integer g.

From this it follows that

$$x_1 + r(a-1) = ga^{n+1} + (-1)^n(s-r)a^n,$$

which takes the form $x_1 + r(a-1) = ga^{n+1}$ for $s = r$, as indeed it must. In order to determine the integer g, we must be given additional information about x_1, for example "as small as possible" or "between two given limits."

V. COMMUTER PUZZLES

241. Simple commuter puzzle. Here is an interesting problem which is well-known: Smith takes the train daily from his work in the city back to the suburb where he lives and is met at the station by his chauffeur, who drives him home. One day Smith finishes his work earlier than usual and arrives in his suburb one hour earlier than usual. He then walks home from the station, and meets his chauffeur on the way. The chauffeur stops and takes Smith home; this brings Smith home 10 minutes earlier than usual. How long did Smith walk? The time for stopping and picking up is to be disregarded, and it is also assumed that the chauffeur arrives at the station at the same moment as the train.

The car has been spared the distance from the meeting point to the station and back. The terms of the problem indicate that it must take 10 minutes to cover that distance, so that the car would need 5 minutes to get from the station to the meeting point. It took Smith 60 − 10 = 50 minutes more than usual to get home from the station. This is because he walked from the station to the meeting point instead of being driven. This shows that it takes Smith 50 minutes more than the car to cover the last-mentioned distance, so that he must have walked for 5 + 50 = **55 minutes.**

242. More difficult commuter puzzle. In §241 it was tacitly assumed that the car travels from the house to the station as fast as it travels on its way back. If this is not the case, we do not have sufficient data to compute the time it took Smith to walk. In the next problem, we assume that the road rises from the station to the house, so that the car travels more slowly from the station to the house than from the house to the station. We do assume, however, that the road rises equally sharply everywhere, so that the car travels equally fast everywhere from the house to the station, and similarly on its way back, but then at a lower speed.

Once again, Smith arrives in his suburb one hour earlier than usual, walks home from the station, and meets his chauffeur. The latter, however, does not notice Smith, and drives on. "What a pity," Smith says to himself. "If the chauffeur had seen me, I would have been home 15 minutes earlier than usual. I'll keep on walking; the chauffeur will catch up with me in 18 minutes, and I'll be home at the usual time." How long had Smith been walking when his chauffeur drove past him on his way to the station? The time the chauffeur needs to ascertain that Smith is not on the train is to be disregarded; so it is assumed that after his arrival at the station the chauffeur immediately returns to join Smith (but without driving any faster than usual).

243. Solution of the puzzle of §242. It takes the car 15 minutes to drive from the meeting point to the station and back, and 18 minutes to drive from the meeting point to the station and continue back to the overtaking point, so it takes 3 minutes to drive from the meeting point to the overtaking point. It takes Smith 18 minutes to cover the last-mentioned distance, so on its way from the station to the house the car must travel six times as fast as Smith walks. Smith takes 60 − 15 = 45 minutes more than the car to cover the road from the station to the meeting point. Since Smith's time is six times as great as the car's time, the difference is five times as great, and 45 minutes must be five times the time the car needs to travel from the station to the meeting point. Hence, that time must be $45/5 = 9$ minutes, so that the time Smith has taken in walking from the station to the meeting point must be equal to $6 \times 9 =$ **54 minutes.**

In connection with this puzzle we ask the following question: How could Smith know that he would have been home 15 minutes earlier if the driver had seen him at the meeting point, and that the car would again catch up with him in 18 minutes? Smith had never before walked home from the station, and he did not know at what speed he walked; nor did he know at what speed the car traveled from his house to the station and back; also, there are no milestones along the road.

What Smith did know was that the car could go from the house to the station $1\frac{1}{2}$ times as fast as in the opposite direction. He also knew that he had been walking for 54 minutes when he met the car. So he knew that the car would be at the station in $60 - 54 = 6$ minutes, and would require $1\frac{1}{2} \times 6 = 9$ minutes to be back at the meeting point; so if the chauffeur had noticed him, $6 + 9 = 15$ minutes would have been saved. The time Smith needs for his walk from the meeting point to the overtaking point is 54/9 times as long as the time that the car needs for that distance, that is, six times as much, so the difference is five times as large; this difference is the amount of time by which the car (on its way back from the station) is later at the meeting point than Smith, so that the car needs $15/5 = 3$ minutes to travel from the meeting point to the overtaking point. Hence, Smith needs $15 + 3 =$ **18 minutes** (we could also say $6 \times 3 = 18$ minutes) to cover that distance.

VI. PRIME NUMBER PUZZLES

244. Prime number puzzle with 16 squares. The problem is to rearrange the 16 numbers in *Figure 106* in such a way that the sum of each adjacent pair of numbers (either horizontally or vertically) is a prime number. We call a number a prime number when it is greater than 1 and has no divisor greater than 1 and less than itself. Hence, the prime numbers are 2, 3, 5, 7, 11, 13, 17, 19, 23, 29, 31, 37, 41, 43, 47, and so forth. Solutions that arise from one another by rotation or reflection are considered to be identical; in this way we obtain groups of eight solutions such that we count each group as a single solution.

Since 2 cannot arise as the sum of two of the numbers 1, 2, . . . , 16, each of the 24 sums (of two adjacent numbers) must be odd. The odd numbers, therefore, have to be placed on the white squares and the even numbers on the black squares in *Figure 107* (or the other way round, which comes to the same thing). From this it follows that we can obtain another solution from any given solution by adding 1 to the eight odd numbers and subtracting 1 from the eight even numbers: since every pair of adjacent numbers consists of an even and an odd number, the 24 sums are left unchanged by the additions and subtractions in question. We call the two solutions conjugate. It may happen that two conjugate solutions correspond by reflection or rotation, but this is not a necessary feature. If this does not happen, we obtain a group of 16 solutions which transform into one another by rotation or reflection, or by additions and subtractions of 1; otherwise we obtain a group of only eight solutions.

1	**2**	**3**	**4**
5	**6**	**7**	**8**
9	**10**	**11**	**12**
13	**14**	**15**	**16**

Fig. 106

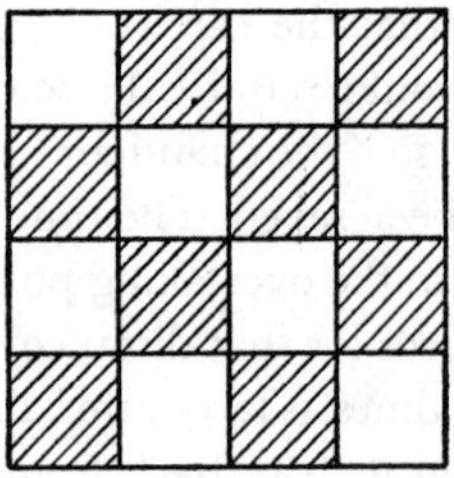

Fig. 107

245. Solution of the puzzle of §244. To solve the puzzle, we first consider how to make the 24 sums all odd and none divisible by 3; for only one of the sums, (1 + 2), can be allowed to be divisible by 3. To achieve this, we replace the 16 numbers by their remainders when divided by 6. By this procedure we obtain two numbers 0 (originating from 6 and 12), three numbers 1 (from 1, 7, and 13), three numbers 2 (from 2, 8, and 14), three numbers 3 (from 3, 9, and 15), three numbers 4 (from 4, 10, and 16), and two numbers 5 (from 5 and 11). We then have to arrange these numbers 0, 1, 2, 3, 4, 5 in such a way that 1, 3, 5 are on the white squares, and 0, 2, 4 on the black squares, while a 0 and a 3 are not allowed to be adjacent, nor a 1 and a 2, nor a 4 and a 5; however, a 1 and a 2 may be adjacent once only.

We begin by placing the three numbers 1 and the three numbers 2 in such a way that the required conditions are satisfied, for example in the manner indicated in *Figure 108*. However, in that case we run into trouble with the seven connected squares in which dots have been placed, since if we place a 3 in the top left-hand corner, we can have only a 4 in the adjacent square, next to that 4 only a 3, and so on. So in the seven connected squares we would require the number 3 four times, whereas we have only three numbers 3 available to be filled in; we reach an impasse even sooner when we place a 5 in the top left-hand corner, next to which we must have a 0 and so on.

•	•	•	**2**
2	•	**2**	
•	•	**1**	
•	**1**		**1**

Fig. 108

The three numbers 1 can be distributed among the white squares in 17 essentially different ways; we leave it to the reader to find schemes for these 17 ways. Since a 1 and a 2 may be adjacent only once, while three numbers 2 have to be filled in, we must be able to fill in at least two numbers 2 which are not adjacent to a 1. This can be done in only nine of the 17 cases and, in two of these nine cases, in three ways. The third 2 now has to be filled in in such a way that it is adjacent to at most a single 1, leaving no set of seven or more connected squares. This leaves the four possibilities in *Figure 109*, if we consider arrangements which arise from one another by interchanging 1 and 2 to be equivalent.

In all these cases it turns out that a 1 and a 2 have to be adjacent

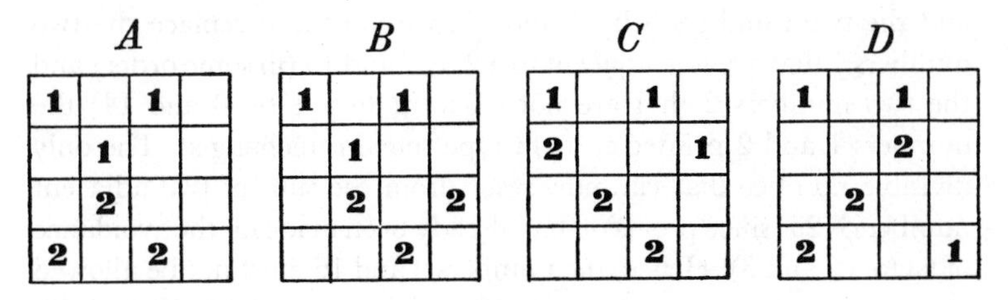

Fig. 109

(since otherwise the number of connected squares becomes too large). By interchange of 1 and 2, diagram *A* (Fig. 109) changes into its mirror image. If we form the conjugates of the solutions resulting from *A*, then these will provide the same solutions reflected about a horizontal line; hence *A* leads to solutions which can be considered as identical in groups of eight, while *B*, *C*, and *D* lead to solutions in groups of sixteen of similar type. In diagram *B*, there are two connected regions of four squares each; we can use the two numbers 0 and the two numbers 5 to fill in whichever of these regions we like. This gives two cases, *B*1 and *B*2. So we can fill in the numbers 0, 1, 2, 3, 4, 5 to have the five possibilities in *Figure 110.*

A

1	0	1	4
0	**1**	4	3
5	**2**	3	4
2	5	2	3

*B*1

1	4	1	4
0	**1**	4	3
5	**2**	3	2
0	5	2	3

*B*2

1	4	1	0
4	**1**	0	5
3	**2**	5	2
4	3	2	3

C

1	4	1	0
2	3	4	1
5	2	3	4
0	5	2	3

D

1	0	**1**	4
0	5	**2**	3
5	2	3	4
2	3	4	1

Fig. 110

Each of these five possibilities leads to a new puzzle, since we have to replace the two numbers 0 by 6 and 12 in some order, the three numbers 3 by 3, 9, and 15, also the three numbers 4 by 4, 10, and 16, and the two numbers 5 by 5 and 11; we must also replace the two numbers 1 that are not adjacent to a 2 by 7 and 13 (in some order) and the two numbers 2 that are not adjacent to a 1 by 8 and 14; the numbers 1 and 2 printed in bold type remain unchanged. The only divisible number that can now result from the sum of two adjacent numbers is 25 (since provision has already been made for the avoidance of factors 2 and 3). Hence, the numbers 9 and 16 must not be allowed to be adjacent, and similarly for 10 and 15, 11 and 14, 12 and 13.

This, of course, produces a large reduction in the number of possibilities.

These restrictions refer only to the adjacency of numbers arising from 0 and 1 (12 and 13), the adjacency of numbers arising from 2 and 5 (14 and 11), and the adjacency of numbers arising from 3 and 4 (9 and 16, or 15 and 10). These three cases are independent. So if each of them leads to a certain number of possibilities (for a specified case *A*, *B*1, *B*2, *C*, or *D*), then we need only multiply these numbers to obtain the total number of solutions provided by the case in question.

For the remaining part of the solution, it is advantageous to have 16 cut-out numbered squares as in Figure 106 (these should be copied on not too small a scale), and to arrange them in such a way that one of the five cases arises. Now we move the numbers arising from 2 and 5 in such a way that 14 and 11 are not adjacent; during this rearrangement the number 2 has to stay where it is, of course. The same is true for the number 1 when we rearrange the numbers arising from 0 and 1.

246. Examination of the five cases. CASE A (*Fig. 111*). The uppermost 0 is next to 13, and thus must be replaced by 6, which requires

A

7	6	13	16
12	1	10	3
11	2	9	4
8	5	14	15

B_1

7	4	13	16
6	1	10	3
5	2	9	14
12	11	8	15

B_2

13	4	7	6
16	1	12	11
3	2	5	8
10	9	14	15

C

1	16	7	6
2	3	10	13
11	8	9	4
12	5	14	15

D

7	12	1	16
6	11	2	3
5	8	9	10
14	15	4	13

Fig. 111

the other 0 to be replaced by 12; from this it appears that the left-hand 1 has to be replaced by 7, and the other 1 by 13. The bottom 5 is adjacent to 14, and hence has to remain 5, after which the numbers 2 and 5 (that is, the true numbers related to these) can be filled in in one way only. The uppermost 3 is adjacent to 10 and to 16, and hence cannot be replaced by 15 or 9; so it has to remain 3. Likewise, the bottom number 4 has to remain 4. The remaining numbers 3 and 4 can then be entered in only two ways, one of which is shown. Hence case *A* leads to $1 \times 1 \times 2 = 2$ solutions.

Case B1. The numbers 0 and 1 can be filled in in three ways (one of which is shown in *Fig. 111*) in such a way that 12 and 13 will not be adjacent. The numbers 2 and 5 can be filled in in three ways, too (with 14 and 11 not adjacent). The filling in of the numbers 3 (3, 9, 15) and 4 (4, 10, 16) in such a fashion that neither 9 and 16 nor 15 and 10 will be adjacent can be done in 14 ways. Hence, case *B*1 leads to $3 \times 3 \times 14 = 126$ solutions in all; one of these is shown.

Case B2. The right-hand number 1 is adjacent to 12, and hence has to be replaced by 7, and thus the left-hand number 1 must be replaced by 13; the filling in of the numbers 0 (that is, of the true numbers related to these) can be done in two ways. The filling in of the numbers 3 and 4 can be done in 14 ways. Hence, case *B*2 leads to $2 \times 1 \times 14 = 28$ solutions, one of which is shown (Fig. 111).

Case C. We can fill in the numbers 0 and 1 in two ways, and the numbers 2 and 5 in one way. The filling in of the numbers 3 and 4 can be done in eight ways, so that case *C* leads to $2 \times 1 \times 8 = 16$ solutions, one of which is shown (Fig. 111).

Case D. Here the situation is exactly the same as in case *C*, so that *D*, too, leads to $2 \times 1 \times 8 = 16$ solutions, one of which is shown (Fig. 111).

Hence, in all, the puzzle has $2 + 126 + 28 + 16 + 16 = 188$ essentially different solutions. If, in cases *B*1, *B*2, *C*, and *D*, we also count conjugate solutions as separate solutions, then there are 374 solutions. If we make distinctions between solutions which arise from one another by rotation or reflection, then there are $8 \times 374 = 2992$ solutions.

247. Puzzle of §244 with a restriction. If, to reduce the number of solutions, we introduce the restriction that each of the 24 sums has to be a prime number less than 31, this reduces to not allowing the numbers 15 and 16 to be adjacent; so the only numbers whose positions are affected are those which arise from 3 and 4. In the five cases there are then 1, 9, 9, 1, 1 possibilities respectively. These five cases then lead to $1 \times 1 \times 1 = 1$, $3 \times 3 \times 9 = 81$, $2 \times 1 \times 9 = 18$, $2 \times 1 \times 1 = 2$, $2 \times 1 \times 1 = 2$ solutions, respectively: amounting to 104 essentially different solutions in all. By distinguishing conjugate solutions, this gives 207 solutions. The solutions given all satisfy the added restriction.

We obtain a considerable reduction of the number of solutions if we also rule out the prime number 7 as a sum of two adjacent numbers. Cases A, $B1$, and $B2$ then give no solutions. C gives the solution already shown, and the one that arises from it by interchanging 7 and 13, while D leads only to the solution already shown. So the puzzle then has only three solutions, or six solutions if we also count the conjugate solutions.

248. Prime number puzzle with 25 squares. We require a rearrangement of the 25 numbers in *Figure 112* such that the sum of every two adjacent numbers (on a white and on a black square) is a prime number. Once again, solutions that arise from one another by rotation or reflection are considered to be identical.

Since this puzzle has a very large number of solutions, we add the restriction that each of the 40 sums be at least 11 and at most 41.

Since the 40 sums are all odd, the 13 odd numbers have to be placed on the 13 white squares, and the 12 even numbers on the black squares. As before, we shall first take heed of the condition that none of the 40 sums can be divisible by 3. To this end, we replace the 25 numbers by their remainders when divided by 6, after which we have to consider four numbers 0, five numbers 1, four numbers 2, four numbers 3, four

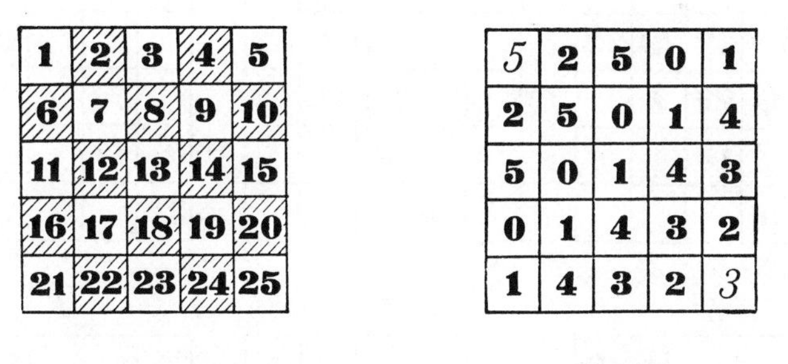

Fig. 112 *Fig. 113*

numbers 4, and four numbers 5. We begin by placing the five numbers 1 on the white squares and the four numbers 2 on the black squares in such a way that a 1 and a 2 are never adjacent. When we do this, we can occupy no region which consists of 9 or more connected squares; for example, if we place a 3, say, on a white square belonging to this region, then a 3 will have to be placed on all the white squares of the region, and a 4 on all its black squares, whereas this would contradict the fact that only four numbers 3 and four numbers 4 are available to be filled in; we also reach an impasse if we start by filling in the numbers 2 and 5 in a region of nine or more connected squares. Such a connected region can be avoided only by placing the numbers 0, 1, 2, 3, 4, 5 as shown in *Figure 113* or by an arrangement in which the italicized 5 in one corner is interchanged with the italicized 3 in the opposite corner. After the numbers 1 and 2 are filled in, the symmetry makes it immaterial whether we place the numbers 4 to the left or to the right of the diagonal.

Now we have to replace the four numbers 0 by 6, 12, 18, 24 (in some order), the five numbers 1 by 1, 7, 13, 19, 25, the four numbers 2 by 2, 8, 14, 20 and so on. Since non-divisibility of the 40 sums by 2 and by 3 has already been taken into account, the only thing we have to do is to ensure that none of the numbers 5, 7, 25, 35, 43, 47, and 49 can appear as a sum.

None of the numbers 1, 19, 25 can be adjacent to 24, while 1 and 19 cannot be adjacent to 6, either, and 7 and 25 cannot be adjacent to 18 (and 13 cannot be adjacent to 12). From this it follows that the numbers 1, 19, 25, 24, and 6 can be filled in only in one of the six

A

			18	1
		12	19	22
	6	25	4	
24	*7*	16		
13	10			

B

			18	19
		12	1	22
	6	25	16	
24	*7*	4		
13	10			

C

			18	1
		12	19	
	24	*7*		
6	*13*			
25				

D

			18	19
		12	1	
	24	*7*		
6	*13*			
25				

E

			12	1
		6	25	
	24	7		
18	13			
19				

F

			12	19
		6	25	
	24	7		
18	13			
1				

Fig. 114

ways shown in *Figure 114*. Also, in each of the diagrams *E* and *F*, there is only one way to fill in the numbers 18, 12, 7, and 13 one after another. In diagrams *A* and *B* we can fill in the numbers 18 and 12 (since 18 cannot be adjacent to 25); here the italicized numbers 7 and 13 can be interchanged. When filling in the numbers 7, 13, 12, and 18 in diagrams *C* and *D*, we must keep in mind that we cannot put 12 adjacent to 13, or 18 to 7; this allows the filling in to occur in two ways, one of which is shown; the other one can be obtained from it by simultaneously interchanging 7 with 13 and 18 with 12.

Diagrams *E* and *F* drop out immediately because of the impossibility of filling in the number 22 (which cannot be adjacent to 13 or to 25). In diagrams *A* and *B*, we can place 22 only against the right-hand edge, and, after that, the number 10 only in the bottom row; this then fixes the positions of 4 and 16 also.

23	**8**	**11**	**18**	**1**
14	**5**	**12**	**19**	**22**
17	**24**	**7**	**4**	**15**
6	**13**	**10**	**9**	**2**
25	**16**	**3**	**20**	**21**

23	**8**	**11**	**12**	**1**
14	**5**	**18**	**19**	**22**
17	**24**	**13**	**4**	**15**
6	**7**	**10**	**9**	**2**
25	**16**	**3**	**20**	**21**

Fig. 115

In each of the diagrams *A*, *B*, *C*, *D*, the number 23 can be only in the upper left-hand corner or in the lower right-hand corner, flanked by 8 and 14 (since 23 cannot be adjacent to any of the numbers 12, 24, 2, and 20). If we place 2 and 20 in the upper left-hand part, then we cannot fit in the number 5 anywhere. Hence, 2 and 20 should be placed in the lower right-hand part, which requires 23, 8 and 14 in the upper left-hand part, and, moreover, places 8 on the top row because otherwise we cannot place 17 anywhere; 17 is then placed in the left-hand row, and 11 in the top row. We have to place 3 next to 20 and 15 next to 2 (against the edge), hence, 9 and 20 have to be on the diagonal. After this, we reach an impasse in diagrams *A* and *B*. In diagram *D*, we cannot place both the numbers 4 and 22, so that only diagram *C* remains. In this diagram, we have to place 20 and 3 in the bottom row (because otherwise we cannot place 22 anywhere), which requires

2 and 15 against the right-hand edge. We can then fill in the numbers 4, 22, 10, 16, 21 and 9 one after another in one way only. We thus obtain the two solutions shown in *Figure 115*, which arise from each other by simultaneously interchanging 7 with 13, and 18 with 12.

249. Puzzle with larger prime numbers. In *Figure 116*, the nine dots have to be replaced by digits in such a way that the horizontal rows contain three numbers which are prime, when read both from left to right and from right to left. Vertically, too, we have to obtain three prime numbers reading both downwards and upwards. Also, for each of the ten diagonal lines, the dots (or dot) on such a line should produce a prime number in either direction.

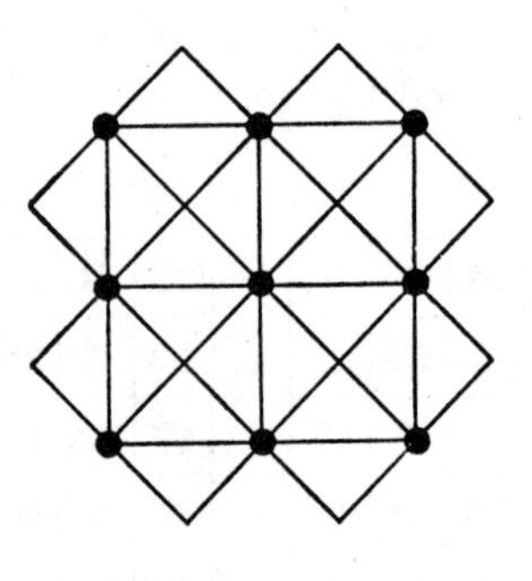

Fig. 116

The digits in the four corners must be prime numbers (consisting of a single digit) but not 2 or 5, or 3-digit numbers divisible by 2 or 5 would result. Hence, in a corner of the square there can only be a 3 or a 7. In the middle of a side of the square, we can only have one of the digits 1, 3, 7, 9; anything else would produce a 3-digit number divisible by 2 or 5. Hence, along a side we must have a 3-digit prime number which is also prime when read from right to left, which has 3 or 7 for its first and for its last digit, and a middle digit which is 1, 3, 7, or 9. The only prime numbers that satisfy these conditions are 313, 337 (or its reverse, 733), 373, and 797. Of these, 337, 733, and 797 should be disregarded, because they would produce a 2-digit number divisible by 3 on a diagonal line. From 313 and 373 we then find the following five solutions:

313	313	313	313	313
151	181	151	757	787
313	313	373	313	313

By rotation, we can write down the third solution in four variants, and the fourth and fifth solutions in two variants each, so that we can also speak of ten solutions. The third solution yields the largest number of different prime numbers (ten, in fact).

VII. REMARKABLE DIVISIBILITY

250. Divisibility of numbers in a rectangle. We imagine a rectangle to be divided into 3×4 small squares, and put a digit in each square. In *Figure 117* these digits have been indicated by a, b, $c, \ldots, l$. The number formed by the digits a, b, c, d will be called A; the number formed by e, f, g, h will be called B, and so on. The number formed by the digits a, e, i (read downwards) will be called D, the number formed by b, f, j will be called E, and so on. Then we have: $A = 10^3 \times a + 10^2 \times b + 10 \times c + d$, and so on; $D = 10^2 \times a + 10 \times e + i$, and so on.

a	b	c	d	A
e	f	g	h	B
i	j	k	l	C
D	E	F	G	

Fig. 117

From this it follows that we have:

$$10^3 \times A + 10 \times B + C = 10^5 \times a + 10^4 \times (b + e) + 10^3 \times (c + f + i) + 10^2 \times (d + g + j) + 10 \times (h + k) + l,$$

with a corresponding result for $10^3 \times D + 10^2 \times E + 10 \times F + G$. Hence we have:

$$10^2 \times A + 10 \times B + C = 10^3 \times D + 10^2 \times E + 10 \times F + G$$

So if the seven numbers A, B, C, D, E, F, G contain six numbers, for instance A, C, D, E, F, G, divisible by 7, say, then it follows that $10 \times B$ is divisible by 7, and that the same is true of B, too. So if all but one of the numbers obtained by reading from left to right or downwards are known to be divisible by 7, then the last number must also be divisible by 7.

This property can be extended to larger rectangles, and to divisibility by 13, or by 17, or in general, by any number that is relatively prime to 10, which means that it must end in 1, 3, 7, or 9. The same property also holds for a figure of arbitrary shape, as long as it consists of equal squares that fit together to form horizontal rows and vertical columns, since by placing zeros before and after the numbers (which has no influence on the divisibility by 7 or 13, and so on), we can reduce this case to that of numbers arranged in a rectangle.

251. Puzzle with multiples of 7. In *Figure 118*, we have to rearrange the 20 digits in such a way as to produce thirteen numbers divisible by 7 when we read the numbers from left to right and downwards; the four digits in the bottom row are to be interpreted as two numbers having two digits each (and not as a single four-digit number). The top number, read horizontally, is a 1-digit number, and hence must be replaced by 7, or by 0 if we admit numbers with 0 as the initial digit.

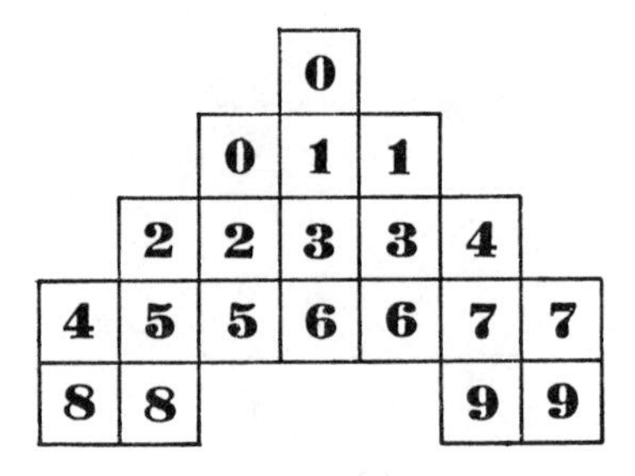

Fig. 118

According to §129, we can arrange the 20 digits in

$$\frac{20!}{(2!)^{10}} = 20^4 \cdot 3^8 \cdot 7^2 \cdot 11 \cdot 13 \cdot 17 \cdot 19 = 2{,}375{,}880{,}867{,}360{,}000$$

ways, that is, in more than 2×10^{15} ways. It is therefore completely impossible to consider all these arrangements individually.

We can simplify the puzzle by noting that divisibility by 7 is not lost when we replace a 7 by a 0, an 8 by a 1, and a 9 by a 2. By doing this we obtain four digits 0, four digits 1, four digits 2, two digits 3, two digits 4, two digits 5, and two digits 6. When we have found a solution with these reduced numbers, it continues to be a solution when we replace two of the four zeros by 7, which can be done in $\frac{4!}{2! \times 2!} = 6$ ways. We also retain a solution when we introduce digits

8 to replace two of the four digits 1, and digits 9 to replace four digits 2. Hence, each solution with the reduced numbers leads to 6 × 6 × 6 = 216 solutions, so that reducing the numbers produces a considerable simplification. Yet this simplification is not so great that we can solve the puzzle by rearranging the reduced numbers in all possible ways, which are still over 10^{13} in number.

In solving the puzzle, we can make advantageous use of the property proved in §250. If the two lower 2-digit numbers are both divisible by 7, then so is the number formed by the lower four digits. This shows that we need not occupy ourselves with the divisibility by 7 of the large 7-digit number; if it has been settled that the thirteen numbers other than this largest number are divisible by 7, then so is the 7-digit number. This, too, produces a considerable simplification.

Among the numbers to be checked for divisibility by 7, there are five numbers having two digits or less and six three-digit numbers (since the middle vertical number begins with 0 or 7—this initial digit can be omitted when divisibility by 7 is being considered) and, furthermore, one 5-digit number (since the 7-digit number has been removed from consideration). To solve the puzzle, it is convenient to write down all three-digit numbers divisible by 7, using reduced digits. These are the following numbers:

000	014	021	035	042	056	063	105	112	126
133	140	154	161	203	210	224	231	245	252
266	301	315	322	336	343	350	364	406	413
420	434	441	455	462	504	511	525	532	546
553	560	602	616	623	630	644	651	665	

252. Multiples of 7 puzzle with the largest sum. The number of solutions of the puzzle with the multiples of 7, from §251, is very large even after reducing the digits. In order to limit this number we require the solution for which the sum of the thirteen numbers is as large as possible. With this aim we look for a solution in which the largest number begins with 99—hence, after diminishing the digits, with 22. The left-hand vertical number is then 21, the lowest number at the left is 14, and the second vertical number 224; hence, the third horizontal number begins with a 2, even when the digits have not been reduced. As the third digit of the largest number we take 8, which becomes 1 after the digits are reduced. The third vertical number from the left ends in 1 (after reduction of digits), and is hence equal to 021, 161, 231, 301, 511, or 651 (because the digit 4 cannot occur twice in this number). The third horizontal number is made largest by taking 511 as the third vertical number and changing each of the two numbers 1 into an 8. In that case the two equal numbers to the right and below are 00, 35, or 63. This gives the following eleven possibilities, when we also fill in the second horizontal number (the three-digit number):

0	0	0	0	0	0
525	532	546	553	504	546
21346	21635	21125	21432	21200	21...
2211630	2214160	2213360	2216610	2211663	221...3
14 00	14 00	14 00	14 00	14 35	14 35

0	0	0	0	0
560	504	525	532	553
21010	21...	21...	21514	21004
2212463	221...6	221...6	2210006	2214106
14 35	14 63	14 63	14 63	14 63

In three of the eleven cases we reach an impasse when filling in the three remaining vertical multiples of 7 (in such a way that each digit occurs the proper number of times); in these cases none of those three vertical numbers has been filled in. They have been filled in for the remaining eight cases. However, the third horizontal number (the 5-digit number) cannot be a multiple of 7; this can be seen at once from the initial digits 21 (in some cases in combination with the list of numbers in §251, which gives the three-digit multiples of 7).

Next we try 301 as the third vertical number (to represent 378 in non-reduced digits), which makes the second digit of the five-digit number attain its largest value but one. The two equal numbers, lower right, are then 14, 56, or 63. We can now fill in the second horizontal number in different ways, which turn out to be nine in number. In all cases but one, the three vertical numbers can be filled in as multiples of 7, in some cases in two ways, so that the number of cases rises to the twelve following possibilities:

0	0	0	0	0	0
336	350	301	301	336	343
20506	20652	20120	20460	20...	20601
2210251	2210631	2214635	2212135	221...5	2212105
14 14	14 14	14 56	14 56	14 56	14 56

0	0	0	0	0	0
364	301	301	315	315	350
20011	20205	20250	20001	20100	20140
2212305	2211546	2211456	2215426	2212456	2211256
14 56	14 63	14 63	14 63	14 63	14 63

Only in the sixth case is the third horizontal number (the five-digit one) divisible by 7. This can be seen quickly from the three-digit multiples of 7 listed in §251; since 203 and 301 are divisible by 7, so is 20601. Hence the fourth horizontal number (the seven-digit number) is also divisible by 7 (see §251).

Thus we find two solutions (*Fig. 119*) for which the sum of the thirteen multiples of 7 will turn out to be as large as possible (see §253), with a value of 10,019,520. The fact that the interchange of the digit 1 at the right and the digit 8 at the right does not produce a change in the sum of the 13 numbers is a consequence of the fact that each of these digits is the units digit in one of the thirteen numbers, and also the hundreds digit in another number.

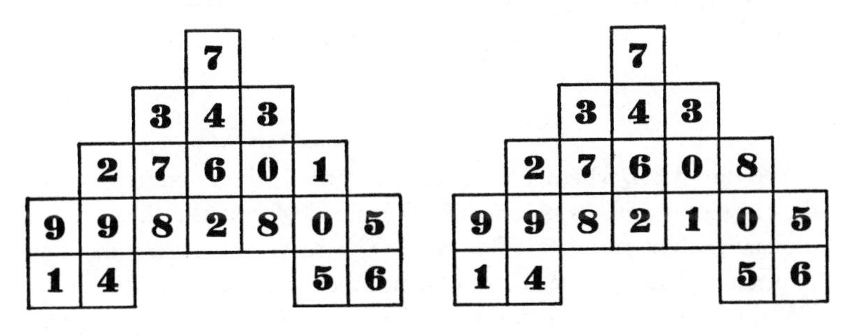

Fig. 119

253. Proof that the solutions found do in fact yield the largest sum. With the solutions found in §252, the unfavorable circumstance arises that the fourth digit of the largest number turns out to be as small as possible, which makes the sum of the two digits in the thousands place only 9, for the horizontal numbers in these solutions. Since the sum of the hundreds digits is 30 for these solutions, while this sum can in no case be made higher than 46 by making a different choice for the thousands digits, we need not examine cases other than those in which the sum of the thousands digits of the horizontal numbers is at least 8.

Hence, we still have to consider the cases in which the third vertical number is 161, 651, 231, or 021. This will be done in the manner indicated in §252. The number 161 gives no solution at all (we again have 221 as initial digits for the seven-digit number), while 651 yields only the two solutions in *Figure 120.* For these, the sum of the thirteen numbers is 10,019,205, which is less than for the solutions found in

§252, as was to be expected in connection with the smaller sum of the thousands digits of the horizontal numbers (8 instead of 9). With 231 or 021 as the third vertical number, it is profitable to note that the fourth digit of the seven-digit number should not be taken to be less than 5 or 6, respectively. This does not lead to any solutions.

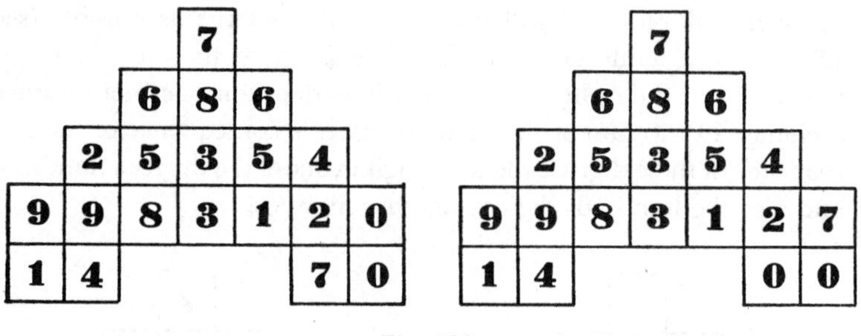

Fig. 120

We might ask whether we could not obtain the maximum sum of the thirteen numbers by choosing 7 instead of 8 as the third digit of the seven digit number. This is not the case, since the sum of the thousands digits of the horizontal numbers is 9 (in the solutions of §252), whereas that sum can never be larger than 8 + 8 = 16 (when the seven-digit number begins with 99).

254. Multiples of 7 with the maximum product. The puzzle requires considerably more work when we do not want the sum, but the product, of the thirteen multiples of 7 to be as large as possible. Then we must pay attention to the initial digits of all numbers, because one low initial digit, even for one of the small numbers, makes a drastic reduction of the product in question. We should pay special

			7			
		9	3	8		
	5	2	0	1	7	
9	0	4	1	2	5	6
8	4				6	3

Fig. 121

care to the initial digits of the horizontal numbers which have 3, 5, or 7 digits, because each of these initial digits is also the initial digit of a vertical number; hence, these three digits in particular should be chosen high (or at least not too low).

Since interchanging two digits (0 and 7, or 1 and 8, or 2 and 9) has a much more radical effect on the product of the thirteen numbers than it has on the sum, it can be stated beforehand that we will find only one solution with the largest product. After considerable work, the solution turns out to be the one in *Figure 121.* The product of the thirteen numbers is over $210{,}583 \times 10^{33}$.

The puzzle with the largest product has been borrowed from the investigations of Dr. S. Kirederf, the well-known Egyptologist and mathematician at Eugaheht, who revealed the secret of a stone found in the picturesque step pyramid at Saqqara, which was built by the famous architect Imhotep by order of the great king Zoser of the Third Dynasty in the twenty-ninth century B.C. The red granite stone, inlaid with white alabaster, gives a symbolic representation of

Fig. 122

the step pyramid; the hieroglyphics on it mean: "Secret of Zoser, King of Upper and Lower Egypt." Since it is known that Zoser had a fervent admiration for mathematics, the inscription and the occurrence of a harp player amid the decoration representing the harmony of mathematics led Dr. Kirederf to conjecture that the mysterious symbols in the 20 squares, which represent the stone blocks, have a mathematical meaning. After years of prolonged study, it appeared to him in 1932 that we have here the original forms of our digital symbols, grouped according to the supposed solution of the puzzle of the largest product, as discussed above. This sensational discovery also showed that our decimal system was already known to the ancient Egyptians, albeit to some adepts only!

By Dr. Kirederf's kind permission, we are able to present a photographic reproduction of the Saqqara stone (*Fig. 122*). It shows that the solution of *Figure 123* has been immortalized on the stone. In this solution, the product of the thirteen multiples is less than $205{,}240 \times 10^{33}$, hence less than in the correct solution given earlier. Dr. Kirederf has not been able to determine whether Imhotep intentionally depicted the largest product but one on the stone, or whether he overlooked something.

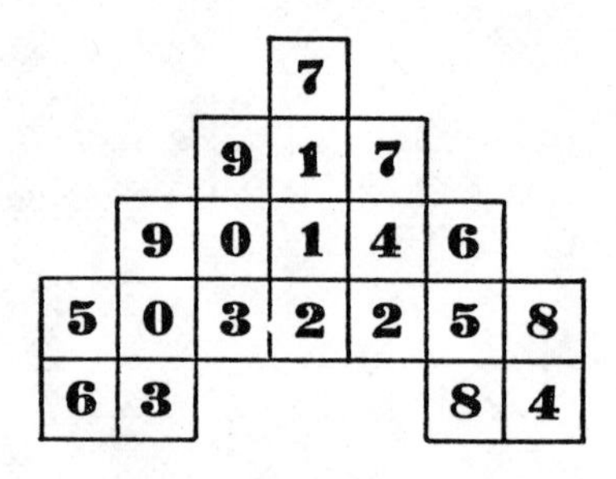

Fig. 123

VIII. MULTIPLICATION AND DIVISION PUZZLES

255. Multiplication puzzle "Est modus in rebus." In the multiplication below, the letters and the dots have to be replaced by digits. Similar letters represent similar digits, but the possibility that two different letters are replaced by the same digit is not excluded. Each of the ten digits corresponds to at least one letter, so there is precisely one case where two different letters are replaced by the same digit. We require the determination of all digits.

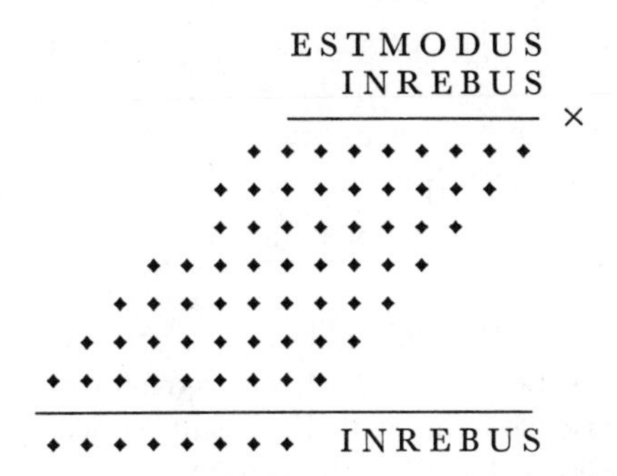

The message "est modus in rebus" which is formed by the multiplicand and the multiplier is the best-known part of the quotation following the preface of this book. We hope that in this puzzle we have not exceeded the limits mentioned in the quotation, although we must admit that the puzzle is far from easy. A perverse translation of the phrase which would suit this book is, "Rebuses are in fashion."

But now down to business. We shall call the multiplicand (the 8-digit number) X, and the multiplier Y. Since there are seven partial products, the digits of Y are all different from 0. The number formed by the last two digits of X (or of Y) will be called z. The product $X \times Y$ has the same last two digits as z^2. According to the letters that occur in the multiplication, z^2 has the same last two digits as z, so that $z^2 - z$, that is, $z(z - 1)$, must be divisible by 100. From this it follows that either z is divisible by 4 and $(z - 1)$ by 25, or z by 25 and $(z - 1)$ by 4; in the first case $z = 76$, in the second case $z = 25$. Since Y and $X \times Y$ have the last seven digits in common, $X \times Y - Y$, that is, $Y \times (X - 1)$, must be divisible by 10^7. The number formed by the first six digits of X will be called x; the number formed by the first five digits of Y will be called y; we then have $X = (100 \times x) + z$ and $Y = (100 \times y) + z$. From the locations of letters E in the multiplicand and in the multiplier, the last digit but one of y is equal to the initial digit of x. From the numbers of digits of the partial products it appears that the last digit of y is smaller than the other digits of y (and also smaller than the digits of z), that the product of the last two digits of y (that is, $E \times B$) is less than 10, and that E (the last digit but one of y) is greater than 2; hence the last digit of y must be either 1 or 2.

The case $z = 76$. In this case, Y is not divisible by 5, and $(X - 1)$ is not divisible by 2; hence Y must be divisible by $2^7 = 128$, and $(X - 1)$ by $5^7 = 78{,}125$. Since $Y = (100 \times y) + 76$ and $(X - 1) =$

$(100 \times x) + 75$, it follows that $(25 \times y) + 19$ must be divisible by 32, and $(4 \times x) + 3$ can be written as $3125 \times t$, where t is a multiple of 4 plus 3 (because 3125 is a multiple of 4 plus 1), so we can put $t = (4 \times v) + 3$, which leads to:

$$x = (3125 \times v) + \frac{(3125 \times 3) - 3}{4} = 3125 \times v + 2343.$$

Since $(25 \times y) + 19$ is divisible by 32, and hence by 4, we find that $(y - 1)$ must be divisible by 4, which makes y a multiple of 4 plus 1. When we associate this with the results concerning the last two digits of y, it follows that these digits must here be 41, 61, or 81. The multiplicand X is then 46 . . . 76 or 66 . . . 76 or 86 . . . 76. In the first case, x (being a multiple of 3125 plus 2343) is one of the numbers 461718, 464843, 467968, in the second case one of the numbers 661718, 664843, 667968, and in the third case one of the numbers 861718, 864843, 867968. In all these cases, too many equal digits occur in X, so that X and Y together would contain less than ten different digits; it may also be noted that in all the numbers found for X the digit 0 is missing, while 0 cannot occur in Y. Hence, the case $z = 76$ does not lead to any solutions.

The case $z = 25$. In this case, Y is not divisible by 2 and $(X - 1)$ is not divisible by 5; hence Y must be divisible by 5^7 and $(X - 1)$ by 2^7. From $Y = (100 \times y) + 25$ and $X - 1 = (100 \times x) + 24$ it follows that $(4 \times y) + 1$ is divisible by 3125 and $(25 \times x) + 6$ is divisible by 32. From the latter fact we conclude that x is even, after which it follows (by trying out $x = 2, 4$, and so on) that x is a multiple of 32 plus 10. According to these results, the number $(4 \times y) + 1$ can be written as $3125 \times w$, where w is a multiple of 4 plus 1, so we can put $w = (4 \times u) + 1$; as a consequence we have:

$$y = (3125 \times u) + \frac{3125 - 1}{4} = (3125 \times u) + 781.$$

Since the last digit of y must be 1 or 2, u must be even, which makes y a multiple of 6250 plus 781. Since Y cannot contain the digit 0 or two pairs of equal digits (we may also save time by noticing that Y cannot here begin with 1), the numbers 6328125, 6953125, and 9453125 become the only numbers which deserve consideration for Y. In the last two cases, X begins with 3, which is incompatible with the number of digits in the second partial product. It follows that we must have $\mathbf{Y} = \mathbf{6328125}$ and $X = 85 \ldots 25$; here the dots represent the digits 0, 4, 7, 9 in some order. Since x is a multiple of 32 plus 10, the last digit of x must be even, hence 0 or 4, and the last digit but one of x must then be odd, hence 7 or 9. Thus, only eight possibilities for x deserve consideration, only one of which turns out to be a multiple of 32 plus 10. This gives X = **85079425.**

256. Multiplication and division puzzle. The working of a multiplication sum is indicated below by dots, together with a check produced by dividing by the multiplier B, to give the multiplicand A as the resulting quotient. The digits of the division, too, are all represented by dots. It is further given that the product is a multiple of 9 plus 2. The problem is to fill in all the digits.

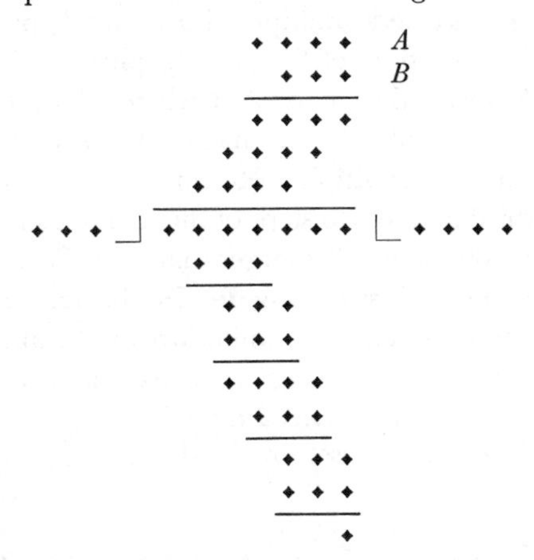

The layout of the multiplication and division shows that the digits of B and of A are all different from 0. The products used in the first and third steps of the division must begin with 9. Hence the division shows that the first and the third digits of the quotient A are equal, that the second digit of A is smaller than this, and that the fourth digit of A is not greater than this; this means that the first digit of A is at least 2. The seven digits of the product $A \times B$ show that the third partial product begins with a 9. Hence, the first digit of A is not a 2, since in that case A would be at least 2121 and at most 2122; the first digit of A is not equal to 4, either, since in that case A would be at least 4141 and at most 4344; and in these cases, A could not provide a partial product beginning with 9. For the same reason, A does not begin with 5, 6, 7, or 8. If A begins with 9, then B = 111, and is thus a multiple of 3; therefore the same is true of $A \times B$, and this is in contradiction to the fact that $A \times B$ is a multiple of 9 plus 2. So the first digit and the third digit of A must be equal to 3, while the second digit must be 1 or 2, and the fourth digit must be 1, 2, or 3. The first digit of B must be 3. According to the multiplication (in which the partial products are four-digit numbers) the second digit and the third digit of B are at most 3. Since $A \times B$ is a multiple of 9 plus 2, we have the following possibilities (9m − 4 stands for "multiple of 9 minus 4," and so on):

$$\begin{array}{lllllll} A: & 9m-4, & 9m-2, & 9m-1, & 9m+1, & 9m+2, & 9m+4, \\ B: & 9m+4, & 9m-1, & 9m-2, & 9m+2, & 9m+1, & 9m-4. \end{array}$$

From the digits of the number A (which is not a multiple of 3) we conclude that A must be a multiple of 9 minus 1, plus 1, or plus 2, which makes B a multiple of 9 minus 2, plus 2, or plus 1, in the respective cases. From the digits of B we conclude that B is not a multiple of 9 plus 2 or plus 1. This means that B must be a multiple of 9 minus 2, and A a multiple of 9 minus 1; this gives $A = 3131$. After this, the third and fourth steps of the division show that $31 \times B$ is larger than 10,000, so that B is larger than 322. Therefore, $B = 331$.

***257. Terminating division puzzle.** In the terminating division shown below, the dots represent unknown digits, and in the entire division the locations of only seven digits 7 are known. However, it is also possible that a dot represents a digit 7.

We shall indicate the divisor by D, the quotient by Q, the digits

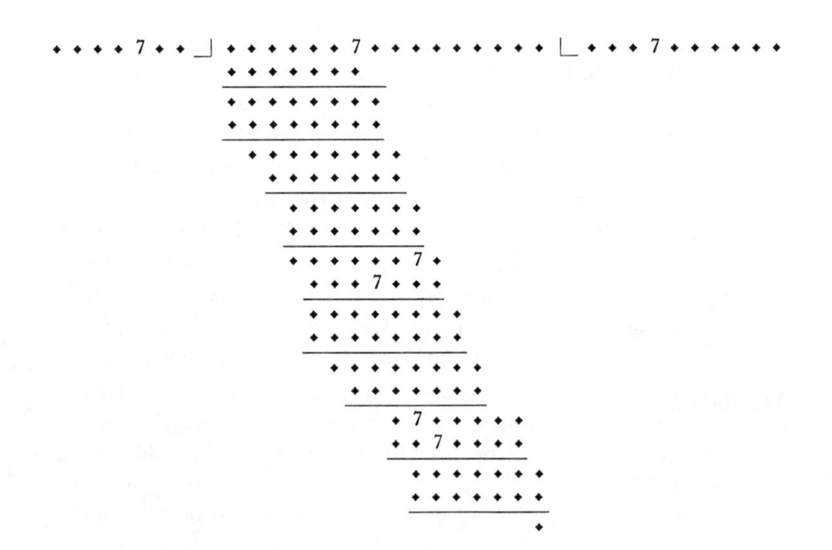

of the divisor by $d_1, d_2, \ldots, d_7$, and those of the quotient by q_1, $q_2, \ldots, q_{10}$. Hence, it is given that $q_4 = 7$, while it is clear at once that $q_8 = 0$. The digits q_3 and q_7 must both be larger than q_4 (hence larger than 7) and less than q_2 and q_6; consequently, $q_3 = q_7 = 8$ and $q_2 = q_6 = 9$. From $q_7 = 8$ it follows that $8 \times D$ is less than 10,000,000 and at least equal to $10{,}000{,}000 - 97{,}999 = 9{,}902{,}001$; hence D must be less than 1,250,000 and greater than 1,237,750, so that $\mathbf{d_1 = 1}$, $\mathbf{d_2 = 2}$, and $d_3 = 3$ or 4. From this it follows further that $q_5 = 8$. Since the fourth digit of $q_5 \times D$, thus of $8 \times D$, is a 7, this shows that $d_4 = 4$ or 9, and that d_6 is at most 4. The assumption that $d_3 = 3$ leads to $d_4 = 9$ (because D is greater than 1,237,750), from which it follows (in connection with the third digit of $q_9 \times D$ being 7) that we must have $q_9 = 2$ or 7. From the thirteenth row of the division sum it is evident that $(800 + q_9 + 1) \times D$ is a ten-digit number. However, $803 \times d_6d_7$ is less than $803 \times 1{,}240{,}000$, hence less than 995,720,000, so that $q_9 = 2$ drops out, and only $q_9 = 7$ remains to be examined. The second digit of the product obtained when $7q_{10}$ (that is, the number written with the digits 7 and q_{10}) is multiplied by D is a 7; but we have

$$7q_{10} \times 1{,}239{,}7d_6d_7 = 86{,}779{,}000 + (q_{10} \times 1{,}239{,}7d_6d_7) + (70 \times d_6d_7);$$

the second digit of this number is not 7 for any of the possible values $1, 2, \ldots, 8$ of q_{10}, so that $q_9 = 7$ is not possible, either, which makes $d_3 = 3$ drop out. So we must have $\mathbf{d_3 = 4}$, and $D = 1{,}24d_4, 7d_6d_7$ (where $d_4 = 4$ or 9). The third digit of $q_9 \times 1{,}249{,}7d_6d_7$ is not a 7 for any of the possible values of q_9, so we must have $\mathbf{d_4 = 4}$. From the fact that the third digit of $q_9 \times 1{,}244{,}7d_6d_7$ is a 7 it follows that $q_9 = 4$. The second digit of

$$4q_{10} \times 1{,}244{,}7d_6d_7 = 49{,}788{,}000 + (q_{10} \times 1{,}244{,}7d_6d_7) + (40 \times d_6d_7)$$

is a 7, from which it follows that $q_{10} = 6$. The seventh digit of $898{,}046 \times 1{,}244{,}7d_6d_7 = 1{,}117{,}797{,}856{,}200 + 898{,}046 \times d_6d_7$ is a 7, from which we can deduce (since d_6d_7 is less than 50) that $d_6d_7 = k \times 11$, where $k = 0, 1, 2, 3$, or 4. Hence, for the dividend $Q \times D$ we find:

$$q_1, 987{,}898{,}046 \times 1{,}244{,}7d_6d_7 = (q_1 \times 1{,}244{,}700{,}000{,}000{,}000) + 1{,}229{,}636{,}697{,}856{,}200 + (k \times 10{,}866{,}878{,}605) + (q_1 \times k \times 11{,}000{,}000{,}000).$$

The seventh digit of this number is the last digit of $6 + k \times (q_1 + 1)$. This has to be a 7, by the terms of the problem, so that $k \times (q_1 + 1)$ must end in 1. In connection with $k = 0, 1, 2, 3$, or 4 (keeping in mind that q_1 is different from 0), it follows that $k = 3$, hence $\mathbf{d_6} = \mathbf{d_7} = \mathbf{3}$, and $q_1 = 6$. Consequently

$$D = 1{,}244{,}733, \qquad Q = 6{,}987{,}898{,}046.$$

The dividend is then found as the product

$$Q \times D = 8{,}698{,}067{,}298{,}491{,}718.$$

All data are satisfied, as is confirmed by performing the division.

***258. Repeating division puzzle.** The division sum below obtains the expansion of a common fraction, reduced to its lowest terms, as a repeating decimal fraction. The dots represent the digits; these are all unknown. The repeating digits have been indicated by a line over their dots. The problem is to determine all the digits.

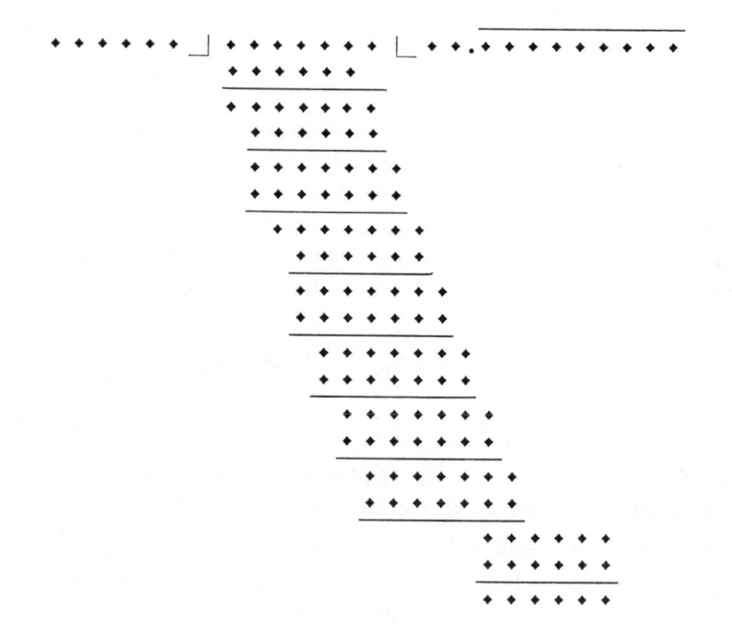

Multiplication of the result by 10 produces a completely repeating fraction which has nine repeating digits. The denominator of the fraction (reduced to its lowest terms) that is equal to this pure repeating fraction must be a divisor of

$$10^9 - 1 = (10^3 - 1) \times (10^6 + 10^3 + 1)$$
$$= 9 \times 111 \times 1{,}001{,}001 = 3^4 \times 37 \times 333{,}667;$$

here 37 and 333,667 are prime numbers. Since that denominator must arise from the original denominator by dividing it by 2, 5, or 10, the new denominator has 5 or 6 digits and must therefore be equal to 333,667. This means that the *denominator* of the original common fraction must have been 2 × 333,667 = **667,334**. If the digits which follow the decimal point of the repeating fraction are denoted by $q_1, q_2, \ldots$, then the repeating fraction is $11.q_1 1 q_3 q_4 q_5 q_6 0001$. In the problem, the third number from the bottom has the form ..0000. If we take the number formed by the first two digits and divide it by 667,334, then we obtain the following expansion of a completely repeating fraction:

```
667334 | ◆ ◆                  | 0.00011q3q4q5q6  (repeating)
         6 6 7 3 3 4
         -----------
         ◆ ◆ 2 6 6 6 0
             6 6 7 3 3 4
             -----------
             ◆ 5 9 3 2 6 0
             ◆ ◆ ◆ ◆ ◆ ◆ ◆
             -------------
               ◆ ◆ ◆ ◆ ◆ ◆ 0
               ◆ ◆ ◆ ◆ ◆ ◆ ◆
               -------------
                 ◆ ◆ ◆ ◆ ◆ ◆ 0
                 ◆ ◆ ◆ ◆ ◆ ◆ ◆
                 -------------
                   ◆ ◆ ◆ ◆ ◆ ◆ 0
                   ◆ ◆ ◆ ◆ ◆ ◆ ◆
                   -------------
                             ◆ ◆
```

The first digit of the number in the fifth row of this last division sum is at least 3, so that q_3 must be at least 5. Hence, the dividend of this division sum must be greater than 667,334 × 0.000115 = 76.74341, and less than 667,334 × 0.00012 = 80.08008; since the dividend (still of the last division sum) is an integer and must be even, it must be 78 or 80. Since 80 leads to the repeating fraction $0.\overline{000119880}$ and q_6 is not equal to 0, the dividend cannot be 80, and therefore it must be 78. With this, the digits of the last division sum have all been found; in particular, the number in the third row is shown to be 112660. For the original division sum we now know:

667334⌋ ◆ ◆ ◆ ◆ ◆ ◆ ◆ ⌊11.$q_1$1 etc.
6 6 7 3 3 4
◆ ◆ ◆ ◆ ◆ ◆ ◆
6 6 7 3 3 4
◆ ◆ ◆ ◆ ◆ ◆ 0
◆ ◆ ◆ ◆ ◆ ◆ ◆
1 1 2 6 6 6 0

So the number in the sixth row must end in 4, which is possible only when $q_1 = 6$ (since q_1 is greater than 1). After this, the division sum can be completed as follows:

667334⌋ 7752341 ⌊11.61 etc.
667334
1079001
667334
4116670
4004004
1126660

and we find **7,752,341** for the numerator of the common fraction. The mixed repeating fraction is $11.\overline{6168830001}$.

IX. DICE PUZZLES

259. Symmetries of a cube. A cube is standing on a table (*Fig. 124*). We move the cube in such a way that it takes up the same position as before, without requiring that every point of the cube, considered as a material body (for example, a block from a child's set, covered with colored pictures), has to be in its original position. In how many ways can we set down the cube?

The answer is quite simple. We can choose any of the six faces as the base. When we have made a choice as to this, we can still give the

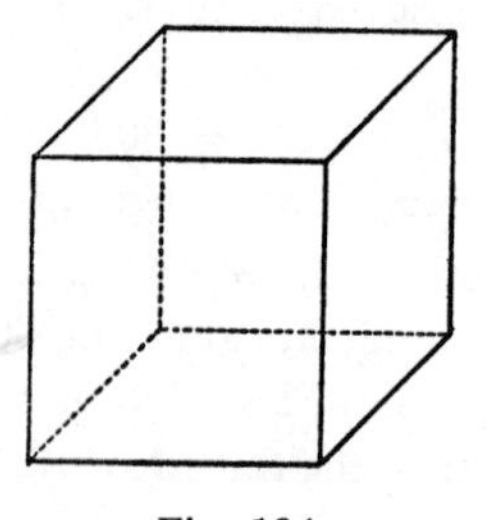

Fig. 124

cube 90° rotations around the line which joins the centers of the base and the top, so that any of four lateral faces can be the front. This gives $6 \times 4 = 24$ ways in all to set down the cube.

This can be done in more than one way because of the fact that the cube possesses various axes of symmetry. A body is said to have an axis of symmetry, or a symmetry-axis, if there is a straight line with the property that the body can be rotated around that line by an angle less than 360°, to bring it into coincidence with its original position.

The smallest angle for which this occurs is the result of dividing 360° by a whole number to obtain, for example, 120° or 90°; we then speak of a symmetry-axis of order 3 or 4, respectively; with a symmetry-axis of order 4, say, the body will allow a rotation by 90° or 180° in either direction. We can now pose the problem of determining all the symmetry-axes for a cube.

The cube has four symmetry-axes of order 3, formed by its diagonals (the lines which connect opposite vertices); three axes of order 4, the lines which connect the centers of opposite faces; and, further, six axes of order 2, the lines which connect the midpoints of opposite edges. This, too, leads to the number of ways found above; by a rotation around an axis of symmetry of order 3, 4, or 2 we can bring the cube into 2, 3, or 1 new situations, respectively, for which the position in space is the same. In all, this yields $4 \times 2 + 3 \times 3 + 6 \times 1 = 23$ new positions. Together with the original position, we thus get 24 situations for the cube from the child's set, such that it occupies the same position in space.

***260. Group of symmetries.** If we successively rotate the cube around two of its axes of symmetry (by appropriate angles), its final position in space is the same as at the start. So the final result could also have been achieved by a rotation around a third axis of symmetry. We express this by saying that the rotations form a group; we also speak of a group of symmetry-axes.

Among the symmetry-axes, there are some that form a smaller group, a so-called subgroup. Two successive rotations around symmetry-axes of that subgroup (again by suitable angles) will jointly amount to a rotation around a symmetry-axis that also belongs to the subgroup. Each symmetry-axis separately can be considered as a subgroup; also a symmetry-axis of order 4 can be used as an axis of order 2, to provide a subgroup.

The foregoing shows that the existence of two symmetry-axes gives an assurance that other symmetry-axes exist. Thus, from an axis l of order 4, and an axis of order 2 which cuts l perpendicularly at a point O, it follows that there are three more axes of order 2 which are also perpendicular to l at O, and which mutually include angles of 45°. Conversely, if there are two symmetry-axes of order 2 which intersect at a point O at an angle of 45°, the line through O perpendicular to both axes is a symmetry-axis of order 4, and there are two more symmetry-axes of order 2 through O (perpendicular to the axis of order 4). These five symmetry axes form a group, a subgroup of all the symmetry-axes of the cube. A smaller subgroup (a subgroup of the subgroup) is formed by three symmetry-axes of order 2 that have one point in common and are pairwise perpendicular; any two of these symmetry-axes imply the presence of the third one.

***261. Symmetries of the regular octahedron.** A regular octahedron (bounded by eight equilateral triangles) has the same group of symmetry axes as the cube, and so has four axes of order 3, three of order 4, and six of order 2 (*Fig. 125*). We leave it to the reader to

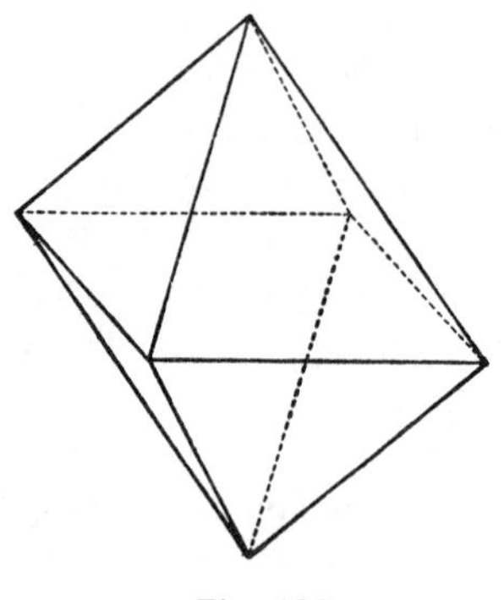

Fig. 125

identify the positions of these symmetry-axes in the body.

The correspondence of the symmetry-axes of the two polyhedra is also exhibited by the fact that the centers of the six faces of a cube are the vertices of a regular octahedron, while the centers of the faces of a regular octahedron form the eight vertices of a cube. We can give such dimensions to the cube and the octahedron that we can imagine the two bodies (interpenetrating each other) to be positioned in such a way that each of the twelve edges of the cube is a perpendicular bisector of some one of the twelve edges of the octahedron.

262. Eight dice joined to make a cube. We have eight completely identical dice, which have the property that the face with 6 spots is opposite the face with 1 spot, with 2 and 5 also opposite each other, and similarly for 3 and 4 (*Fig. 126*). The eight dice are glued together to form a cube which has an edge twice as long as that of each of the dice. We require the number of ways in which this can be done. Interchanging the dice, and at the same time preserving the positioning of the spots, is not to be considered as a different assembly (since the dice are completely identical). If the cube is rotated in such a way that it resumes its position with an interchange of faces, then we do consider this as a different form of assembly.

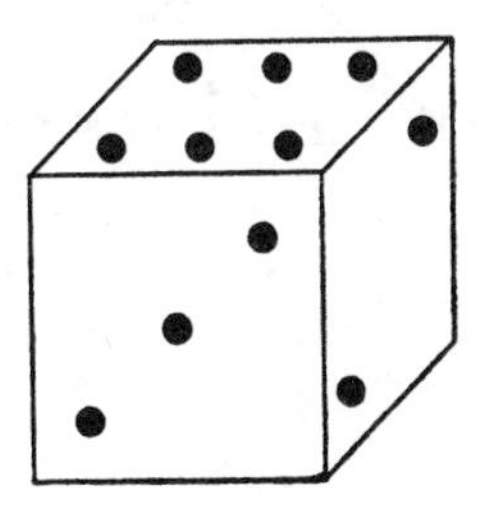

Fig. 126

We can put down a die in 24 ways such that every time it has the same position considered as a body, although the spots are differently placed (see §259). This holds for each of the eight dice, which shows the number of assemblies to be 24^8 = **110,075,314,176**, that is, over 110 billion (in the American sense).

***263. More difficult puzzle with eight dice.** The puzzle of §262 becomes more attractive, but also substantially harder, when we consider solutions to be identical if they arise from each other through a rotation of the cube. Since we can put down the cube 24 ways, the required number would be 24 times as small as the number of §262, provided that no symmetrical cubes existed, that is, cubes that are transformed into themselves by rotation, even when regard is also paid to the spot numbers (which means that any question of reflectional symmetry with respect to a plane can here be disregarded). There are six types of cubes which show symmetry of this kind, as shown in *Figure 127*.

Each of the six types is characterized by its symmetry-axes, and thus by the relevant subgroup of the group of the cube's symmetry-axes (see §260). Since the three spot numbers in a corner are always different, symmetry-axes of order 3 are never present. In regard to the invisible spot numbers in the figures, the reader is referred to §264.

TYPE I (the one with the highest symmetry): In the position shown, the line joining the centers of the front and the back is a symmetry-axis of order 4. The line joining the centers of the top and the base, the line joining the centers of the left-hand and the right-hand faces, and the two lines joining the centers of opposite edges perpendicular to the front and back are symmetry-axes of order 2; these four lines are perpendicular to the symmetry-axis of order 4. We can set down the

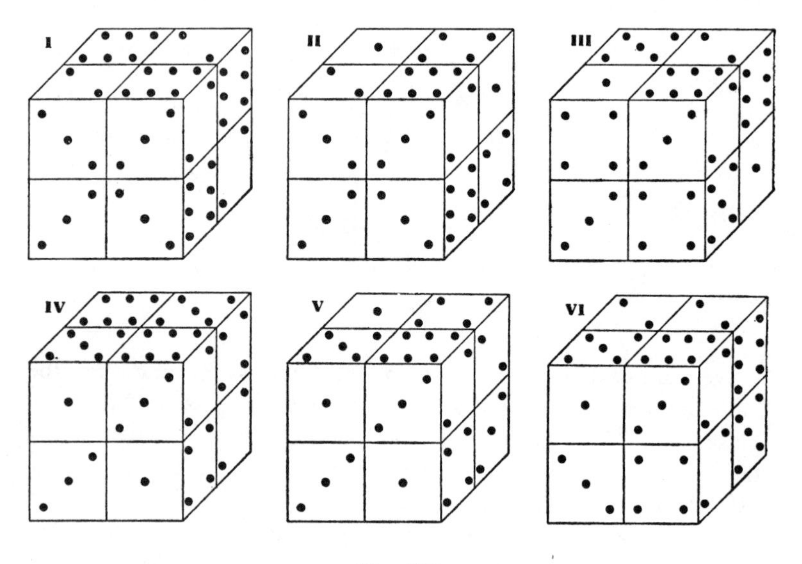

Fig. 127

cube in 8 ways such that the spot numbers, too, remain the same, and in $24/8 = 3$ ways such that the spot numbers are different every time. The latter statement can immediately be verified by considering that the symmetry-axis of order 4 can be any of the three lines which join centers of opposite faces.

Disregarding the position of the cube, there are **24** solutions of type I, for when we have made a choice of the faces for the equal spot numbers (which fixes the position of the symmetry-axis of order 4), we then can position one of the dice in 24 ways, after which the positions of the remaining dice follow from rotations around the symmetry-axes.

Type II: The line joining the centers of the front and the back is a symmetry-axis of order 4. We can set down the cube in four ways such that the spot numbers are unchanged, and in six ways such that the spot numbers are differently placed every time.

If we choose the faces with equal spot numbers as the front and the back, then we can position the die in the lower right front corner arbitrarily, as well as the die in the lower right back corner, after which the positions of the remaining dice are determined by rotations around the symmetry-axis of order 4. This gives 24×24 positions. Among these, there are the 24 solutions of type I, so that 23×24 solutions remain. These are identical in pairs for 180° rotations around the line joining the top and the base. Hence, there are $23 \times 12 =$ **276** solutions of type II (disregarding the position of the cube).

Type III: The line joining the centers of the front and the back, and the two lines joining the centres of opposite edges perpendicular to front and back, are symmetry-axes of order 2. We can set down the cube in 4 ways such that the spot numbers are the same, and in 6 ways such that the spot numbers are different every time.

We can place the lower right front and lower left front dice arbitrarily, after which the position of the remaining dice is determined. This gives 24×24 ways, among which are the 24 solutions of type I. Again the remaining solutions are identical in pairs for rotations by 180° around the line joining the centers of the top and the base, so that there are **276** solutions of type III.

Type IV: The three lines joining the centers of opposite faces are symmetry-axes of order 2. As with type III, these are pairwise perpendicular, and we can set down the cube in 4 ways such that the spot numbers remain the same, and in $24{:}4 = 6$ ways such that the spot numbers are altered; the latter statement can easily be verified directly.

If we set down the lower right front and upper right front dice arbitrarily, which can be done in 24^2 ways, then the position of the remaining dice follows as a result of the required symmetry. However we then also obtain the 24 solutions of type I, each of them three times (with three positions of the symmetry-axis of order 4). This gives $24^2 - 3 \times 24 = 21 \times 24$ positions. The foregoing can also be worded as follows: we can give an arbitrary position to the lower right front die, with a choice of 24 ways; when this choice has been made, as in Figure 127, for example, we can position the upper right front die in 21 ways, since we have to avoid the three positions where the spot numbers 1, 4, and 5 meet at the upper right front corner (this would lead to type I in some position). For the remaining 21×24 ways, the position of the cube has been taken into account, so that there are $21 \times 24/6 =$ **84** solutions of type IV.

We now proceed to the less regular types.

TYPE V: The line joining the centers of the front and the back is a symmetry-axis of order 2. We can set down the cube in two ways such that the spot numbers remain the same, and in 12 ways such that the spot numbers are different every time.

We can place the four lower dice in 24^4 ways, after which the position of the remaining dice is determined. Among these, the 24 solutions of type I occur three times each (with different positions of the axis of symmetry of order 4), the 23×12 solutions of type II twice each, the 23×12 solutions of type III twice each, and the 7×12 solutions of type IV six times each. This gives

$$24^4 - 3 \times 24 - 23 \times 24 - 23 \times 24 - 21 \times 24 = 48 \times 6877$$

solutions of type V, where the line joining the front and the back is the symmetry-axis, and where the position of the cube has been taken into account. We thus obtain every solution four times (namely, rotated by 90° around the line joining the centers of the front and the back, and both positions thus obtained rotated by 180° around the line joining the centers of the top and the base), so that there are $12 \times 6877 =$ **82,524** solutions of type V.

TYPE VI: The line joining the center of the upper right edge to the center of the lower left edge is a symmetry-axis of order 2. We can place the cube in 12 ways such that the spot numbers are different every time.

If we leave the position of the symmetry-axis unchanged, but if, for the rest, we take into account the position of the cube, then we obtain (as with type V) 24^4 placings. These include the 24 solutions of type I, once each, and the 23×12 solutions of type III, twice each, so that

$$24^4 - 24 - 23 \times 24 = 24^2 \times 575$$

placings remain. In this way, we obtain every solution twice (namely, also rotated by 180° around the line joining the centers of the front and the back), so that there are $288 \times 575 =$ **165,600** solutions of type VI.

If we also take into account the position of the cube, then, as appears from the foregoing, the number of solutions with one or more axes of symmetry is:

$$(3 \times 24) + (6 \times 23 \times 12) + (6 \times 23 \times 12)$$
$$+ (6 \times 7 \times 12) + (12 \times 12 \times 6877) + (12 \times 288 \times 575)$$
$$= 72 \times (1 + 23 + 23 + 7 + 13754 + 27600)$$
$$= 24^2 \times 8 \times 647 = 2{,}981{,}376.$$

Hence, the number of solutions without an axis of symmetry, when we take into account the position of the cube is:

$$24^8 - 24^2 \times 8 \times 647 = 24^2 \times 8 \times (23{,}887{,}872 - 647)$$
$$= 24^2 \times 8 \times 23{,}887{,}225,$$

so that the number of solutions without a symmetry-axis, without taking into account the placing of the cube, is:

$$24 \times 8 \times 23{,}887{,}225 = \mathbf{4{,}586{,}347{,}200}.$$

Hence, without taking account of the position we find the total number of solutions to be:

$$24 + 12 \times 23 + 12 \times 23 + 12 \times 7 + 12 \times 6877 + 288$$
$$\times 575 + 192 \times 23{,}887{,}225 = 48 \times 95{,}554{,}083$$
$$= \mathbf{4{,}586{,}595{,}984}.$$

***264. Which are the invisible spot numbers?** In §263, six types of cubes have been drawn, while the symmetry-axes of these cubes have been given. We now pose the problem: to derive from these data the spot numbers on the invisible faces.

For types I, III, and IV this takes little trouble.

With type II, the symmetry-axis of order 4 shows us which spot numbers are on the base and on the left-hand face, and we learn further that on the back there are four equal spot numbers. From the upper right back die it follows (in connection with the placing of the spot numbers on the dice indicated in Figure 126) that these are four faces each with a 5.

With type V, the symmetry-axis indicates which spot numbers are on the base and on the left-hand face, and we learn further that on the back the diametrically opposite spot numbers are equal. The upper right-hand number, hence also the lower left-hand number, on the back is 1. From the direction of the line joining the 3's on the right-hand face, it follows (in connection with Figure 126) that the lower right-hand number and the upper left-hand number on the back must each be a 2 or must each be a 5. Hence, there are two possibilities here.

With type VI, the symmetry-axis indicates which spot numbers are on the back, and we learn further that on the base the same spot numbers occur as on the left-hand face. The right-hand front number on the base is a 6, hence this spot number also occurs at the back of the top of the left-hand face. Since on the back there is a 1 at the lower right, there is a 4 on the base at the back of the right-hand face, hence also on the top front of the left-hand face. On the base, there is a 2 or a 5 at left front, and on the left-hand face, bottom front, there is a 1 or a 6, correspondingly, so that on the bottom face there is also a 1 or a 6, respectively, at the left back. Hence there are two possibilities here, also.

Chapter XIV:
PUZZLES OF ASSORTED TYPES

I. NETWORK PUZZLE

265. Networks. By a network we mean a collection of line segments. These are called the edges of the network and their endpoints are called its vertices. It is assumed that any two edges have nothing else in common, or only a single vertex in common. The edges of a network need not be the edges of a polyhedron. We can form a network, for instance, by taking the twelve edges of a cube, together with the four body-diagonals (each of which connects a pair of opposite vertices to the center of the cube, and can be assumed to provide two edges; *Fig. 128*), or else a network could consist only of four edges that radiate from a single point.

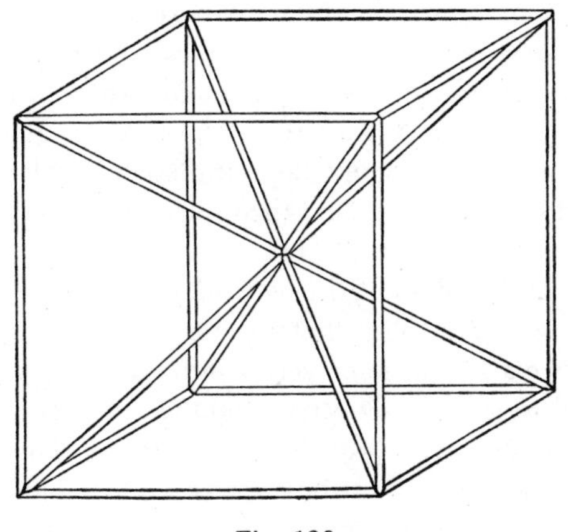

Fig. 128

We shall call a network connected when it does not fall apart into two or more disjoined networks. Two networks are called disjoined when no edge of one network has a point in common with any edge of the other network.

266. Puzzle on open and closed paths. We imagine that a point moves along one or more edges of the network, starting at a vertex and ending in a vertex, but in such a way that no edge is traversed twice or more. The moving point may pass through the same vertex twice or more often. Such an assembly of edges will be called a path. A path will be called closed or open according as its endpoints coincide or do not coincide. Each of the vertices of a closed path can be considered as a starting point or an endpoint.

We shall say that a network has been divided into paths when these paths have no edge in common, and jointly form the whole of the network. This raises the problem of dividing a given network into the smallest possible number of paths. In particular, we ask how we may determine from a network whether it can be traversed in a single path, and, if so, whether that path is open or closed, or whether the network can equally well be traversed by an open path or a closed one. For these purposes we can naturally assume that the network is connected.

267. Relation to the vertices of the network. The number of edges which meet at a vertex will be called the order of that vertex. If this number is even (or odd), we shall speak of an even (or odd) vertex. The sum of the orders of all vertices of a network must amount to twice the total number of edges, since each edge is counted twice in that sum, by reason of the two vertices which lie on it. So the sum of the orders taken for all the vertices must be an even number. This implies that the network must contain an even number of odd vertices (possibly zero). A network will be called even if all its vertices are even vertices (as in Figure 128); if there are at least two vertices which are odd vertices, we shall then refer to the network as an odd network.

Suppose that the edges of the network can be traversed in a single closed path (which implies that the network must be a connected network). If we arrive at some vertex in traversing the path, we then must proceed to leave it by way of another edge of the network. In this way, the edges meeting at the vertex are associated in pairs, and consequently the vertex must be an even vertex. This shows that an odd network can certainly not be transformed into a single closed path.

Next we assume that the edges can form a single open path. In this event, each vertex of the network must be an even vertex, except for the two endpoints of the path. These must be odd vertices, since the edges that meet at one of these endpoints must again correspond in pairs, apart from the edge which forms a beginning or end for the path. This means that the network must have exactly two odd vertices;

the endpoints of the path cannot be any vertices other than the two odd vertices. From this it also follows that no network can ever be considered both as a single open path and also as a single closed path.

If a connected network has $2 \times k$ odd vertices (at least 4 in number), it then follows, in the same way, that the network cannot be divided into fewer than k paths. If a division into k paths is possible, then these must all be open paths; the endpoints of these paths must be the $2 \times k$ odd vertices; in this case, no two of the k paths can have an endpoint in common.

268. Splitting a network. At a vertex A of some connected network N, suppose that we start a path which makes use of an edge AB. The network N, which is produced by deletion of the edge AB from N, can be called the *residual* network. Now it may happen that this residual network N_1 splits up into two disjoined parts, in which case we can distinguish a *terminated* network N_2 (containing the vertex A but not the vertex B) from a *continuation* network N_3 (which contains the vertex B but not the vertex A). Each of N_2 and N_3 is then a connected network. If B is a first-order vertex of N (which means that AB runs to a dead end at B), then N_1 is necessarily a connected network which has A as a vertex, but not B; and then the residual network consists *only* of a terminated network.

Let us assume that deletion of AB has split off a terminated network N_2. If B is an even vertex of the continuation network N_3 (or if N_3 is missing, because AB goes to a dead end at B), then B must be an odd vertex of N. If, however, B is an odd vertex of N_3, then N_3 must have yet another odd vertex, and this must be an odd vertex of the original network N. This means that in either case N must have at least one odd vertex C (either B itself, or a vertex other than B) which is not a vertex of N_2 (but which *is* a vertex of the combination of AB and N_3).

If we now decide to start a path from A, using some edge AD which is different from AB, and find that deletion of AD will also cause a splitting of N, AB must then become a part of the terminated network (which then contains A but not D), and the like must apply to the continuation network N_3 of AB, if deletion of AB gives rise to one.

This implies that the terminated network, for deletion of AD, must include the odd vertex C, and must thus be an odd network. So we now know that if we start a path from a vertex A, among the edges radiating from A there is at most one edge whose use can split off a terminated network which will be an even network. If only one edge of N passes through A, then this edge cannot cause a split of the type considered.

This makes clear that the edge AB with which we start a path (at A) can always be chosen in such a way that deletion of this edge from N does not produce the splitting-off of an even network. When we draw a second edge of the path, the continuation network N_2 (connecting with AB) plays the same role as the complete network N did originally. Hence, we can choose this second edge in such a way that no even network is split off from N_2. Continuing in this way, we can entirely avoid the splitting-off of any even network.

It may be noted that these results imply a further consequence: if an even network splits off when the edge AB is deleted, then no splitting-off will be caused if we start with another edge passing through A (if such is present). However, this is immaterial in our further argument.

269. Case of an odd network. Let N be a connected odd network with $2 \times k$ odd vertices (at least two). We start a path at an odd vertex, and continue it as far as possible, always avoiding the splitting off of an even network; the path must then end at some other odd vertex. If N has at least four odd vertices, then the remaining network N' has at least $2 \times k - 2$ odd vertices, because the endpoints of the completed path are transformed into even vertices by deletion of the path whereas the other vertices are still of unaltered kind (even or odd), if any vertices still remain in N. We now form a second open path, from an odd vertex of N' to another odd vertex of N', again avoiding the splitting-off of an even network, continuing in this way as long as we can. We repeat this until every odd vertex of N has occurred as an endpoint of an open path; k open paths have then resulted. Together, they must constitute the entire network N, since otherwise the remaining network would be even, which would be in contradiction to the facts that no even network has split off and that the original network was connected.

We now have a division into the smallest number of paths, since a division into fewer than k paths is not possible (as we showed in §267).

270. Case of two odd vertices or none. As a special case ($k = 1$) in the results of §269, we have the result that a connected network with two odd vertices can be traversed in a single open path.

If the connected network N is an even network, and if we delete from N an arbitrary edge AB, this causes no splitting-off (since, according to §268, splitting-off would require that an odd vertex was present in N), so the remaining network N_1 must here be a connected network. N_1 then has two odd vertices, A and B. It follows that we can traverse N_1 in a single open path from B to A. Addition of the edge AB then produces a single closed path. So a connected even network can be described in a single closed path.

271. Another determination of the smallest number of paths. If the connected network N has $2 \times k$ odd vertices, then we begin by forming k open paths, from an odd vertex to an odd vertex every time (paying no attention to splitting-off). If this does not exhaust the network, one or more even networks must then remain. From these, we form closed paths, starting at an arbitrary vertex and ending at the same vertex (again, without paying attention to splitting-off). If N is even, we form closed paths exclusively.

If N is odd, we can incorporate the closed paths (if present) into the k open paths and, by so doing, get rid of them, since any closed path must have a vertex in common with at least one other path, because N is connected. If A is this common vertex, and if a point is traversing that other path, we can arrange on reaching A to traverse the closed path first, before continuing that other path (if it has not yet been traversed completely). The path that arises from the union of two paths is closed or open, according as the other path is closed or open; in the latter case the path that arises from the union has the same endpoints as the odd path with which the even path has been united. In the same way, we can incorporate the other closed path (or closed paths that arise from unions) into other paths, until eventually k open paths remain.

If N is even and if more than one closed path has resulted, we can use this same method to unite some closed path (which must have a vertex in common with at least one other path) with another closed path, to produce a single closed path, and continue this until finally there is only a single closed path.

It is true that this proof is shorter than the one given in §§267–270, but the earlier proof has the advantage of also indicating considerations which require attention in forming the paths, so that it is not necessary to make errors first, and then correct them at the end, so to speak. Every time it is proposed to add an edge to a path, there is need only to watch whether this would cause an even network to split off; if so, the continuation can be made by using any other edge, and this is discontinued only when it becomes impossible to extend the path.

II. BROKEN LINES THROUGH DOTS

272. Broken line through nine dots. Nine dots have been arranged in a square which has been divided into four smaller squares. A broken line, consisting of four joined straight-line segments, has to pass through the nine dots, starting at a dot, and ending at another dot. The broken line is allowed to intersect itself.

This problem has only one solution, provided that broken lines which correspond by rotation or reflection are considered to be equivalent. We shall omit the argument, and confine ourselves to drawing the solution (*Fig. 129*).

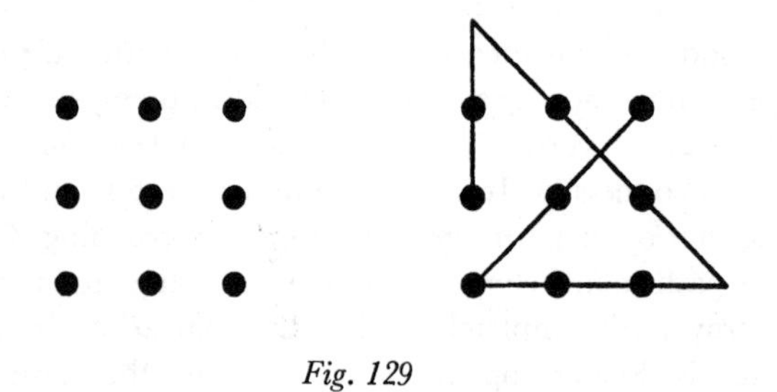

Fig. 129

273. Broken line through twelve dots. We replace the nine dots of §272 by twelve dots, which are arranged in a rectangle divided into six squares, and we require a broken line, consisting of the smallest possible number of straight-line segments, to pass through the twelve dots and pass through each dot once only. To keep the number of solutions finite, we further require that each segment of the broken line must pass through at least two of the dots.

The smallest number of line segments to produce a broken line that satisfies the requirements is five. Five line segments allow a single closed broken line through the twelve dots (*Fig. 130*). By deleting

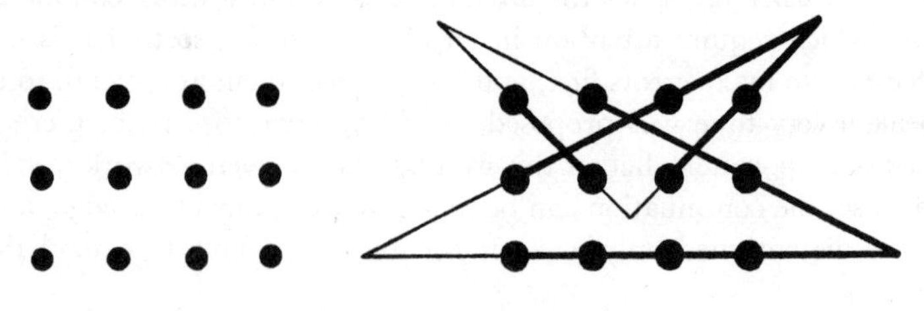

Fig. 130

one of the three lightly drawn sections of the closed broken line, we get an open broken line which satisfies the condition. Apart from these three solutions, which begin at a dot and end at a dot, we have the sixteen solutions shown in *Figure 131* which cannot be completed to form a closed broken line without increasing the number of line segments.

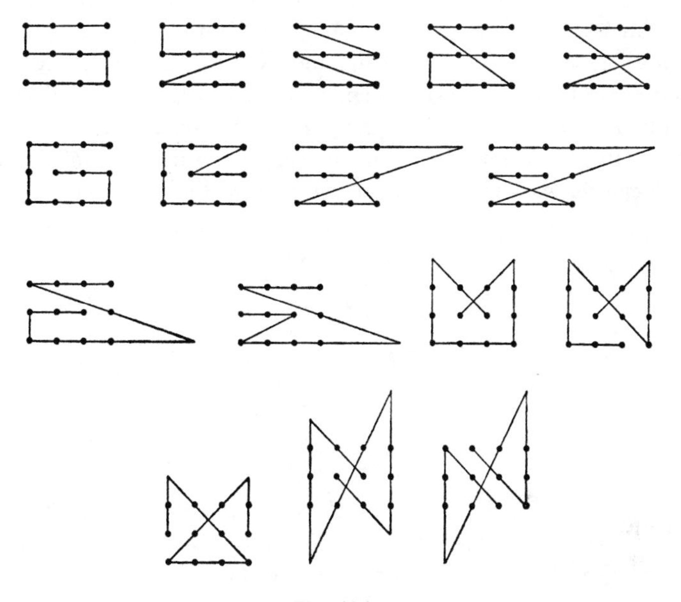

Fig. 131

It is easy enough to find a broken line of five segments that runs through the twelve dots, because some of the solutions are very obvious. The difficulty consists entirely in finding the whole set of solutions. The argument has been omitted, since it will be given for the corresponding puzzle with 16 dots.

274. Broken line through sixteen dots. The puzzle becomes more difficult when extended to sixteen dots arranged in a square divided into nine smaller squares (*Fig. 132*). First we take the case in which the broken line through the sixteen dots consists of five line segments. Without discussion it is clear that the broken line cannot contain four horizontal line segments. But then it cannot contain three horizontal segments, either, since otherwise each of the four dots of the remaining horizontal row would require a segment of the broken line, and all four segments would have to be different (since there is no fourth horizontal segment). Together with the three horizontal line segments, this gives $3 + 4 = 7$ line segments, which is too much. From this it follows further that the broken line cannot contain two horizontal line segments, either, since $2 + 4 = 6$ line segments is too much, also. Of course, there cannot be two vertical line segments, either.

Since 5×3 is less than 16 (the number of dots), at least one of the five line segments has to contain four dots. If it is a diagonal, then we reach an impasse with the remaining twelve dots, which have to be placed on four straight-line segments. Hence, the broken line must contain a horizontal segment, say, on which four dots lie. Since there is no other horizontal line segment, no four of the remaining twelve dots can lie on a single segment. So each of the four non-horizontal

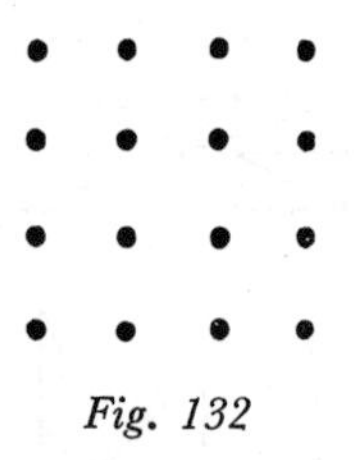

Fig. 132

segments would have to contain three of those twelve dots; however, this is impossible, since there cannot be two vertical line segments. From this we conclude that a broken line of five line segments running through sixteen dots is impossible.

275. Broken line with six line segments through 16 dots. As in §274, we argue that the broken line cannot contain three horizontal or three vertical line segments. From this it is evident that among the sixteen dots only two separate sets of four (that is, sets of four having no dot in common) can occur which between them account for two line segments. The two line segments can be the two diagonals (case A in *Figure 133*), or two line segments that are both horizontal or both vertical—for example, both horizontal (case B in Figure 133). In case B the remaining eight dots cannot include three dots that lie on the same line segment, since this would require a third horizontal segment; hence, the remaining eight dots must be able to be divided into four pairs such that each pair lies on a segment of the broken line.

It is also possible that there are no two separate sets of four on two

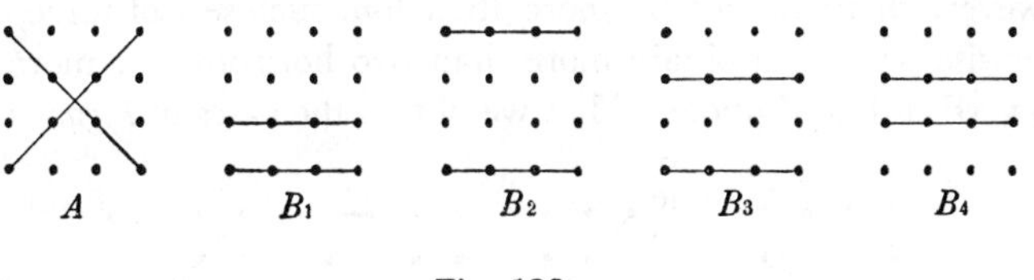

Fig. 133

line segments, but that there is a line segment on which four dots lie. This line segment can be a diagonal (case C) or run horizontally (case D). Among the twelve remaining dots, there can be no three separate sets of three that jointly occupy three line segments, since otherwise we would obtain three horizontal or three vertical line segments and thus reach an impasse. However, two separate sets of three jointly occupying two segments must occur among the remaining twelve dots, in view of the relation $4 + 3 + 3 + 2 + 2 + 2 = 16$. We obtain the possibilities in *Figure 134*.

If there is no line segment on which four dots lie (case E), then there must be at least four separate sets of three jointly occupying four line segments (in view of the relation $3 + 3 + 3 + 3 + 2 + 2 = 16$).

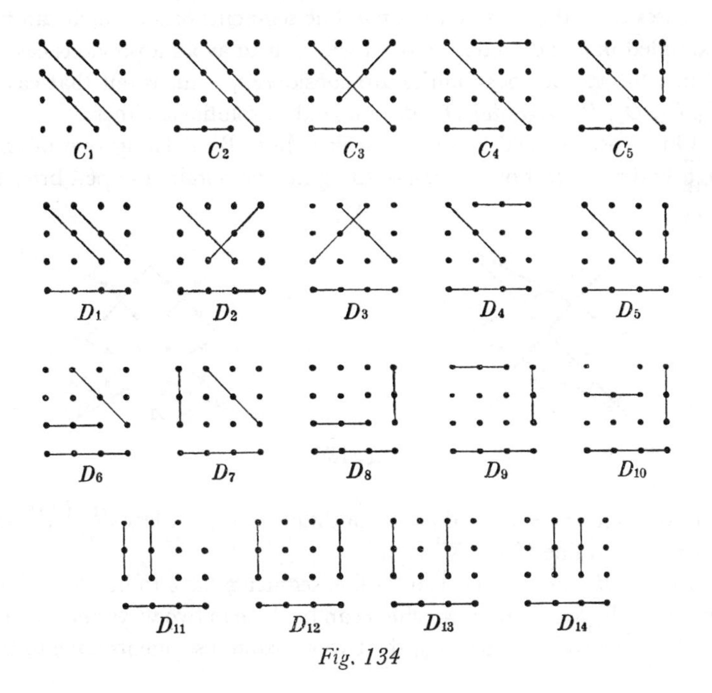

Fig. 134

However, there cannot be more than four such sets of three, since otherwise we would obtain more than two horizontal or more than two vertical line segments. Thus we obtain the cases in *Figure 135*.

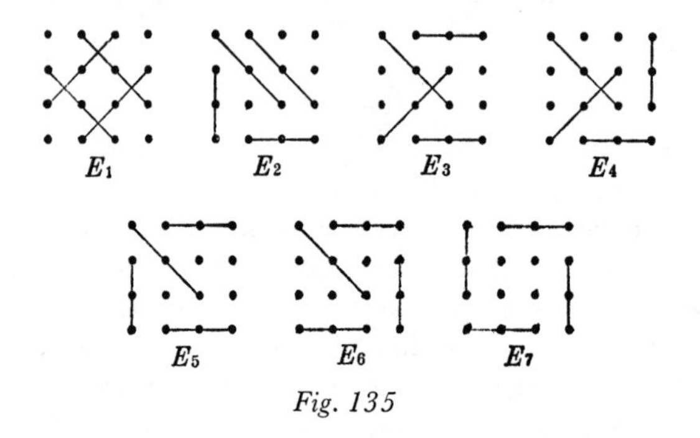

Fig. 135

In cases B_3, B_4, C_1, D_{10}, D_{13}, D_{14}, and E_1, there is a line segment which cannot be extended in either direction, and in cases E_2, E_4, E_5, and E_6, there is a line segment for which this is not possible without obtaining a line segment with four dots (and hence a previous case). In cases E_3 and E_7, more than two line segments occur which can be extended in one direction only (if we want to avoid a previous case). When the various possibilities are considered, it turns out that cases C_3, C_5, D_2, D_5, D_8, and D_{11} do not lead to solutions, either.

Only case *A* yields a closed broken line. By deleting one of the lightly drawn portions in *Figure 136* (left) we obtain an open broken

Fig. 136

line through the sixteen dots. In addition to this, *A* leads to another open broken line (Fig. 136, right).

In case B_1, the two horizontal line segments have to be connected by two or by four segments, which can be done in two ways and in one way, respectively. In case B_2, the two horizontal segments have to be connected by one segment or by three segments. This gives the six solutions shown in *Figure 137*.

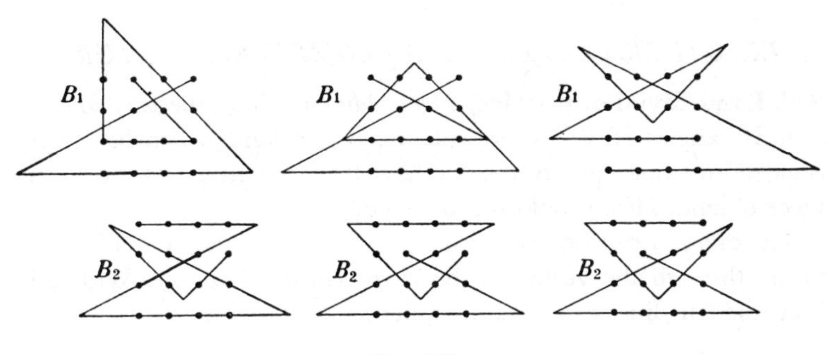

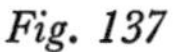
Fig. 137

Case D_1 gives the four solutions shown in *Figure 138*, in which the two diagonal segments are connected to each other by two or three line segments. Cases C_2, C_4, D_4, D_9, D_{12}, E_2, E_5, and E_6 also lead to solutions, which, however, all occur among the solutions resulting

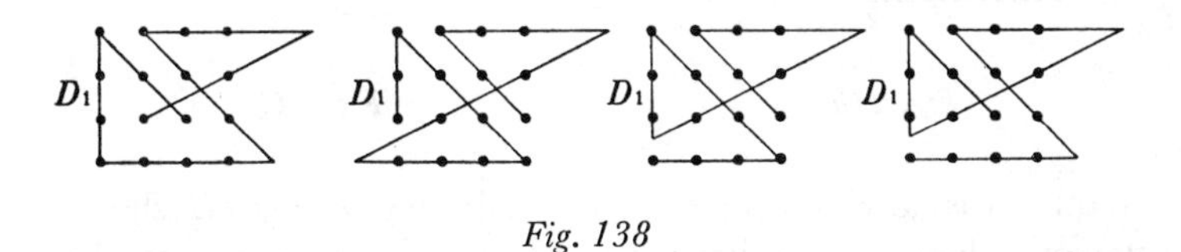

Fig. 138

from D_1. In addition, D_3 gives two further solutions (*Fig. 139*). D_6 and D_7 also give solutions, but no new cases.

In all, we obtain fifteen essentially different open broken lines that satisfy the condition, among which there are two symmetrical ones (corresponding to case A). If we make distinctions between solutions that arise from each other by rotation or reflection, then every solution counts as eight solutions, except the two symmetrical ones,

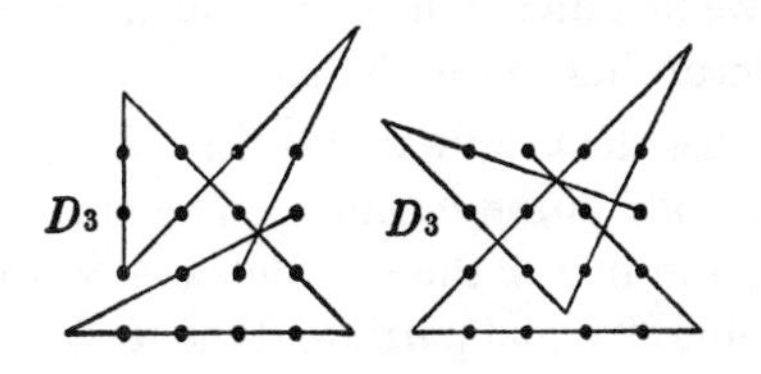

Fig. 139

which count as four solutions each; the number of solutions is then $(8 \times 13) + (4 \times 2) = 112$.

III. OTHER PUZZLES OF A GEOMETRICAL NATURE

276. Land division puzzles. *Figure 140* (in which the tiny squares indicate angles of 90°) represents a piece of land which has to be divided into four equal pieces that have the same shape as the original piece of land. How should this be done?

The division can be produced by four straight lines, one of which passes through the vertex in the lower left-hand corner. With this hint, the four lines can be easily found.

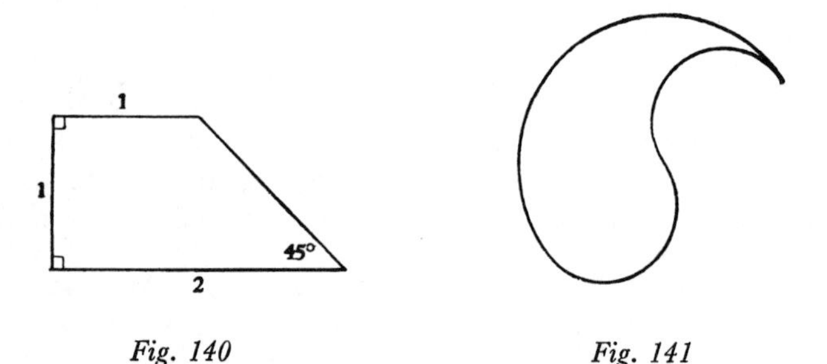

Fig. 140 *Fig. 141*

Another easy division puzzle is the following. *Figure 141* shows a region of a plane which is bounded by three semicircles, two of which are the same size, while the third is twice this size. This region has to be divided into two pieces of identical size and shape.

It will be evident that this can be done by drawing a semicircle of the same size as the two smaller bounding semicircles. By using two such semicircles, the region can be divided into three equal parts of identical shape, and so on.

277. Another division puzzle. The region shown in *Figure 142* has to be divided into the smallest possible number of identical parts.

To achieve this we first divide the figure into a right-angled triangle and a rectangle. According to the Pythagorean theorem, the hypotenuse of the triangle has the length $\sqrt{9^2 + 12^2} = 15$, which is also the length of the longer side of the rectangle. The area of the triangle is $\frac{1}{2} \times 9 \times 12 = 54$, and that of the rectangle is $4 \times 15 = 60$. So if we divide the triangle into 9 equal parts, and the rectangle into 10 equal parts, then these 19 parts will all have the same area. They will also have the same shape if we make them right-angled triangles with perpendicular sides of lengths 3 and 4.

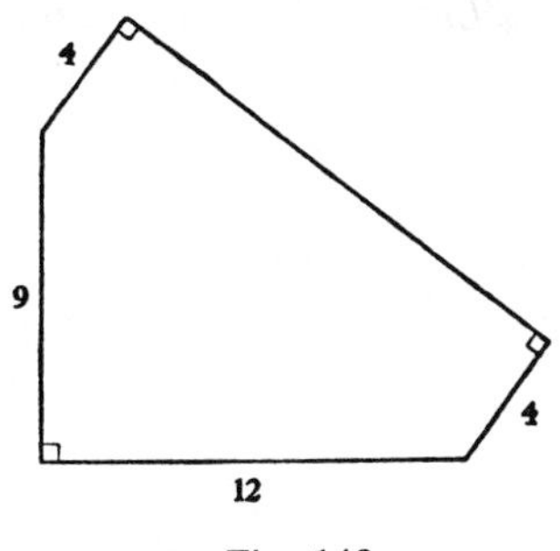

Fig. 142

278. Polyhedron puzzles. A body bounded by planes (a polyhedron) is projected onto a plane. Visible edges (or rather their projections) are indicated by continuous lines; hidden edges that do not coincide with visible edges (in the projection) are shown as dotted lines. The visible faces are hatched. The diagrams shown in *Figure 143* are drawn in this way.

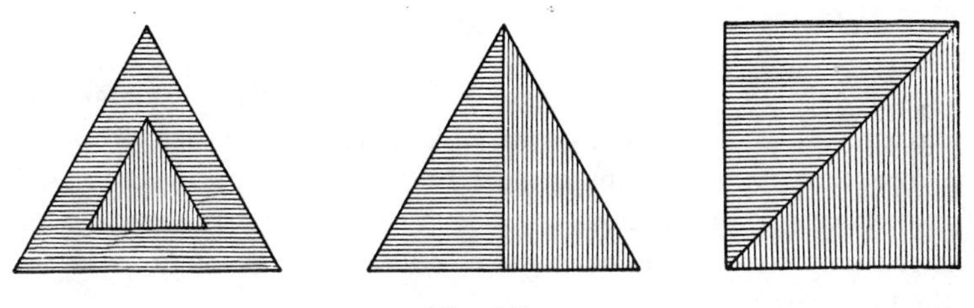

Fig. 143

No dotted lines have been given here, because there are no lines that have to be dotted. The number of faces of the body is as small as possible. How large is this number, and what shape can the bodies have?

The smallest numbers of faces are 7, 5, and 6, in the successive cases.* We leave it to the reader to imagine the polyhedra that satisfy the question.

* [Thus the original. But the second and third numbers are both one unit too large!—T.H.O'B.]

The following presents a similar problem: To cut off from a cylindrical cork, by as few plane cuts as possible (for instance, with a flat knife), enough to produce a body which can just pass through each of the three openings shown in *Figure 144*. The intention is that the body, in an appropriate position, should entirely fill the hole in each of the three cases.

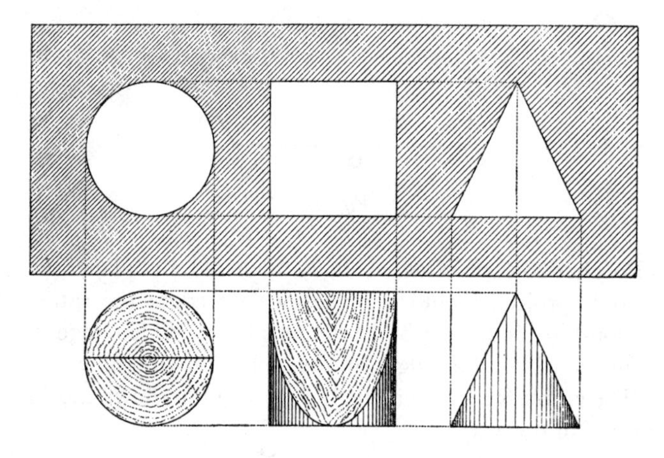

Fig. 144

It is possible to achieve this by two plane cuts. They touch the circumference of the base of the cork at two diametrically opposite points, and make the same angle with the axis of the cork. Below each of the three holes, the body has been drawn in the position in which it can pass through the hole; the middle diagram shows the shape of the body clearly.

IV. QUEENS ON THE CHESSBOARD

279. Eight Queens on the chessboard. A well-known problem, formulated and solved in 1850 by Nauck, and of interest to no less a person than the great German mathematician Gauss (1777–1855), is the following:

Place eight Queens on a chessboard in such a way that none of the Queens is attacking another Queen. All solutions are required.

This problem has no solution with symmetry of reflection, since two Queens that were mirror images of one another diagonally, vertically, or horizontally, would attack each other. Moreover, it is soon observed that there is no solution that becomes self-identical after a 90° rotation.

A solution that becomes self-identical after a 180° rotation gives rise, with its reflections, to four solutions (which can be considered to be equivalent). A solution that does not become self-identical through a 180° rotation gives rise with its reflections to eight solutions (which can be considered as one solution).

The solutions of the puzzle can be easily found by noting that each of the vertical files (*a*-file, *b*-file, and so on) must contain a Queen (this also shows immediately that it is impossible to place more than eight Queens without an attack situation being present). You can begin by placing a Queen on the *a*-file, on *a*1 first of all. Then you place a Queen on the lowest unguarded square of the *b*-file; next, on the lowest unguarded square of the *c*-file, and so on. In this way, you reach an impasse with the Queen on the *f*-file and so you make a minimum upward displacement of the Queen on the *e*-file, to an unguarded square. You continue in this way; if the Queen can no longer be moved upwards in the *e*-file, then you move the Queen on the *d*-file, and so on.

In this way you will find two solutions with a Queen on *a*1; each of these solutions is found in two variants which arise from each other by reflection in the diagonal from *a*1 to *h*8. You can use this as a check, but better still you can use it to shorten the solution. For example, if you have put Queens on *a*1 and *b*5, you need only consider the squares *f*2, *g*2, and *h*2 of the second rank; you then find each of the two solutions once only. If you have put a Queen on *a*1 and one on *b*7, then you put the Queen of the second rank on *h*2; locating that Queen reduces your solving time considerably. The case with Queens on *a*1 and *b*8 does not have to be investigated, since it is equivalent to having Queens on *a*1 and *h*2.

If you next consider the case in which the first Queen is placed on *a*2, then you should not place a Queen on *h*1 or *h*8, since this does not give any new solutions (that is, solutions unrelated by rotation or reflection to solutions already found). By placing one Queen on *a*2 and another on *b*8, *g*1, or *h*7, you can obtain two variants of a single solution. When the case of a Queen on *a*2 has been dealt with, then you place the first Queen on *a*3. Here you should not place a Queen on any of the squares *b*1, *b*8, *g*1, *g*8, *h*1, *h*2, *h*7, and *h*8, since again this gives no new solutions. With the one Queen on *a*3 and another on *f*1, the solution has two variants; a solution also may appear in three variants. If you have placed a Queen on *a*3, and other Queens on the *b*-, *c*-, and *d*-files, with no Queen as yet located on the eighth rank, then the Queen of the eighth rank can only be put on *e*8, or else you will obtain solutions you have already found; this, too, reduces solving time. If you place the first Queen on *a*4, then you should not place a Queen on any of the squares *b*1, *b*8, *c*1, *c*8, *f*1, *f*8, *g*1, *g*8, *h*1, *h*2, *h*3, *h*6, *h*7, and *h*8. Hence, you can immediately place Queens on *d*8, *e*1, and *h*5; after that you cannot place a Queen on the *c*-file. This completes the entire investigation, since a Queen on *a*5 comes to the same thing as one on *a*4.

We recommend that you go through this procedure not on paper, but on the chessboard, using the eight white Pawns as Queens. When you have appreciated the principles of the system to be followed, then your reasoning will be fast and sure, and you will quickly find all the solutions. Of these, twelve are essentially different, namely those shown in *Figure 145* (where the dots represent the Queens); they have been arranged in the order in which you will find them when you use the method described.

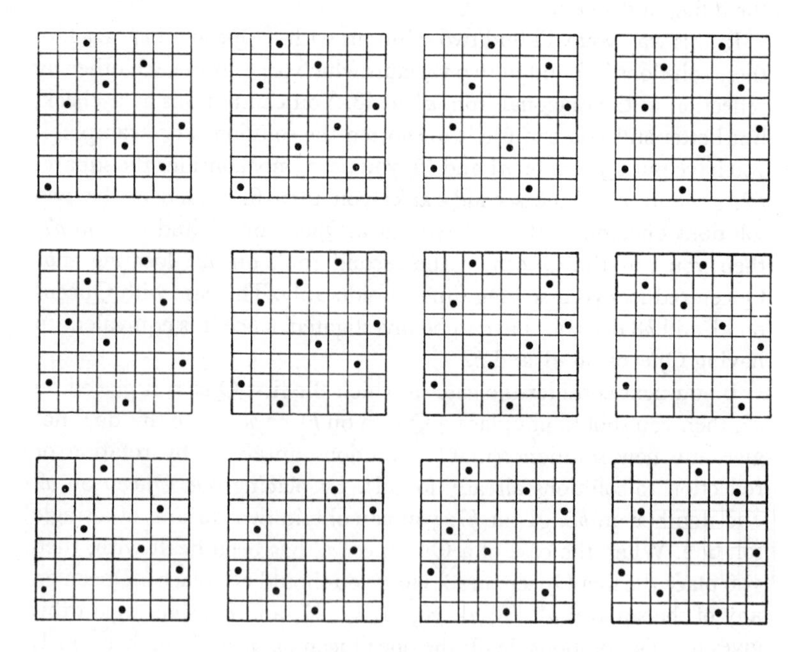

Fig. 145

There is only one solution (the tenth) that becomes self-identical after a rotation through 180°. If you count rotated and reflected solutions as different, then the number of solutions is $(8 \times 11) + (4 \times 1) = 92$. There is only one solution (the eleventh) in which no three Queens lie on any straight line (irrespective of its direction).

If the problem is to determine the solutions that become self-identical after a 180° rotation, then this can be done quickly. Here you should not place a Queen on a diagonal, and each location of a Queen determines the location of another Queen.

280. Queens on a smaller board. If we replace the regulation chessboard by a square board of arbitrary dimensions, the corresponding problem would be to place on it a number of non-attacking Queens equal to the number of ranks on the board (thus, for example, five Queens on a board with 5×5 squares). The smallest board on which this is possible is one of $4 \times 4 = 16$ squares. For 16 squares there is one solution; this becomes self-identical after a rotation through 90°. On a board of 25 squares, there are two solutions, one of which does not become self-identical when rotated through 180°, whereas the other one does become so when rotated through 90°. On a board with 36 squares, there is one solution; it becomes self-identical when rotated by 180°. On a board with 49 squares, there are six solutions, two of which become self-identical when rotated through 180°. The solutions for boards with 4^2, 5^2, 6^2, or 7^2 squares are shown in *Figure 146.*

If you make distinctions between the solutions that arise from each

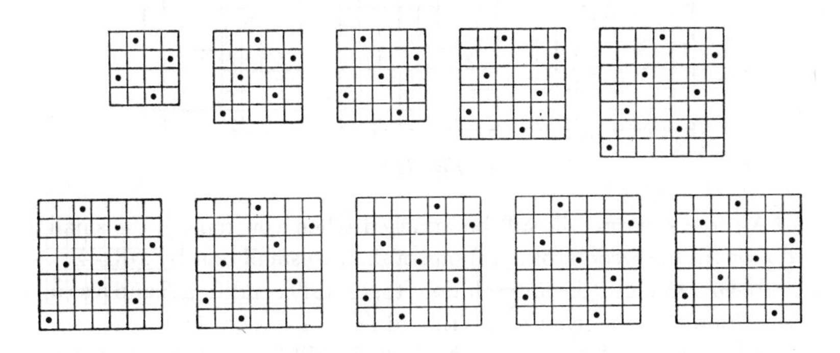

Fig. 146

other by reflection or rotation, then there are $2 \times 1 = 2$ solutions with 4^2 squares, $(8 \times 1) + (2 \times 1) = 10$ solutions with 5^2 squares, $4 \times 1 = 4$ solutions with 6^2 squares, and $(8 \times 4) + (4 \times 2) = 40$ solutions with 7^2 squares. With 5^2, 6^2, or 7^2 squares there are no solutions which avoid placing three or more Queens in a straight line.

281. Five Queens on the chessboard. Here the problem is to place five Queens in such a way that each free square (a square on which there is no Queen) is attacked by at least one Queen. One of the following two restrictions can be added to this requirement: (*A*) None of the Queens is to be attacked by another Queen. (*B*) Each of the Queens has to be attacked by at least one other Queen.

This problem, which is old, perhaps even very old, has many more solutions than that of the eight Queens (91 essentially different solutions with restriction *A*) and it is also more difficult to find all solutions. It is true that some of the solutions are very obvious. In one of these, you place a Queen on each of the borders (and not on a diagonal) in such a way that all border squares and all squares adjacent to the border squares are controlled by a Queen; for example, placing Queens on *a*2, *b*8, *g*1, and *h*7; in this case only four squares *c*3, *c*6, *f*3, and *f*7 (which are the corners of a square consisting of 16 small squares) are free from being controlled. When a Queen is placed on one of these four squares, the other three squares are attacked. In this way, restriction *A* is satisfied. This solution is represented by the diagram at the left in *Figure 147*; the two other diagrams show other solutions.

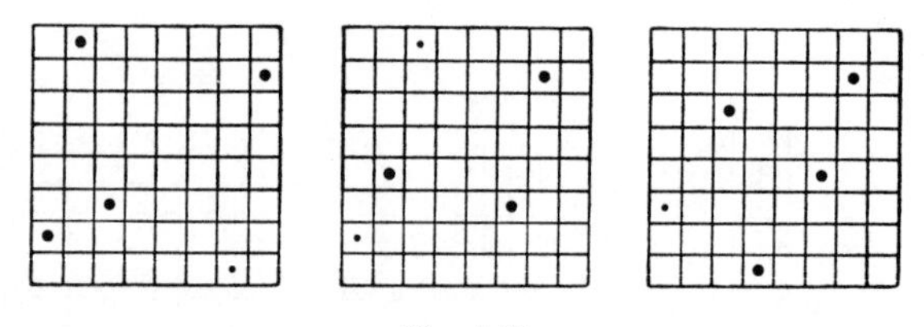

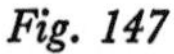

Fig. 147

All free squares will still be attacked when you replace the small dot or (in the second solution) one of the two small dots by a Rook.

Figure 148 shows the four solutions that satisfy condition *B* and also a condition that the five Queens should lie in a straight line.

All squares and also the pieces used will still be attacked when you replace the small dot by a Rook in the first two cases.

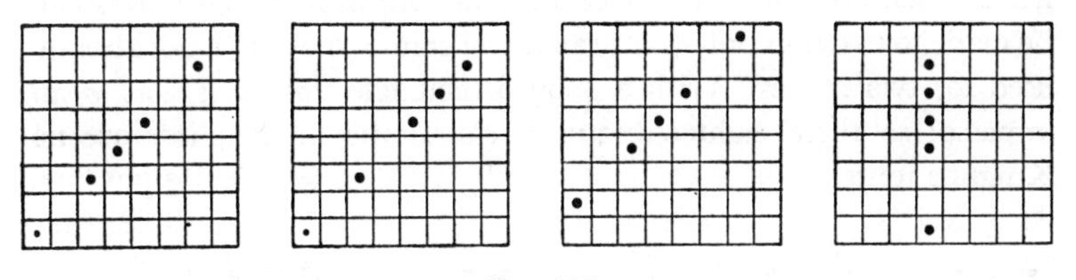

Fig. 148

V. KNIGHT ON THE CHESSBOARD

282. Knight's tour. As you know, the Knight moves two squares horizontally and one square vertically, or one square horizontally and two squares vertically. By a closed Knight's tour we mean a broken line, formed by Knight's moves, which ends at the starting point and which does not enter any square more than once. Now the problem is to divide the 64 squares of the chessboard into a number of closed tours which have no square in common, and in particular to pass through the 64 squares in one closed tour. It is not too difficult to find such a closed tour. Essentially different solutions exist in large number, and it would be extremely laborious to find them all.

Instead of the entire chessboard we can take some set of the squares of this board and require a division of these squares into closed tours. Since the Knight jumps from a white square to a black square, or vice versa, a closed Knight's tour must contain an equal number of white and black squares. Hence, a given set of squares certainly cannot be divided into closed tours (or be passed through by one tour) when the numbers of white and black squares are different (in particular, this shows that the number of squares has to be even). This does not mean, however, that a division into closed tours is always possible when there are as many white squares as black.

283. Simple examples of closed Knight's tours. In a 3 by 4 rectangle (that is, a rectangle divided into 3×4 squares), or in a 4 by 4 or a 4 by 5 rectangle, a division into closed tours is possible in one way only (see *Fig. 149*).

The centers of the squares are represented by dots. In the first diagram we obtain two closed tours of six moves each, in the second diagram four tours of four moves each, and in the third diagram two tours of ten moves each. Drawing the tours is extremely simple. You start in the four corners; from each of the corners, the two moves can be made in one way only. Then you turn your attention to other dots from which no move has yet been made and from which moves can be made to two dots only (naturally, you cannot move to a dot already used for two moves). With a 3 by 6 rectangle this leads you to an impasse; so these eighteen squares cannot be divided into closed Knight's tours.

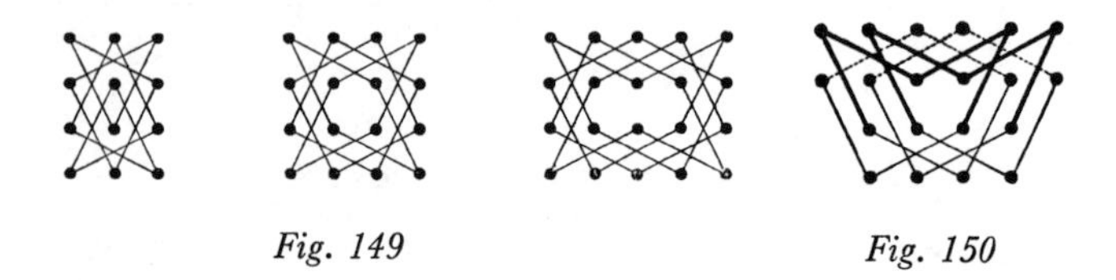

Fig. 149 *Fig. 150*

The next case, with twenty dots (or squares), is slightly more complicated. We start with the moves, indicated by heavy lines in *Figure 150*, from four dots of the uppermost row. After that, we can draw the thin lines from the dots of the bottom row; from this the dotted moves follow. We obtain two closed tours of ten moves each.

The figures described in this section are all symmetrical and produce no other solutions by reflection.

284. Other closed Knight's tours. In the case of a 3 by 8 rectangle, you have to distinguish between several cases after having drawn the sixteen moves that are indicated by heavy lines in the diagrams of *Figure 151*. The distinction is made by the thin lines; the dotted lines

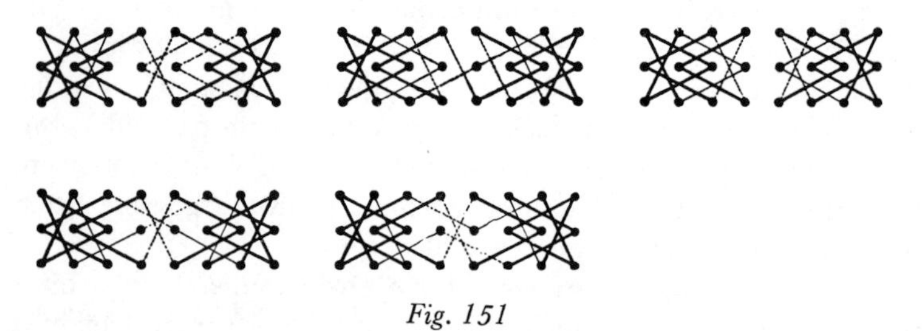

Fig. 151

are drawn last. The division into cases has been arranged in such a way that figures which arise from each other by rotation or reflection appear only once. In the first diagram we have one tour of eight moves, two tours of six moves each, and one tour of four moves. In the third diagram we have four tours of six moves each, while in each of the three diagrams we have one tour of eighteen moves and one tour of six moves.

The case of a 3 by 10 rectangle requires more work. A restriction to pass through the 30 squares in one closed tour makes for a considerable simplification in the distinction of different cases (again indicated by thinly drawn lines in *Figure 152*).

You can divide the 30 squares into more than one closed tour in seven essentially different ways: in two ways by tours of twenty, six, and four moves; in two ways by tours of eighteen, eight, and four moves; in two ways by two tours of eight, one of six, and two of four moves; and in one way by one tour of ten, two of six, and two of four moves. We leave it to the reader to draw these closed Knight's tours in the various cases.

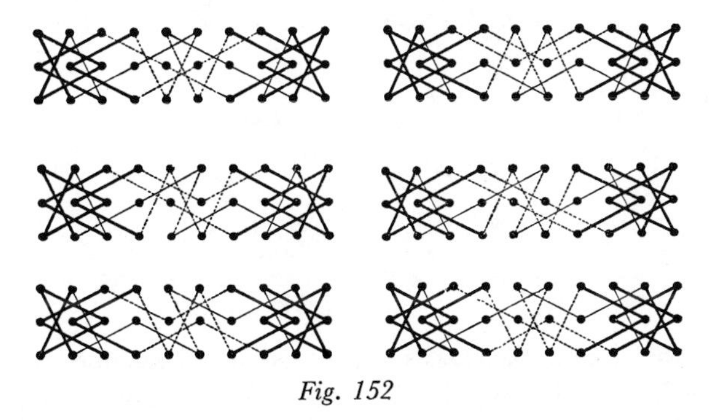

Fig. 152

***285. Angles of a closed Knight's tour.** In a Knight's tour, angles of different magnitudes can occur. In *Figure 153*, they all occur, and in increasing magnitude. Angle 3 is a right angle; angle 6 is a straight angle. Angle 1 (the smallest) will be called k, assuming a right angle as the unit. Then angles 1, 2, 3, 4, 5, and 6 are equal to k, $1 - k$, 1, $1 + k$, $2 - k$, and 2, respectively. Angles 1 and 2 are acute; angles 4 and 5 are obtuse. In a closed Knight's tour, let r be the number of right angles, and g the number of straight angles. The numbers of angles with magnitude k, $1 - k$, $1 + k$, and $2 - k$ will be called w, x, y, and z, respectively. Since the total number of moves of a closed tour is even (see §282), it follows that $w + x + r + y + z + g$ is even.

Let r' be the number of right angles at which the Knight turns right, and r'' the number of right angles at which it turns left (with $r' + r'' = r$); w', w'', x', x'', y', y'', z', and z'' will have an analogous meaning. On making an angle k, the Knight deviates by an angle $2 - k$ (the supplement of k) from its original direction, clockwise or counterclockwise according as the Knight jumps to the right or to the left. On making an angle $1 - k$, the change in direction is $2 - (1 - k) = 1 + k$ to one side or the other, and so on. Starting in the middle of a Knight's tour, until the Knight has returned to its starting point, the total change of direction of the Knight is

$$(w' - w'')(2 - k) + (x' - x'')(1 + k) + r' - r'' + (y' - y'')(1 - k) + (z' - z'')\,k = 2(w' - w'') + x' - x'' + r' - r'' + y' - y'' - k(w' - w'' - x' + x'' + y' - y'' - z' + z''),$$

(where the clockwise sense is taken to be positive).

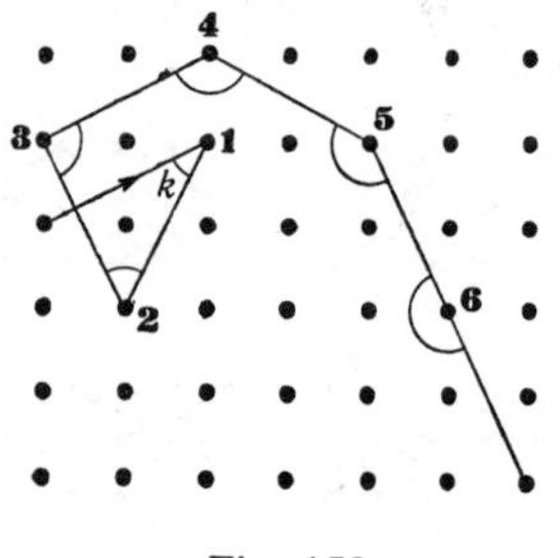

Fig. 153

Since the Knight, back at its starting-point, has regained its original direction, the total change in direction is a multiple of 4, hence $4 \times v$, where v is an integer (positive, negative, or 0). Hence we have

$$k(w' - w'' - x' + x'' + y' - y'' - z' + z'') = 2(w' - w'') + x' - x'' + r' - r'' + y' - y'' - (4 \times v).$$

With the aid of advanced mathematics, it can be proved that k is an irrational number, that is, it cannot be written as the quotient of two integers; one can also express this by saying that an acute or obtuse angle of the Knight's tour, and a right angle, are not in the proportion of two integers. This can be reconciled with the relation last found only if both the equated numbers are 0. Consequently

$$w' - w'' - x' + x'' + y' - y'' - z' + z'' = 0,$$
$$2(w' - w'') + x' - x'' + r' - r'' + y' - y'' = 4 \times v.$$

This shows that the numbers $w' + w'' + x' + x'' + y' + y'' + z' + z''$ and $x' + x'' + r' + r'' + y' + y''$ are even. Because of $w' + w'' = w$, and so on, this means that the numbers $w + x + y + z$ and $x + r + y$ are even. This expresses the fact that the numbers r, $x + y$, and $w + z$ have the same parity, that is, they are all even or else all odd. Since $r + (x + y) + (w + z) + g$ is also even, g too has the same parity as r.

Hence, for a closed Knight's tour, the number of jumps straight ahead must have the same parity as the number of jumps at a right angle and as the number of jumps at one of the angles $1 - k$ and $1 + k$ (the largest acute and the smallest obtuse angle, which are supplementary to each other) and as the number of jumps at one of the angles k and $2 - k$ (smallest acute and largest obtuse angle).

That r and $w + r$ have the same parity can be proved more easily as follows. We pass through the Knight's tour by going from the middle of a jump to the middle of the next jump every time. When this is done, the Knight moves along a distance which is represented by an integral number in the horizontal direction, if we take the smallest distance between the dots to be 2. The horizontal displacement is even when an angle $1 - k$, $1 + k$, or 2 is formed, and odd when an angle k, 1, or $2 - k$ is formed. When the Knight is back at its starting point, the total horizontal displacement is 0, and thus even. Hence, the number of odd horizontal displacements (for the individual jumps) is even, so that $w + r + z$ is even, and so r and $w + z$ have the same parity. In connection with the fact that $r + (w + z) + (x + y) + g$ is even, it further follows that g and $x + y$ must also have the same parity; however, this reasoning does not show that this parity is the same as that of r and $w + z$.

286. Knights commanding the entire board. We now pose the problem: Place as few Knights as possible on a chessboard in such a way that each square is controlled (by at least one Knight), including the squares on which there is a Knight.

This puzzle presents a striking example of the advantage to be gained from partitioning a puzzle into two equal smaller puzzles (see §18), since the Knights that command the white squares have to be placed on black squares, and conversely. Hence we can first demand that as few Knights as possible (to be called black Knights) are placed on the black squares in such a way that all white squares are controlled. Starting from the white border squares, which have to be controlled, it soon becomes evident, after some trials, that the smallest number of Knights is seven, and that with seven Knights only two essentially different positions are possible (see *Figure 154*).

Two other solutions arise from this by reflection in the white diagonal (from *a*8 to *h*1). The solutions for Knights on the white squares are found by reflection in a horizontal or vertical line.

By superposing the two solutions (for the black and for the white Knights) we obtain the solutions of the original problem. Here 14 is

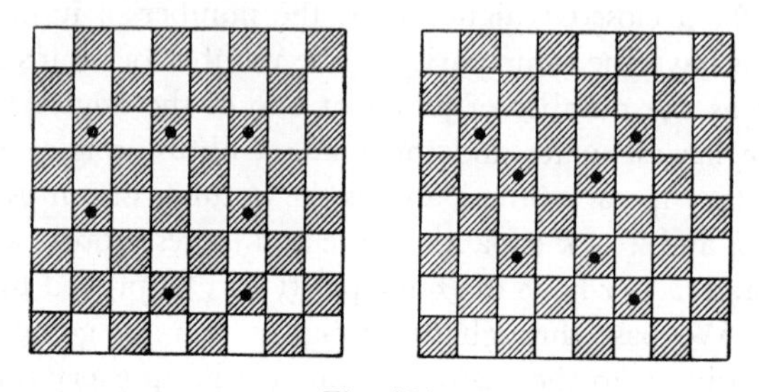

Fig. 154

the smallest number of Knights, while there are three essentially different solutions: you can choose the first solutions both times, or the second solution both times, or you can combine the two solutions. The solutions of the original problem thus obtained are shown in *Figure 155.*

No essentially different solutions are obtained by reflecting the white Knights in the black diagonal.

If you count reflected or rotated solutions as different, then the first and the second diagrams lead to four solutions each, while the third diagram yields eight solutions. This amounts to sixteen solutions in all.

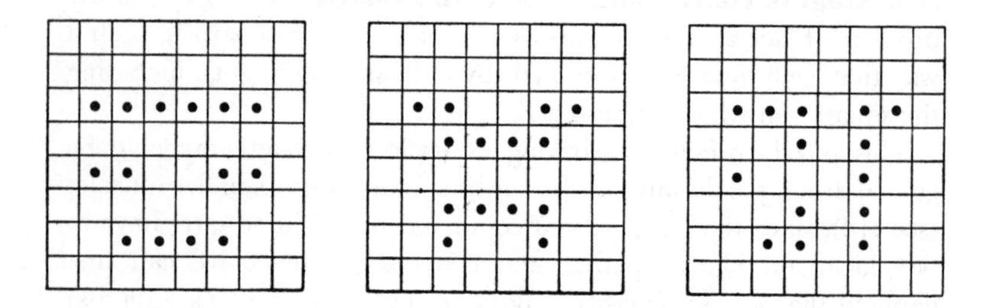

Fig. 155

VI. PROBLEM ON NAMES

***287. Who wins what?** Ten players from a bridge tournament find themselves in the same train, five being passengers and five being railway employees. The passengers are called Smith, Jones, Clark, Stone, and Black, while the five railway employees—an engineer, a fireman, a guard, a chief guard, and a mail clerk—bear the same five names in some order. It is further known that:

1. The passenger Smith won \$10.60 more than the chief guard did.
2. There is a difference in age between the passenger Black and the engineer, of 5 years to the day.
3. The passenger Clark is not related to the guard's namesake, even by marriage.
4. Last week, the fireman was punished for failure to report to work. In the newspaper paragraph the offender was identified only as a railway employee and by his initials, but it was impossible to imagine that the chief guard could have been the culprit.
5. The passenger Black is precisely as old as the engineer was when the passenger Jones was as old as the railway employee Stone was when the engineer was as old as the railway employee Stone now is.
   Moreover, the passenger Black is precisely as old as the railway employee Stone was when the passenger Jones was as old as the engineer was when the passenger Jones was as old as the engineer now is.
6. One of the passengers is the son of one of the railway employees. The mail clerk is married to the mother of the passenger in question.
7. The engineer is a neighbor of the fireman's namesake.
8. The guard and his maternal grandfather celebrate their birthdays on the same day.
9. The fireman differs in age by less than $2\frac{1}{2}$ years both from his namesake and from the engineer.
10. The chief guard's eldest grandson is married to a sister of the passenger Black, while the chief guard's youngest grandson is married to the mother of the engineer's namesake.
11. The passengers Jones and Stone live in Bridgeport.
12. The mail clerk had winnings exactly one quarter of his namesake's winnings, and the chief guard had winnings exactly one fifth of his namesake's winnings.
13. The chief guard's namesake celebrates his birthday one week before the railway employee Stone.
14. The guard is married to a daughter of the passenger Smith.
15. The railway employee Stone's mother's father was born on the same day as the passenger Clark.
16. One of the passengers is the father of one of the railway employees.
17. The engineer differs less in age from the mail clerk than from any of the eight other persons.

18. The railway employee Stone won \$2.80 more than the railway employee Smith did; the railway employee Clark won \$3.25.
19. The railway employee Smith lives in Stamford.
20. Two of the railway officials won \$6.60 each.

What are the names and the winnings of the railway employees (engineer, fireman, and so forth)? Which piece of information is superfluous?

***288. Solution of the names puzzle.** For brevity, persons will be referred to as pSm (passenger Smith), pJ (passenger Jones), rSt (railway employee Stone), e (engineer), c (chief guard), and so on. From datum 5 (first part) we know that e is older than rSt, hence e is not named Stone. The ages of pB, e, pJ, and rSt (in years) will be called B, e, J, and S, respectively. It was $e - S$ years ago that e was as old as rSt is now. At that time, rSt's age must have been $S - (e - S) = (2 \times S) - e$; this was pJ's age some $J - (2 \times S - e) = J + e - (2 \times S)$ years ago, so that at that time e's age was $e - (J + e - 2 \times S) = (2 \times S) - J$; and this was pJ's age some $J - (2 \times e - J) = (2 \times J) - (2 \times e)$ years ago, so that rSt's age then was $S - (2 \times J - 2 \times e) = S + (2 \times e) - (2 \times J)$. From datum 5 (second part) we then have $B = S + (2 \times e) - (2 \times J)$. From datum 5 (first part) we know that e is older than pB, which in connection with datum 2 establishes that $e = B + 5$. This makes the last-found relation transform into $2 \times J = S + B + 10$. Taking this together with $B = (2 \times S) - J$ leads to the results $J = S + 3\frac{1}{3}$, $B = S - 3\frac{1}{3}$. In connection with datum 13, this shows that c's name cannot be either Jones or Black. We also derive the results $e = S + 1\frac{2}{3}$ and $J = e + 1\frac{2}{3}$, so that e has the same difference in age from rSt as from pJ; this shows that m's name cannot be Stone (in view of datum 17). Since datum 15 shows that pC is much older than rSt, and hence also much older than e, and since the latter is five years older than pB, it follows from datum 9 that f's name cannot be Clark or Black. According to datum 15, pC's mother is certainly older than 72, so that c's youngest grandson cannot be married to that mother. This, together with datum 10 (second part), shows that e's name is not Clark.

If e's name were Smith, datum 19 would imply that he lived in Stamford, so in that event, according to datum 11, he could be a neighbor neither to pJ nor to pSt. According to datum 7, f's name could not then be either Jones or Stone, which would mean that in this case there would then no longer be any name available for f. Consequently, e's name cannot be Smith.

From what we have shown, it follows that c's name must be Clark, Stone, or Smith. If c's name is Clark, then datum 1 and datum 18 imply that pSm has won \$13.85; since \$13.85 cannot be divided by 4, it follows from datum 12 that m's name cannot then be Smith. If c's name is Stone, it follows from datum 18 that c has won \$2.80 more than rSm, so that pSm has won \$13.40 more than rSm in that case (in view of datum 1); since \$13.40 cannot be divided by 3, rSm's winnings cannot here be a quarter of pSm's winnings, which means (in view of datum 12) that m's name cannot then be Smith. Since this is equally true when c's name is Smith, we now know that m's name is not Smith.

From datum 4 it follows that the names of f and c cannot jointly be Smith and Stone, in either order. Since it has already been established that neither e nor m has either of the names Stone or Smith, it follows that g's name must be Stone or Smith. From datum 9, it follows that f is neither the son nor the father of a passenger. Since what we have already found shows that e's name must be either Jones or Black, and since e's age differs little from those of pJ and pB (because of $J = e + 1\frac{2}{3}$ and $e = B + 5$), e is neither the son nor the father of a passenger. According to 17, m's age, too, can differ little from those of pJ and pB; hence, if m's name is Jones or Black, he is neither the son nor the father of a passenger. If we assume that g's name is Smith, it would follow from 14 that g is neither the son nor the father of a passenger; but then, since datum 6 and datum 16 indicate that railway employees are the son of a passenger and the father of a passenger (so these railway employees would here be m and c), this would now show that m's name could not here be Jones or Black so it would have to be Clark, which would require c's name to be Stone. In view of pC's age (as implied by datum 15), it would follow that pC would have to be m's father, and that pSt would have to be a son of c; it would further follow from datum 6 (second part) that m would be married to pSt's mother, so that pC would be the father of the stepfather of pSt, that is, of c's namesake. Since this would be contrary to 3, it follows that g's name cannot be Smith, so that the guard must have the name Stone. In connection with datum 8 and datum 15, this then shows that rSt has the same birthday as pC, and datum 13 then shows that c's name is not Clark. As a consequence, the chief guard must have the name Smith. From this it successively follows that the fireman must have the name Jones, that the engineer must have the name Black, and that the mail clerk must have the name Clark.

Datum 18 then shows that it was the mail clerk who won \$3.25. From datum 1 and datum 12 it then follows that the winnings of pSm are \$10.60 more than, and also five times as large as, c's winnings; this shows that the chief guard's winnings are \$2.65—a fourth of \$10.60. From datum 18 it then follows that the winnings of the guard are \$5.45. Finally, datum 20 shows that the fireman and the engineer must have won \$6.60 each.

The first part of 10 has not been used, and indeed cannot be used.

VII. ROAD PUZZLES

289. Simple road puzzle. A pedestrian has to go from *A* to *B* in *Figure 156*, and in so doing, he also has to go round all eight of the circles. He is not allowed to go over any part of a road twice, but he is allowed to arrive at the same point more than once. In how many ways can he do this?

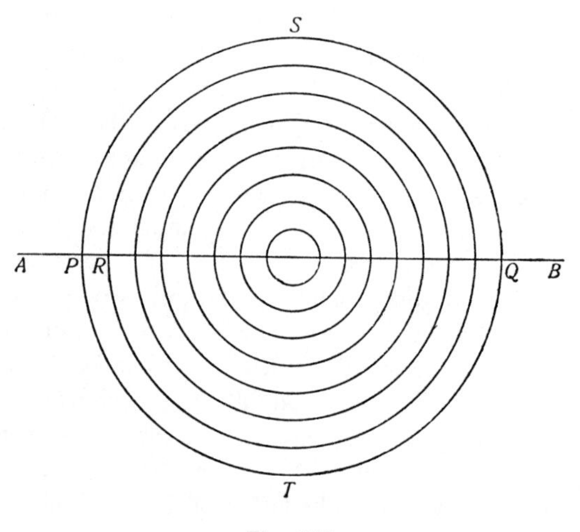

Fig. 156

If he had one circle to traverse (instead of eight), he could meet the requirements in six ways; on his first arrival at the circle, he could continue in three ways. When he then came to the second point of intersection of the line and the circle, he would have a choice between two roads, so that in all there would be $3 \times 2 = 6$ possibilities (this is immediately evident also by considering the number of permutations of the three roads: $3! = 6$). The six possibilities are shown in *Figure 157*.

Now we take two circles (*Fig. 158*). If we count only one possibility for the trip from C to D along the straight line and the smaller circle, our last result shows that there would be six possibilities to consider (because of three choices to be made at C, and two at D). In each of

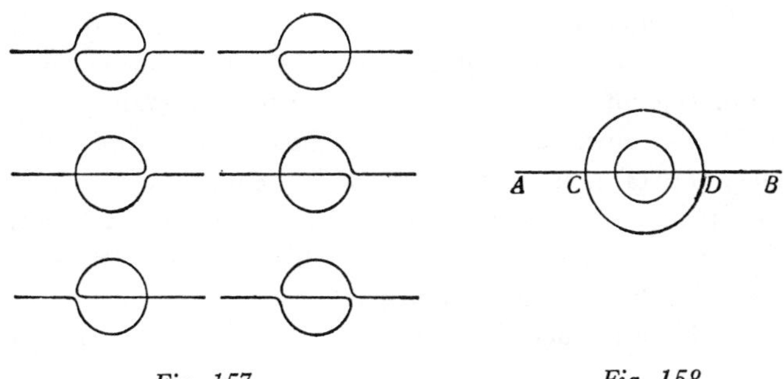

Fig. 157 *Fig. 158*

these six cases, there is still the need to make the trip from C to D (or back) along the straight line and the smaller circle, which makes each of the six cases split into six sub-cases. This produces $6 \times 6 = 6^2 = 36$ possibilities in all. This shows that every additional circle makes the number of possibilities six times as large, so that with 3 circles there are 6^3 possibilities, and with 8 circles $6^8 =$ **1,679,616** possibilities.

It may be remarked that the results of §§265–271 give immediate proof that it is possible to traverse all the roads as required, since the figure can be interpreted as a network (with some of the edges curved) for which all vertices, except A and B, are even—and, indeed, all of order 4. With *Figure 159*, for example, the problem would involve something impossible.

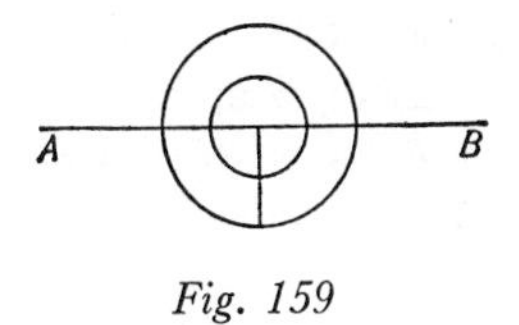

Fig. 159

290. More difficult road puzzle. We pose the same problem as in §289, but this time with respect to *Figure 160*. Among the points of contact of the circles, there may be one or more at which a movement from left to right changes into a movement from right to left. After arrival at such a point of contact for the second time, the movement must continue from left to right. Later, the pedestrian must return to the point of contact in question, and then a movement from right to left must change into a movement from left to right. However, points of contact where this occurs (turning points) need not be present, since it is quite possible for the man first to go continually to the right, from A to D, then continually to the left, until he is back at C, and then again continually to the right, from C to B. The points of contact where the movement is reversed, that is, the turning points, can be chosen in 2^5 ways, since each of the five points of contact may either be or not be one of these turning points. As soon as a choice of the turning points has been made, this then fixes the order in which the points of contact are encountered, after which the only remaining question is to decide which of three roads the man takes from one point of contact to the next, each time. He must first go from C to the first

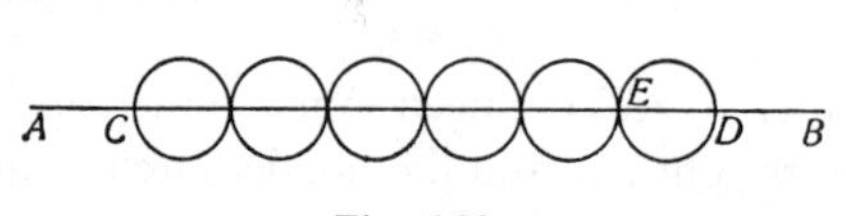

Fig. 160

turning point (numbered from left to right), then back to C, next to the second turning point, then back to the first turning point, from there to the third turning point, back to the second turning point, and so on. The trip made from any point of intersection of the circle and the line AB to any adjacent point of intersection has to be made three times over (twice from left to right, and once from right to left), along a different road every time; so these trips can be done in $3! = 6$ ways. This means that the total number of ways in which the entire trip from A to B can be executed is $2^5 \times 6^6 =$ **1,492,992.**

This number can be obtained more quickly as follows: A trip from E (the last point of contact) to D and back can be made when the pedestrian has arrived at E for the first time, but also after he has reached E for the second time, without yet having been at D (two cases). The three roads between E and D (or back) can be chosen in $3! = 6$ ways. This means that each addition of a circle makes the number of ways in which the entire trip can be made become $2 \times 6 = 12$ times as large. Since this number is 6 with one circle, it must be equal to $6 \times 12^5 =$ **1,492,992** with six circles.

291. Still more difficult case. Again we ask the same question as in §289, but this time for the upper diagram in *Figure 161.* This diagram can also be replaced by the lower one, if we assume that each of the line segments *CD*, *DE*, *EF*, *FG*, and *HG* then has to be traversed three times, and the arc *CJH* twice. In the lower diagram, we can disregard the question of the order in which the three roads between *F* and *G* or the two semicircles between *C* and *H* are to be traversed. We then can multiply the number of possibilities for this simplified problem by $(3!)^5 \times 2! = 2 \times 6^5$, to obtain the result for the original problem.

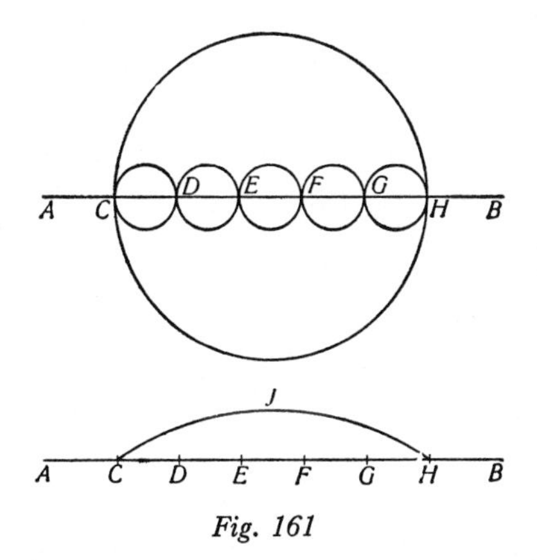

Fig. 161

The pedestrian can traverse the road *CJH* or back in four orders, namely *CJH–HJC*, *HJC–CJH*, *CJH–CJH*, and *HJC–HJC*. First we shall consider the case *CJH–HJC*. Arriving at *C*, the man can first cover the road *CJH* there and back, and then go from *C* to *H* in 2^4 ways (via *D*, *E*, *F*, *G*), since he can choose turning points (see the first solution of the puzzle of §290) in 2^4 ways (making *D* either a turning point or not, and the same for *E* and the rest). He can also make the trip in one of the two following ways: *CDCJHJC*__DH__, *CJHGHJ*__CG__*H*, where he can make the trips *DH* and *CG* in 2^3 ways each (as appears from the possible turning points *E*, *F*, *G* in the first case, and *D*, *E*, *F* in the second case). As well as this, he can make a trip in one of the three following ways: *CECJHJC*__EH__, *CDCJHGHJC*__DG__*H*, *CJHFHJ*__CF__*H*, where the road indicated in bold type can be covered in 2^2 ways, and we continue in this fashion. For the final case, we find that the trip can be made in one of the following six ways:

CHCJHJCH, *CGCJHGHJCH*, *CFCJHFHJCH*,
CECJHEHJCH, *CDCJHDHJCH*, *CJHCHJCH*.

Hence, in the simplified problem, in the case *CJH–HJC*, the trip can be made in $2^4 + 2 \times 2^3 + 3 \times 2^2 + 4 \times 2 + 5 \times 1 + 6 = 63$ ways. The same is true for the case *HJC–CJH*, which can be reduced to the previous one by imagining the trip to be made in the opposite direction, from *B* to *A*.

In the case *CJH–CJH*, too, there are 63 ways to get from *C* to *H* (still in the simplified puzzle), namely:

*CJ***HC***JH* (2^4), *CDCJ***HD***CJH* (2^3), *CJH***GC***JHGH* (2^3), *CECJ***HE***CJH* (2^2), *CDCJH***GD***CJHGH* (2^2), *CJH***FC***JHFH* (2^2), *CFCJ***HF***CJH* (2), . . ., *CHCJHCJH*, *CGCJHCJHGH*, . . ., *CDCJHCJHDH*, *CJHCJHCH*;

the bold-type parts of the road can be traversed in the numbers of ways indicated in parentheses. In the case *HJC–HJC* the trip can be made in one way only, *CHJCHJCH*. Thus, the simplified puzzle has a total of $(3 \times 63) + 1 = 190$ solutions. For the original puzzle (the one with the six circles) this gives $2 \times 6^5 \times 190 =$ **2,954,880** possibilities.

For the reader who is familiar with arithmetico-geometric series, we give the extension to the case of n small circles spanned by one large circle. In the simplified problem the number of ways in each of the cases *CJH–HJC*, *HJC–CJH*, and *CJH–CJH* is then equal to:

$$2^{n-1} + 2 \times 2^{n-2} + 3 \times 2^{n-3} + \cdots + (n-2) \times 2^2 + (n-1) \times 2 + n + (n+1) = 2^{n+1} - 1,$$

which equals 63 for $n = 5$, as is proper. In the simplified problem the number of ways is $3(2^{n+1} - 1) + 1 = (3 \times 2^{n+1}) - 2$, hence in the original problem it is $4 \times 6^n \times (3 \times 2^n - 1)$.

***292. Modification of the puzzle of §289.** We take Figure 156, and again we require the pedestrian to go from *A* to *B* without passing through the same section twice. However he need not pass over all the roads. In how many ways can he do this?

The required number for eight circles will be called X_8, for seven circles X_7, and so on. To find X_8 we note that the man has two ways to get from *A* to *B* without passing through the point *R* (by way of *PSQ* or *PTQ*). Furthermore, he can get from *A* to *B* in X_7 ways without setting foot on the outer circle *PQ*, and finally in $6 \times X_7$ ways via *R*, making a trip round the whole of the other circle. For if a trip from *R* to *Q* or back, avoiding *P* (which can be completed in X_7 ways) is counted as a single possibility, then we obtain six ways to get from *P* to *Q* via *R*, traversing the entire outer circle (*PSQRPTQ*, *PSQTPRQ*, *PRQSPTQ*, *PRQTPSQ*, *PTQSPRQ*, *PTQRPSQ*). Hence, we find that $X_8 = (7 \times X_7) + 2$, hence $X_8 + \frac{1}{3} = 7(X_7 + \frac{1}{3})$. Likewise,

$X_7 + \frac{1}{3} = 7(X_6 + \frac{1}{3})$, $X_6 + \frac{1}{3} = 7(X_5 + \frac{1}{3})$, and so on. Thus $X_8 + \frac{1}{3} = 7^8(X_0 + \frac{1}{3})$. Now $X_0 = 1$, since here there is nothing but the straight road AB. Consequently, $X_8 + \frac{1}{3} = 7^8 \times \frac{4}{3}$. From this it follows that $X_8 = \frac{1}{3}(4 \times 7^8 - 1) = \frac{1}{3}(4 \times 5764801 - 1) = \frac{1}{3} \times 23059203 =$ **7,686,401**. Hence the trip can be completed in over $7\frac{1}{2}$ million ways.

Of course, the number X_8 can also be found by successively computing X_1, X_2, X_3, and so on, from $X_1 = 7X_0 + 2$, $X_2 = 7X_1 + 2$, and so forth.

VIII. SOME PUZZLES ON SUMS

293. Puzzle with seven sums. In *Figure 162* the six numbers have to be rearranged in such a way that the seven sums for pairs of numbers in adjacent squares (adjacent either horizontally or vertically) are all different. All solutions are required. Solutions which arise from each other by reflection or rotation are considered to be identical.

1	**2**	**3**
4	**5**	**6**

Fig. 162

A solution remains a solution if we replace each of the six numbers by its difference from 7, that is, if we interchange 1 and 6, 2 and 5, and 3 and 4. When this is done, each of the seven sums is replaced by its difference from 14, so that different sums remain different. The solution thus obtained (which may be the same as the original one) will be called the complementary solution. We can make use of this to reduce the solving time, since we can assume that the sum of the two middle numbers is at most 7. In order to avoid getting the same solution twice, we assume an upper location for the smaller of the two middle numbers, and we take the top left-hand number to be less than the top right-hand number. We arrange the solutions according to their top center numbers, in increasing magnitude, and, in case of equality, according to their lower center numbers. Cases which agree even in these numbers will be subdivided according to the top left-hand numbers, and, in cases of equality, again according to the top right-hand numbers.

Solutions which are the same as their complements can arise only when 1–6, 2–5, or 3–4 are taken as the middle two numbers. These middle numbers do not lead to solutions which differ from their complements. Solutions of this latter type can be found by taking 1–2, 1–3, 1–4, 1–5, 2–3, and 2–4 successively as the central numbers; of these cases, the first, the third, and the fifth come to a dead end.

It is a good idea to draw squares on cardboard, cut them out, and number them; then you will quickly find all solutions by shifting the squares according to the classification system indicated. This will give the following solutions:

214	213	216	123	126
536	654	354	645	345

and their complements, and also the following solutions, which transform into their complements after a rotation:

213	315	124	326	132	135
465	264	356	154	546	246

If we also require six different sums for pairs of numbers that either are diagonally adjacent, or are connected by a Knight's move, this allows only the solution $\begin{matrix}216\\354\end{matrix}$ and its complement $\begin{matrix}561\\423\end{matrix}$

294. Modified puzzle with seven sums. We take the same puzzle as in §293, but now we require that the seven sums of adjacent numbers have as few different values as possible (see *Figure 163*).

a	b	c
x	d	y

Fig. 163

Since the sums $a + b$, $b + c$, and $b + d$ are to be different, we need to have at least three different sums. We obtain not more than three different sums only if x and y satisfy the relations:

$$a + x = b + d = c + y, \quad x + d = b + c,$$
$$d + y = a + b;$$

other combinations of equal sums prove to be incompatible with the condition that a, b, c, and d are all different. It follows easily from the above relations that we must have $2 \times d = a + c$. Since we can assume that a is less than c, this can also be written as $a = d - v$, $c = d + v$.

This leads further to $x = b + v, y = b - v$. Since we can assume b to be less than d, the six numbers $b, b - v, b + v, d, d - v$, and $d + v$ are mutually different only when $b = 2, d = 5, v = 1$, or $b = 3, d = 4, v = 2$ (since we must have $b - v = 1$ and $d + v = 6$). This gives the two solutions $\begin{matrix}426\\351\end{matrix}$ and $\begin{matrix}236\\541\end{matrix}$. These are identical to their complements (when 1 and 6, 2 and 5, and 3 and 4 are interchanged).

295. Puzzle with twelve sums. In *Figure 164,* the numbers 1, 2, . . . , 9 have to be rearranged in such a way that the twelve sums of pairs of adjacent numbers (horizontally or vertically) are all different.

1	2	3
4	5	6
7	8	9

Fig. 164

One solution changes into another solution (which may or may not be essentially different)—once again we shall call this a complementary solution—if we replace each number by its difference from 10, which replaces 1 by 9, 2 by 8, and so on. In at least one of two complementary solutions, a number 1, 2, 3, 4, or 5 must occupy the central square. So we can confine ourselves to finding solutions for which this applies. If 5 is the number in the central square, then we can further assume that the sum of the four border numbers (numbers situated at mid-points of sides of the square) is at most 20, since this sum is replaced by its difference from 40, in the complementary solution.

To find all solutions by systematic trial, we fill in the squares in a definite order, starting with the central square, in which we first enter a 1. We can assume that the smallest border number is in the top row, and that the left-hand border number is smaller than the right-hand one. After selecting the number in the central square, we enter the top border number, then the left-hand one, next the right-hand one, and then the lower border number. We shall also assume a definite order for inserting the corner numbers. When there are two cases, we shall give priority to the one for which the first available number to be entered is the smaller. As soon as a sum appears which is equal to a sum already present, we increase the last number entered by 1; if it is 9, then we give a unit increase to the next-to-last number entered and reduce the last-entered number as much as possible. It requires a great deal of perseverance and shifting of cut-out squares to find all solutions. The system has been constructed to avoid the production of complementary solutions and of solutions which arise from one another by rotation or reflection, equivalent to solutions already found.

We obtain four solutions which transform into their complements after a rotation through 180°, namely:

412	312	124	231
357	456	357	456
896	897	689	978
diag.			

It turns out that there are no solutions which transform into their complements by reflection. If we look exclusively for solutions that are identical to their respective complements (so that 5 is in the central square, of course), then all solutions can be found rather quickly. There are 66 solutions which differ from their complements; arranged in the manner indicated above, these are:

423	324	324	317	413	314	517	139	219	217	214
618	619	718	429	528	528	829	527	436	439	536
975	578	965	586	769	679	436	864	758	586	987
	(diag.)						(succ.)			

214	821	721	128	713	512	512	215	318	126	127
537	435	435	536	249	348	349	649	749	345	349
869	769	698	749	568	679	687	387	526	789	568
		(diag.)	(succ.)	(succ. diag.)				(Kt)		

321	321	321	132	132	231	216	216	218	218	312
546	546	549	546	546	648	354	359	359	359	456
987	798	687	987	798	597	987	748	467	476	978
				(diag.)						

and their complements. In all, there are 70 solutions.

If we count complements as identical, there is only one solution in which the four corner numbers are even. If we require the eight "diagonal" sums (sums of diagonally adjacent pairs of numbers) to be different also, then there are five solutions; these have been indicated by a caption "diag." If we require not only these twelve to be different, but also the eight "Knight's sums" (sums of pairs of numbers that are connected by a Knight's move), then there is only one solution; this has been indicated by the caption "Kt." For three of the solutions, eleven successive numbers occur among the twelve sums; these solutions have been indicated by the caption "succ." There is one solution for which eleven among the twelve different sums are successive and for which the eight diagonal sums are different as well.

***296. Another puzzle with twelve sums.** We modify the puzzle of §295 to the effect that we require the twelve sums of adjacent numbers to have as few different values as possible.

$$\begin{matrix} w & b & x \\ c & a & d \\ y & e & z \end{matrix}$$

The number of different sums can certainly not be less than four, since the sums $a + b$, $a + c$, $a + d$, and $a + e$ must all be different. To find a solution with four different sums, we determine the numbers w, x, y, z such that:

$$w + c = a + e = x + d, \quad c + y = b + a = d + z.$$

For w, x, y, and z we then find the relations:

$$w = a + e - c, \quad x = a + e - d, \quad y = b + a - c, \quad z = b + a - d.$$

If the equation $w + b = a + d$ is also to be satisfied, it is necessary to have $b + e = c + d$. This then requires also that $y + e = a + d$ and $b + x = c + a = e + z$. For the sum of the nine numbers we find $5a + 3b - c - d + 3e$, and hence $5a + 2(b + e)$. Since this sum is also equal to $1 + 2 + \cdots + 9 = 45$, we have $5a + 2(b + e) = 45$. Consequently, $b + e$ must be a multiple of 5. Since a transfer to the complementary solution replaces $b + e$ by its difference from 20, we can assume that $c + d = b + e = 5$, $a = 7$, or $c + d = b + e = 10$, $a = 5$. Since we can assume that b is smaller than c and d, and c smaller than d, we obtain the following cases:

1	1	1	1	2	2	3
273	258	357	456	357	456	456
4	9	9	9	8	8	7

From the relations $w = a + e - c$, and so on, we find the corner numbers. Only the first and sixth cases produce the numbers $1, 2, \ldots, 9$.

These cases give the solutions

918 927
273 456
645 381

and the complement of the first one.

There are no other solutions, since if we use a different scheme to equate the sums $w + b$, $w + c$, $b + x$, $x + d$, and so forth, to the numbers $a + b$, $a + c$, $a + d$, and $a + e$ (that is, with altered correspondence), we always reach an impasse.

***297. Knight's move puzzle with eight sums.** We pose the following problem: To rearrange the nine numbers in Figure 164 in such a way that the eight Knight's sums (sums of pairs of numbers that are connected by a Knight's move) have as few different values as possible.

A solution remains a solution when we interchange 1 and 9, 2 and 8, and so on (so we have complementary solutions). The eight numbers on the border or in the corners can be traversed by a closed Knight's tour. Hence, we have to place these eight numbers in a cycle in such a way that the eight sums of pairs of numbers that are adjacent in the cycle have as few different values as possible. From such a cycle we then find two solutions, since a given number in the cycle can be placed either in a corner or in the middle of a side.

Since there are eight different numbers in the cycle, two sums that are adjacent in the cycle cannot be equal. So among the eight sums there can be no five equal sums. It is impossible for the eight sums to take only two different values, since this would mean that alternate sums in the cycle would have to be equal, and this would require the numbers in alternate sets to be increased (or decreased) by the same amount progressively; this is not possible because the cycle is closed.

However, it turns out that with three different values of the sums, there is a solution. Here we have to distinguish several cases.

CASE I: Four of the eight sums are equal, and so are three of the other sums. Then the eight numbers of the cycle can be written as indicated in the diagram. The four equal sums have been indicated by •, the three equal sums by ○, and the eighth sum by –. The numbers a, b, and v have to be chosen in such a way that eight of the nine

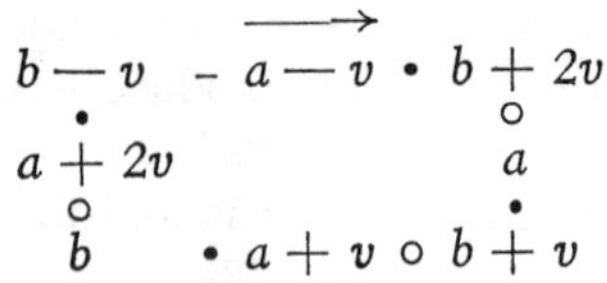

$$\begin{array}{ccccc} b-v & - & \overrightarrow{a-v} & \bullet & b+2v \\ \bullet & & & & \circ \\ a+2v & & & & a \\ \circ & & & & \bullet \\ b & \bullet & a+v & \circ & b+v \end{array}$$

numbers 1, 2, . . . , 9 arise, in some order. We can assume that v is positive (and thus 1 or 2), because otherwise this can be achieved by passing to the complementary solution. Furthermore, we can take a to be less than b, because otherwise we can obtain this by running through the cycle in the opposite direction. Thus, we arrive at the following five cycles of the first type:

18273645 19283746 29384756 18365472 29476583;

the square at which the cycle starts has been indicated by an arrow in the diagram. Each cycle yields four solutions, including the complementary solutions. The solutions resulting from the first cycle

$$\begin{array}{ccc} 174 & \quad & 617 \\ \text{are } 698 & \text{and} & 294 \\ 253 & & 538 \end{array}$$

and the two complementary solutions.

Case II: Four equal sums, and two equal sums twice. In the cycles shown the four equal sums have been indicated by •, the two other sums by ○ and –. In cycle A, the two equal sums are pairwise opposite

$$A.\quad \begin{array}{ccccc} a-x & \bullet & b+x & \circ & a \\ - & & & & \bullet \\ b+z & & & & b \\ \bullet & & & & - \\ a-z & \circ & b+y & \bullet & a-y \end{array}$$

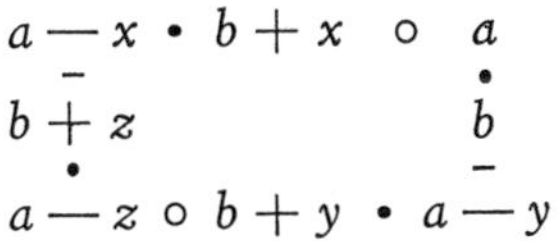

$$B.\quad \begin{array}{ccccc} a-x & \bullet & b+x & \circ & a \\ - & & & & \bullet \\ b+z & & & & b \\ \bullet & & & & \circ \\ a-z & - & b+y & \bullet & a-y \end{array}$$

$$C.\quad \begin{array}{ccccc} a-x & \bullet & b+x & \circ & a \\ \circ & & & & \bullet \\ b+z & & & & b \\ \bullet & & & & - \\ a-z & - & b+y & \bullet & a-y \end{array}$$

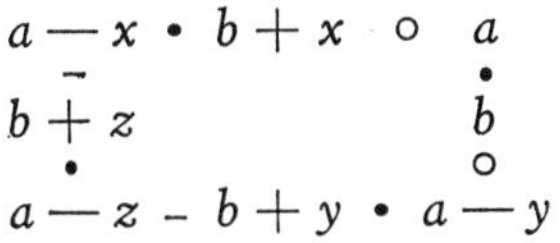

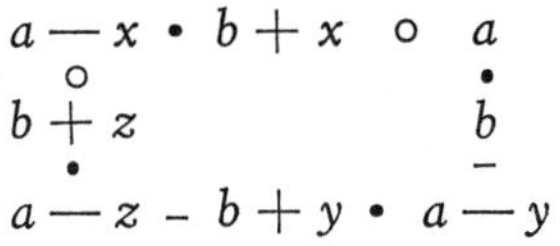

in the cycle. If this is not the case, the equal sums have the positions indicated in cycles B and C. In cycle A we have:

$$(b + x) + a = (a - z) + (b + y),$$
$$(a - x) + (b + z) = b + (a - y),$$

hence $y = x + z$, $x = y + z$, and so $z = 0$, which leads to equal numbers. In cycle B we have:

$$(b + x) + a = b + (a - y),$$
$$(a - x) + (b + z) = (a - z) + (b + y),$$

whence $z = 0$ (and equal numbers). Cycle C leads to $x = y$, and thus to equal numbers. Thus, case II does not give any solutions.

CASE III: Three equal sums twice and two more equal sums. The six sums which are equal in groups of three cannot all be successive in the cycle, since otherwise the two equal sums would be adjacent,

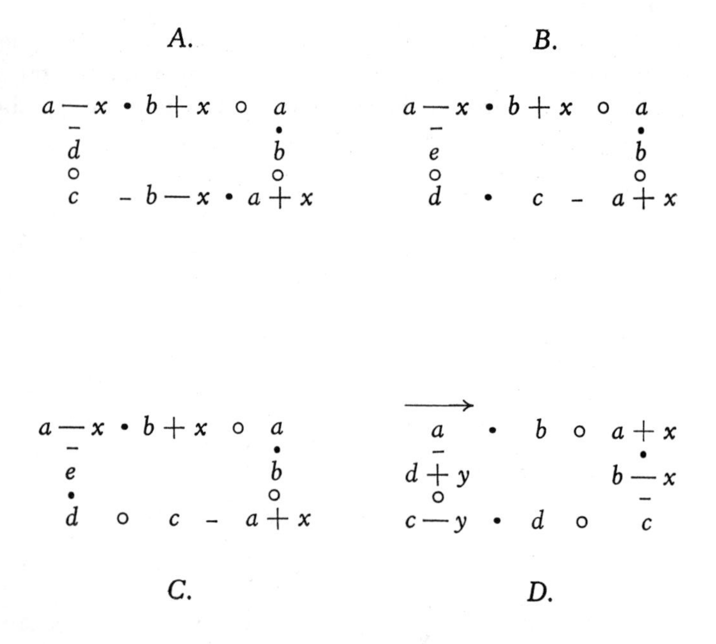

which is impossible. If the six sums are placed in the cycle in such a way that five of these sums succeed one another (taking one from one of the triplets and one from the other alternately), then we obtain cycle A, which leads to:

$$(b + x) + a = d + c, \quad (a - x) + d = c + (b - x),$$

whence we have $2a = 2c - x$. Consequently, x is even, so that we can put $x = 2z$, $c = a + z$, $d = b + z$, to make the eight numbers:

$$a - 2z, \quad b + 2z, \quad a, \quad b, \quad a + 2z, \quad b - 2z, \quad a + z, \quad b + z.$$

No set of values of a, b, and z makes these eight numbers different.

If the six sums which are equal in triplets are placed in the cycle in such a way that only four of the sums are successive, then we obtain the possibilities B and C. For cycle B we have:

$$a + b = d + c, \quad a + b + x = e + d, \quad e + a - x = c + a + x,$$

whence $x = 0$ (and equal numbers). For cycle C we have:

$$a + b = e + d, \quad a + b + x = d + c, \quad e + a - x = c + a + x,$$

which also leads to $x = 0$.

This leaves only the case where the six sums which are equal in triplets are placed in a cycle in such a way that three of these sums are successive, and likewise the other three sums. We then obtain cycle D. In this case we have:

$$a + b = d + c - y, \quad b + a + x = d + c, \quad a + d + y = c + b - x,$$

whence $x = y$, $2c = 2a + 3x$, so that x is even, and thus we can put $x = 2z$. From this it follows that $c = a + 3z$, $d = b - z$. Hence, the eight numbers in the cycle (starting at the arrow) are:

$$a, \quad b, \quad a + 2z, \quad b - 2z, \quad a + 3z, \quad b - z, \quad a + z, \quad b + z.$$

Here we can assume z to be positive, since otherwise this can be achieved by interchanging the roles of a and c (as well as of b and d, and so on). This gives us $z = 1$ or $z = 2$. A cycle transforms into a cycle of the same type (with the same value of z) when we start at the fourth number (hence at $b - 2z$) and read in the opposite direction. So we can assume that a is less than $b - 2z$. In this way we find the following five cycles of the second type:

17354628 18364729 28465739 16527438 27638549,

the second of which coincides with its complement, while the other four are complements of one another. Each of the five cycles yields two solutions (by putting the first number in a corner or in the middle of a side).

If we consider the complementary solutions to be different, but solutions that arise from one another by rotation or reflection to be equivalent, then according to the foregoing, there are **thirty** solutions, twenty of the first type and ten of the second type. There is no solution which coincides with its complement; the two solutions that arise from the second cycle of the second type are complements of one another.

IX. COUNTING-OFF PUZZLES

298. Simple counting-off puzzle. Twenty-five persons are standing in a circle. Someone starts to count off the persons, going round the circle and pointing to a person every time. The verbal counting runs from 1 through 7, then resumes with 1. Persons who are indicated by 7 drop out, and leave the circle. When the entire circle has been covered, the count off continues with the person at whom it started. Finally, only one person remains. With whom should one start in order that a given person should be the last one left?

To solve this, we start the count-off with an arbitrary person, let us say with person A, who has to be the last one left. It then turns out that the person who is the fifteenth in the circle is the last one left if the persons are numbered in the usual way, starting with A as number 1. If the count-off had started with number 2 or 3, and so on, then number 16 or 17, and so on, respectively, would have been the last one left. Hence, to leave A, who can also be considered as number 26, one has to start with number 12 $(= 26 - 15 + 1)$.

299. More difficult counting-off puzzle. Between 40 and 100 persons are arranged in a circle. A count-off of the type 1, 2, 1, 2, 1, and so on, is begun, starting with one of the persons whom we shall call A; the persons indicated by 2 drop out. It is known that this makes A the last one to remain. What is the number of persons originally in the circle?

The number of persons must be even, since otherwise *A* would drop out as soon as the count-off completed the first round. When the count returns to A for the first time, half the persons have dropped out. At that moment, the situation is exactly as it was originally, but with a circle of half the original size. If A is not to drop out after the second round either, half the initial number of persons must then be even, too. This applies after every round, so that the number of persons must be a power of 2, and therefore 2 or 4 or 8 or 16, and so on. Since the number lies between 40 and 100, it must be **64.**

300. Modification of the puzzle of §299. We number the persons in the circle in the usual way, beginning with person A, with whom the count-off started, and continuing in the direction of the count-off. Into the puzzle of §299 we now introduce the following modification: the last person B of the circle (the one having the highest number) is the last one to remain, and again we require to know the number of persons in the circle. We ask the same question about the case where number 7 is the last one to remain. As before, the number of persons lies between 40 and 100.

In the first question, the number of persons must be odd, since otherwise B would drop out in the first round. When B is pointed to for the first time (on a count of 1) half of the others have by then dropped out. So the same situation has arisen as in §299, but with the difference that the count-off now resumes with B. So when B is pointed to for the first time, the number of remaining persons must be a power of 2, and therefore 2 or 4 or 8 or 16, and so on. The number of persons who have already dropped out is one unit less, and therefore 1 or 3 or 7 or 15, and so on, correspondingly, so the original number of persons must be 3 or 7 or 15 or 31, and so on, in the respective cases, which makes it always one less than a power of 2. Since the number lies between 40 and 100, it must be 63.

If number 7 is to be the last person to remain, three persons will have been eliminated by the time that the count first arrives at number 7, and once again, as in §299, there will be the same situation as at the beginning of the count-off. This shows that the number of persons who then remain must be a power of 2, so that the original number must be three units more than a power of 2, which makes it 67.

301. Counting-off puzzle for 1–2–3. We modify the puzzle of §299 to the effect that the count-off is of the type 1, 2, 3, 1, 2, 3, 1, and so forth, beginning with A; every person indicated by 3 drops out. The number of persons in the circle lies between 4100 and 13,000. How large is this number if it is given that A (who is numbered 1 in the circle) is the last one to remain? We ask the same question for the case where number 2 is the last one left; we assume that persons are numbered in the usual way, starting with A.

We proceed with no regard to whether the last remaining person is to be A (number 1), as in the first question, or whether the survivor is to be number 2, as in the second question. The number of persons in the circle will be called m; the number of persons who are left when the count returns to A for the first time will be called n.

If m is a multiple of 3, then $m/3$ persons have dropped out when the count returns to A. Since A then receives a 1 again, the original situation will have arisen again, but with a smaller circle. We then have $n = 2m/3$, and hence $m = 3n/2$, where m and n both refer to the first question (in which number 1 is the survivor) or both to the second question (with number 2 as the survivor).

Suppose now that m is one unit more than a multiple of 3, and that the count-off has come to B, the last person in the circle (so that B will be indicated by 1). By this time, $(m - 1)/3$ persons will have dropped out; in this case it can only be A who is the last one left, since number 2 drops out in the second round, and we are assuming that either number 1 or number 2 will remain to the end. It follows that we here have $m = \frac{m-1}{3} + n$, and hence $m = \frac{3n}{2} - \frac{1}{2}$. When the count-off comes round to B, the original situation repeats, but with a smaller circle; however, we now have a number n concerned with our second question, associated with a number m which refers to our first question.

Suppose now that m is one unit less than a multiple of 3, and that the count-off has returned to A. A then drops out, and a total of $(m + 1)/3$ persons have been eliminated; in this case we assume that number 2 is now to be the last one to remain. This then gives us a result $m = \frac{m+1}{3} + n$, and hence $m = \frac{3n}{2} + \frac{1}{2}$. When the count-off reaches number 2 for the second time, we have the same situation as originally, but with the difference that the circle is smaller and that the number n is one related to our first question (while m is here a number which refers to our second question).

Our arguments show that the required numbers for circles of various sizes (disregarding for the present the limits between which the number of persons has to lie) can be found by repeated multiplication by 3/2. If the multiplicand is even, then the case concerned (no matter whether number 1 or number 2 is to be the last one left) remains the

TABLE 18

No. 1	No. 2	No. 1	No. 2	No. 1	No. 2	No. 1	No. 2
	2	355		69127		13453488	
	3		533		103691	20180232	
4		799		155536		30270348	
6			1199	233304		45405522	
9		1798		349956		68108283	
	14	2697		524934			102162425
	21		4046	787401		153243637	
31			6069		1181102		229865456
	47	9103			1771653		344798184
70			13655	2657479			517197276
105		20482			3986219		775795914
	158	30723		5979328			1063693871
	237		46085	8968992		1595540806	

same. If the multiplicand is odd, a transfer from one case to the other is then involved. If we replace number 2 (as the last survivor) by number 1, a new number has to be rounded downwards to the next lower integer after the multiplication. If we change from number 1 to number 2, then we have a new number which has to be rounded upwards to the next higher integer.

In Table 18 we show some of the numbers concerned. At the top of the column we have indicated whether number 1 or number 2 is the last survivor. In constructing the table we have thought it best not to exceed the total population of the world.

When regard is paid to the given limits, the table shows that the answer to our first question is **9103**, and the answer to our second question is **6069**.

X. SOLVING COUNTING-OFF PUZZLES BY REVERSAL

302. Solution of the puzzle of §298 by reversal. We imagine that the original count-off takes place in the clockwise direction and that the persons face the center of the circle. The numbers on the inside of the circle in *Figure 165* indicate the ordinary clockwise numbering starting at A. We now proceed to perform the count-off in the opposite direction, that is, counterclockwise, as indicated by an arrow

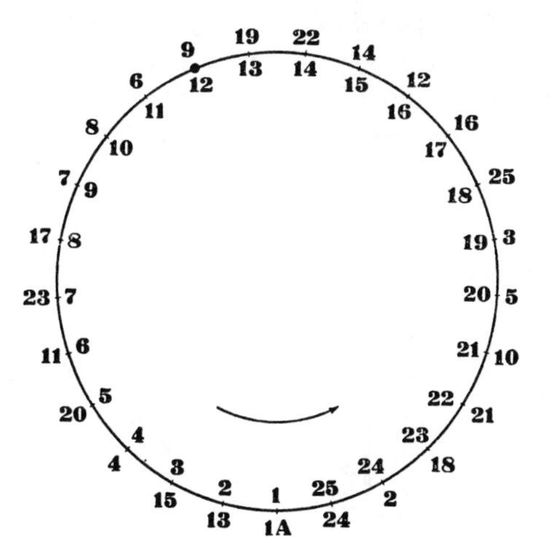

Fig. 165

in the figure. We begin with a circle which contains only A. In the reversed count-off, persons are admitted to the circle in the order and in the places indicated by the numbers on the outside; since A is placed first, the number 1 has been written next to him on the outside. If any number on the outside is subtracted from 26, the difference here indicates the total number of persons who have dropped out thus far in the ordinary count-off, so that, for instance, number 8 (without 17 on the outside) was the ninth to drop out. The reversed count-off begins with 7 every time, starting with the person admitted last, and we continue by 6, 5, 4, 3, 2, 1, after which a new person is placed to the right of the person last indicated (which puts the newcomer between the latter person and the person who previously followed him in the reversed count-off). Beginning with the new person, we continue with 7, and so on. Of course, placing the person after A can occur arbitrarily. The reversed count-off shows that the next person (3 on the outside) has to be put to the right of the person with 2 on the outside. The reversed count-off further shows that between the last-admitted person and A it is necessary to place a person (with 4 on the outside), and so on. The persons already admitted move up, if necessary, to make room. After the last person has been placed, the reversed count-off is continued with 7, 6, 5, 4, 3, 2, 1 once more; when the 1 is reached, it indicates the person (see the dot in the figure) with whom the ordinary count-off has to start, if A is to be last person left.

303. Solution of the puzzle of §299 by reversal. The method of §302 can also be used to solve the puzzle of §299. The numbers in *Figure 166* indicate the order in which the persons are admitted to the circle in the reversed count-off, which starts with the person A, who

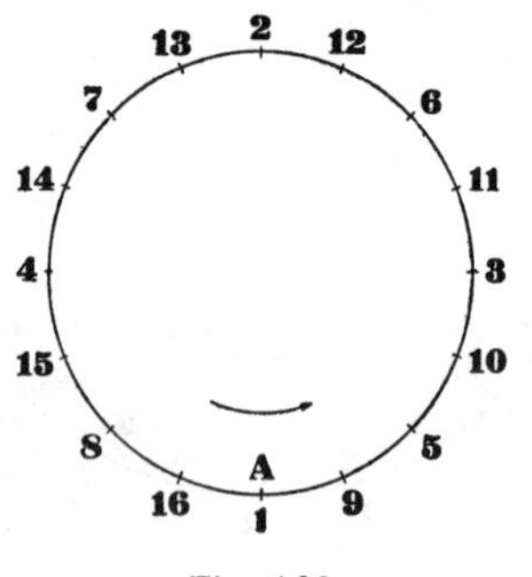

Fig. 166

is the last survivor in the ordinary count-off. To find cases for which the ordinary count-off can start with A, the reversed count-off should be continued until the person admitted last is put directly to the left of A (assuming that all the persons face the center), since in that event the reversed count-off indicates A to be the person with whom the ordinary count-off should start.

Counting off in the reverse direction requires going round the circle a certain number of times. In every round the number of persons is doubled, since a new person is placed between every two previously adjacent persons. It follows that the number of persons must be a power of 2.

304. Solution of the puzzle of §301 by reversal. To solve the puzzle of §301 by reversal, we start with the person B, who is the last survivor in the ordinary count-off. We continue to insert persons until B or his right-hand neighbor is indicated as the person with whom the ordinary count-off should start; hence we make a complete circuit around the circle when placing the persons, but we can make an arbitrary number of such circuits. This will be clarified by the diagrams of *Figure 167*, in which each time the number of circuits is increased by 1. The numbers indicate the order in which the persons are admitted to the circle, and A indicates the person with whom the ordinary count-off starts. In the second, third and fourth diagrams, A and B are the same person, and hence the requirements of our first question in §301 are satisfied. In the first and fifth diagrams, A is the

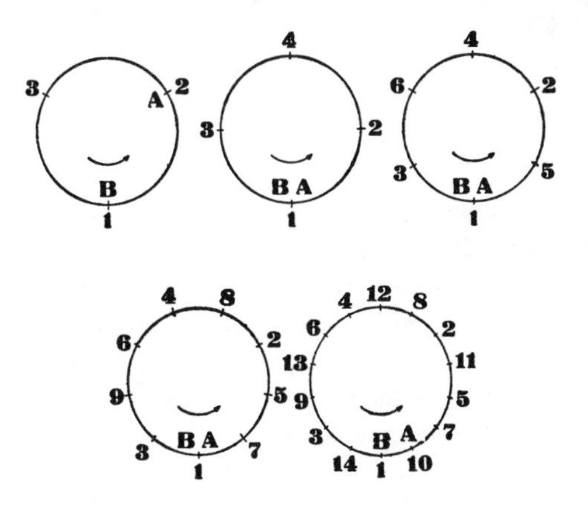

Fig. 167

right-hand neighbor of B; hence the person B who is the last survivor in the ordinary count-off is number 2 if the persons are numbered clockwise, starting with A; this means that in these cases the requirements of our second question in §301 are satisfied.

In each circuit, new persons are placed in every alternate interspace (between alternate pairs of adjacent persons). This shows that the number of persons inserted in any circuit is equal to half the number n of persons already present, rounded off as necessary to the next higher or lower integer. So the numbers that satisfy our first and second question in §301 are found each time from the preceding number by multiplying by 3/2 and rounding off as appropriate.

Four cases are possible: before the new circuit the number n of persons present can be even or odd, while A can be the same person as B or the right-hand neighbor of B. In *Figure 168* the dashes indicate

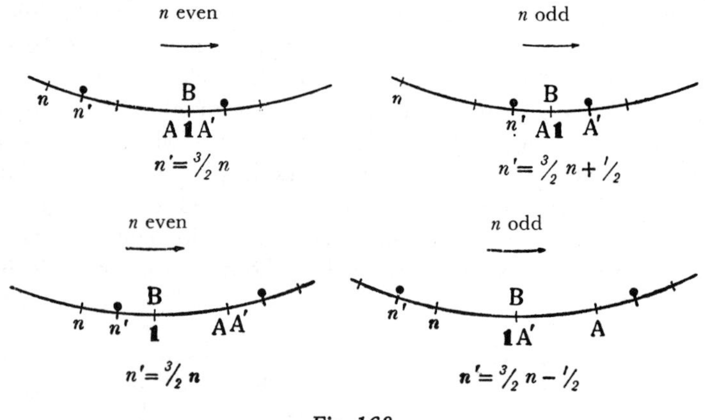

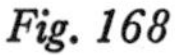
Fig. 168

the persons present before the new circuit, and the dots the persons who are introduced in the new circuit (the right-hand dot first, the left-hand dot last); n' is the number of persons after the new circuit. The letters n and n' also indicate the person introduced immediately before or after the new circuit, respectively. In the diagrams, which are self-explanatory, A′ is the person with whom the ordinary count-off should start, after the new circuit, if B is to be the last survivor; in the diagrams (if they are completed to form whole circles) the number of dots is $\frac{n}{2}, \frac{n}{2} + \frac{1}{2}, \frac{n}{2}$, and $\frac{n}{2} - \frac{1}{2}$, in the successive cases; the resulting formulae appear beneath the diagrams. This leads to the method of determining possible numbers of persons which we also found in §301.

XI. STOCKINGS PUZZLES AND PROBABILITY CONSIDERATIONS

305. Stockings puzzle. The following puzzle is well known: A lady is going out for the evening, and wishes to put on new stockings for the occasion. In her wardrobe she has a pile of stockings, of two different shades, with light and dark stockings all in a heap. It makes no difference to her whether she wears light or dark stockings. In the room where the wardrobe is, she is unable to put on a light, which prevents her from knowing which stockings are light and which are dark. Now, what is the least number of stockings she has to take at least to be sure that among them there will be a pair either of light or of dark stockings? Obviously, the answer is three.

Let us make the problem a little more complicated. Two sisters intend to wear identical costumes to a masked ball. They have forgotten to put on the matching stockings; these can be red, yellow, or blue, but the four stockings have to be of the same color. At the last moment, one of the sisters takes from a pile of red, yellow, and blue stockings a number of stockings that guarantees that she has four stockings of the same color among them. Again the question is: what is the least number of stockings she has to take from the pile? If she takes $3 \times (4 - 1) = 9$ stockings, there is the possibility that she has taken three stockings of each of the three colors. Hence, if she takes ten stockings from the pile, she ensures that she will have four (or more) stockings of the same color; she may even obtain two usable colors, for instance four or more red stockings and four or more yellow ones (with 0, 1, or 2 blue stockings), but this is not certain.

If seven millipedes were in the same situation, also with a pile containing red, yellow, and blue stockings, then at least

$$3 \times (7 \times 1000 - 1) + 1 = (3 \times 6999) + 1 = 20{,}998$$

stockings would have to be taken from the pile to enable them to have 7000 or more stockings of a single color (for at least one of the colors). With 20,997 stockings it might happen (though it would be unlikely) that 6999 red, 6999 yellow, and 6999 blue stockings were taken from the pile.

306. Probability problem on the stockings puzzle. In the first puzzle of §305 we assume that the pile contains light and dark stockings in equal numbers, and we require the probability that the lady will not achieve her object when she takes two stockings at random from the pile. Evidently, this probability is $\frac{1}{2}$, provided that the pile is sufficiently large, since the probability that she takes a second stocking different from the first is as large as the probability that she takes one of its mates. Because the pile is not infinitely large, the probability of a failure is somewhat larger, since if the pile contains n pairs (and thus $2 \times n$ stockings), and if the stocking taken first is a light one, say, then $(2 \times n - 1)$ stockings remain, n of which are dark, so that the probability is $n/(2 \times n - 1)$ that she will take an odd pair of stockings, which gives (for instance) 0.6 if $n = 3$, and 0.52 if $n = 13$.

In the other puzzles of §305 we also assume that the pile contains the same number of stockings for each of the colors, and further that the pile is very large, so that the probability of obtaining a given color is 1/3 even after some stockings have been taken from the pile. We think of the stockings as being taken single from the pile; this obviously has no influence on the probability. In the second puzzle of §305, if one of the sisters takes nine stockings, that is, one less than she should, the probability that she takes three red, three yellow, and three blue stockings in a definite order is then $\frac{1}{3^9}$. However, according to §129, these three red, three yellow, and three blue stockings can be selected in $\frac{9!}{(3!)^3}$ different orders, so that the probability of failure (that is, of having no four stockings of a single color) is:

$$\frac{9!}{(3!)^3} \times \frac{1}{3^9} = \frac{560}{3^8} = \frac{560}{6561} = 0.08535.$$

With three sisters to be dressed alike, sixteen stockings would have to be taken from the pile. In the case of fifteen stockings the probability of failure (with five red, five yellow, and five blue stockings) is:

$$\frac{15!}{(5!)^3} \times \frac{1}{3^{15}} = \frac{28028}{3^{12}} = \frac{28028}{531441} = 0.05274.$$

In the puzzle with the seven millipedes, the probability of the case that we said was unlikely, in §305, when 20997 stockings are taken from a considerably larger pile, is:

$$\frac{20997!}{(6999!)^3} \times \frac{1}{3^{20997}}.$$

The direct reduction to a decimal fraction is extremely laborious However, with the aid of a certain formula, it is easy (see §307).

***307. Approximations to the probabilities of §306.** We extend the last puzzle of §306 by letting a be the number of p-pedes, and k the number of colors of the stockings. To be sure of having at least ap stockings of the same color, $k(ap - 1) + 1$ stockings have to be taken from the large pile. If we take $k(ap - 1)$ stockings, then the probability that success will not be attained (by reason of having $ap - 1$ stockings of each of the k colors) is:

$$\frac{(kap - k)!}{\{(ap - 1)!\}^k} \times \frac{1}{k^{k(ap-1)}}.$$

The Scottish mathematician James Stirling (1692–1770) found a truly remarkable formula, which gives an approximate value for $m!$ (that is, for m factorial), namely:

$$m! = \sqrt{2\pi m}\left(\frac{m}{e}\right)^m.$$

Here π is the well-known ratio of the circumference of a circle to its radius, something already of interest to Archimedes, who gave the approximation 22/7, which is correct to two places of decimals. The ratio 355/113, given by Adriaen Metius (1527–1607), can be easily recalled from a scheme like

```
      3.1...
113 )355
     339
     ---
      160
      113
      ---
```

(where you start by writing down successively two digits 1, two digits 3, and two digits 5, and then divide the number formed by the first three digits into the number formed by the last three digits). This gives a value of π correct to six decimal places, namely: $\pi = 3.141593$.

To return to Stirling's formula, e is the remarkable number 2.7182818285 (continuing with other digits), which serves as the base of the logarithms of John Napier (1550–1617), who was the inventor of logarithms.

In some particular case, for example, that of the seven millipedes, the appropriate method is not to apply Stirling's formula for direct approximation of the two factorials (20997! and 6999!) which appear in the expression for the probability. Instead, you should first reduce the general expression for the probability (the one which uses the letters a, p, and k) by substituting the approximate values of the factorials. Moreover, it is remarkable how much simplification this substitution produces in the formula. The factors e cancel completely, while the factors m^m (for the cases $m = k(ap - 1)$ and $m = ap - 1$) are partially cancelled by each other and partially by the power of k, so that only the factors arising from $\sqrt{2\pi m}$ remain. In this way, we find the following approximation to the probability:

$$\frac{\sqrt{k}}{\{\sqrt{2\pi(ap - 1)}\}^{k-1}}.$$

For the probabilities 0.5, 0.08535, and 0.05274 found in §306 and for the probability which we there refrained from converting into a decimal fraction, the formula gives, in the respective cases:

$$\frac{1}{\sqrt{\pi}} = 0.5642, \quad \frac{1}{2\pi\sqrt{3}} = 0.09189,$$

$$\frac{\sqrt{3}}{10\pi} = 0.05513, \quad \frac{\sqrt{3}}{13998\pi} = 0.0000393863.$$

308. Accuracy of Stirling's formula. Mathematicians have succeeded in evaluating the accuracy of Stirling's formula. The approximate value of $m!$ still requires multiplication by about $1 + \frac{1}{12m}$, that is, it should be increased by about $25/3m\%$. Hence, relatively (that is to say, proportionally, which is what really matters) Stirling's formula becomes still more accurate as m increases.

The effect on the probability found in §307:

$$\frac{\sqrt{k}}{\{\sqrt{2\pi(ap - 1)}\}^{k-1}}$$

is that it should be multiplied by a little more than

$$1 - \frac{k^2 - 1}{12k(ap - 1)},$$

that is, it should be diminished by nearly $\frac{25(k^2 - 1)}{3k(ap - 1)}\%$. When this is done, the probabilities computed with the aid of Stirling's formula in §307 become:

0.4937, 0.08508, 0.05268, 0.000039385.

The first three approximative values differ only slightly from the exact values of the probabilities as found in §306: 0.5, 0.08535, 0.05274. It is true that the approximation procedure is not especially advantageous in these cases, since the probabilities (especially the probability $\frac{1}{2}$) can more easily be computed directly, but it is precisely by applying the approximation procedure to these cases, which can be checked, that we have a clear proof of its usefulness in more complicated cases. In the last case, which is difficult to check, the probability found is correct to the last decimal (that is, to nine decimal places), whereas the value of that probability found in §307 (that is, without the correction) still turns out to be correct to eight decimal places.

***309. Further applications of the results.** The probability determined in §307 and the corresponding approximation formula can also be applied to various other cases. Let us imagine a die with k numbered faces, each of which has the same probability of turning up. With this die, we make kq throws, and we require the probability that each of the k numbers will appear exactly q times. This is entirely the same question as the one on the $k(ap - 1)$ stockings. We need only replace $ap - 1$ by q, after which we find $\frac{(kq)!}{(q!)^k k^{kq}}$ for the exact value of the probability, and

$$\frac{\sqrt{k}}{(\sqrt{2\pi q})^{k-1}}$$

for the approximate value.

According to §308, we still have to multiply this probability by approximately $1 - \frac{k^2 - 1}{12kq}$, from which one can appreciate how large a q we should have if the expression given for the probability is to be reliable. If we make $6q$ throws with an ordinary die ($k = 6$), then the probability of q ones, q twos, q threes, q fours, q fives, and q sixes is approximately equal to:

$$\frac{\sqrt{6}}{(\sqrt{2\pi q})^5} = \frac{0.02475}{(\sqrt{q})^5}$$

provided q is not too small: say, greater than 5.

If a roulette wheel is spun $37q$ times, then the probability that each of the 37 numbers will appear exactly q times is approximately equal to:

$$\frac{\sqrt{37}}{(2\pi q)^{18}} = \frac{1}{383 \times 10^{11}\, q^{18}}$$

provided q is large; say, greater than 30.

XII. MISCELLANEOUS PUZZLES

310. Keys puzzle. A certain board of directors consists of four persons. They agree to provide their safe with such a number of locks (each with a different key), and every member of the board with such a number of keys, to ensure that no pair of members of the board can open the safe with their keys jointly, but that every three members of the board will be able to do this. How many locks should be put on the safe and how many keys should every member of the board receive?

To solve this puzzle, you should pay attention not to the keys that a member of the board receives, but to the keys that he does not receive. Denoting the keys by 1, 2, 3, and so on, we prepare slips of paper, some bearing the number 1, some the number 2, and so forth. These slips are distributed among the members of the board. If a person obtains a slip with the number 2, say, this will mean that he does not receive key number 2. Each member receives all keys for which he does not have the corresponding slips. Among the four members of the board A, B, C, D, any two—let us say A and B—must receive at least one pair of equally numbered slips, say the slips numbered 1, since otherwise they would jointly be able to open the safe. A third member of the board, say C, should not then receive a slip numbered 1, since this would make A, B, and C jointly unable to open the safe, for lack of key 1. If as few locks as possible are to be installed on the safe, then every pair of members of the board must receive only a single pair of like-numbered slips, for example, the slips numbered 1 go to A and B, the slips number 2 to A and C, 3 to A and D, 4 to B and C, 5 to B and D, and 6 to C and D. This means that the safe should have at least **six locks**. Each member of the board receives three (differently numbered) slips, since he has three colleagues on the board. This means that each member of the board receives three of the six keys; the distribution is as follows:

A: 4, 5, 6;   B: 2, 3, 6;   C: 1, 3, 5;   D: 1, 2, 4.

It is clear what the solution will be when the total number of members of the board and the number of members who can open the safe are different, let us say six members, no three of whom are to be able to open the safe, whereas this is to be possible for every four of them. If as few locks as possible are to be installed on the safe, then every set of three members of the board must receive a set of three like-numbered slips. It follows that the number of locks on the safe must be equal to the number of combinations of the six members of the board taken in groups of three, and so is equal to $\frac{6 \times 5 \times 4}{3!} = 20$ (see §128). This means that each member of the board fails to receive a number of keys equal to the number of combinations of his five fellow members of the board in groups of two; that is, there are $\frac{5 \times 4}{2!} = 10$ keys that he does not receive, and therefore $20 - 10 = 10$ keys that he does receive. With eight members of the board, no five of whom can open the safe, though any six can do this, the safe should have $\frac{8!}{5!3!} = \frac{8 \times 7 \times 6}{3!} = 56$ locks and each member of the board should receive

$$56 - \frac{7!}{4!3!} = 56 - \frac{7 \times 6 \times 5}{3!} = 56 - 35 = 21$$

keys. This number of keys is also equal to the number of combinations of his seven colleagues in groups of five, that is, to $\frac{7!}{5!2!}$; this can also be seen directly, since the member of the board has to possess one key for each of these groups, to enable a group of six to be formed who can jointly open the safe.

If there are d members of whom no group of a can open the safe, whereas every group of $a + 1$ can, then the safe should have $\frac{d!}{a!(d-a)!}$ locks, and each member of the board should receive

$$\frac{(d-1)!}{a!(d-a-1)!} \text{ keys.}$$

311. Puzzle on numbers. We seek to determine a number such that together with its multiple of 3, it contains all ten digits once.

If A is the desired number, then $A + (3 \times A)$ has the same remainder when divided by 9 as the sum of the ten digits. Consequently, $4 \times A$ is a multiple of 9, hence so is A, and so is the sum of the digits of A. The number A has five digits. The smallest digit of A is 2 at most, since otherwise the multiple of 3 would contain 6 figures. The number A should not contain a digit that leads to a digit of $3 \times A$ which already occurs in A. Hence, questions of order apart, the digits of A can only form one of the following eleven groups:

(a) 01368 (b) 01467 (c) 02358 (d) 02367 (e) 02457 (f) 13689
(g) 14589 (h) 14679 (i) 23589 (j) 23679 (k) 24579

The group 01359, for example, has not been included here, since the digit 3 would lead to the digit 9, 0, or 1 in $3 \times A$, which already occur in A. 01458 has been omitted, too, since the digit 8 leads to the digit 4 or 5 in $3 \times A$. In all, 15 groups have been omitted for this reason.

If we refuse to accept any digits occurring in A as digits for $3 \times A$, then in case (a), both the digit 3 and the digit 6 must lead to the digit 9 in $3 \times A$, so that case (a) drops out. For the same reason, cases (b), (d), (e), (f), and (j) drop out; in each case the pairs of important digits have been italicized.

Case (g) is impossible, since the digit 9 in A must produce a digit 7 in $3 \times A$ (since 8 and 9 occur in A); this then requires that the digits 5 and 8 of A will both produce a 6 in $3 \times A$. Case (i) is also impossible because the digit 9 of A must produce a digit 7 in $3 \times A$, which shows that the digits 2 and 5 must each produce a 6. Case (k) is impossible; the digits 5 and 9 of A must produce the digits 6 and 8, respectively, in $3 \times A$, and then the digit 2 of A must again produce a digit 6 or 8 in $3 \times A$.

Case (c) turns out to be possible. The digit 0 of A produces the digit 1 in $3 \times A$, which shows that the digit 3 of A produces the digit 9 in $3 \times A$. This means that the digit 0 of A must be followed by 5, and this 5 by 8. If we allow 0 as an initial digit, this gives the following six solutions:

A	05823	05832	20583	23058	30582	32058
$3 \times A$	17469	17496	61749	69174	91746	96174

Case (h) is possible, too. The first digit of A is 1, since otherwise $3 \times A$ becomes a six-digit number. The digit 9 of A gives the digit 8 of $3 \times A$, so that the digit 6 of A must produce the digit 0 in $3 \times A$. The last digit of A must be 4, because it cannot be 6, 7, or 9. The digit last but one in A must be 9. This gives the following two further solutions:

A	16794	17694
$3 \times A$	50382	53082

312. Adding puzzle. The sum of two numbers both formed from the same set of four digits (not necessarily all different) is a third number again formed from this same set of four digits. An initial digit 0 is not permitted. All solutions are required; interchange of the two numbers should not be considered to produce a new solution.

If A and B are the two numbers, and if $A + B = S$, then A, B, and S must differ by a multiple of 9 from the number s which is the sum of the four digits. Consequently, $2s$ and s differ by a multiple of 9, which shows s to be a multiple of 9. The digits of A cannot all be odd, for the last digit of S would then be even (and unable to be a digit of A). The sums of digits of A and B, in the units place, tens place, and so on, cannot all be less than 10, since otherwise S would have a higher digit-sum than A; hence, in the addition, a carry of 1 must occur at least once, which implies that the digits of A cannot all be even, and also that they cannot all be less than 5. Since the smallest positive digit of A is at most 4 (because otherwise S would have more than four digits), only the following sets of four digits deserve consideration, when we disregard the matter of their order:

0018	0027	0036	**0**045	0117	01**2**6	**0**135	**0**1**8**9	**0**225	0279
0369	**0**378	**04**59	0**4**77	1116	1125	1188	**12**69	1278	**1**3**6**8
1449	1458	**14**67	**1**566	1**8**99	2259	2277	**23**49	2**3**58	**23**67
2457	**2**556	2799	2**8**89	3348	3366	3**4**47	3456	3699	3**78**9
3888	4455	**4**599	**468**9	**4**779	**4**788				

Since there is a carry of 1 at least once in the addition, at least one of the four digits, after being increased by 10, has to be equal to the sum of two of the digits, or to the double of a digit. A digit with this property has been printed in boldface; thus, in **12**69 the digits 1 and 2 are bold because of the relations $11 = 2 + 9$, $12 = 6 + 6$. We need consider only those sets of four in which at least one bold-type digit occurs. The first digit of S arises from two digits, different from 0, which have a sum less than 10: because of this we reach an impasse in the cases **0**045, 01**2**6 (since the 2 must arise from $6 + 6$), 0**4**77, 2**8**89. In the cases **0**378, **4**788 the second digit leads to a dead end, and in the cases **1**3**6**8, **1**566, 1**8**99, **23**67, **24**57, **2**556, 3**4**47, 3**78**9 this is true of the third digit. Here, we can also pay attention to the bold digits and, in the cases **1**566, **2**556, 3**4**47, to the fact that the digits 5, 6, 3, respectively, cannot occur in the sum (so that these three cases drop out at once). Together, the eleven remaining cases give the following twenty-five solutions:

1503	1530	1089	1089	2502	4095	4095	4590
3510	3501	8019	8091	2520	4950	5409	4950
5013	5031	9108	9180	5022	9045	9504	9540

1269	2691	1467	1467	1476	1746	2439	4392
1692	6921	6147	6174	4671	4671	2493	4932
2961	9612	7614	7641	6147	6417	4932	9324

2385	2538	2853	3285	4599	4959	4698	4896	4797
2853	3285	5382	5238	4995	4995	4986	4968	4977
5238	5823	8235	8523	9594	9954	9684	9864	9774

In three cases, a double addition is possible, each time involving five numbers which consist of the same four digits:

$$\begin{array}{r}4671\\1476\\\hline 6147\\1467\\\hline 7614\end{array}\qquad\begin{array}{r}2493\\2439\\\hline 4932\\4392\\\hline 9324\end{array}\qquad\begin{array}{r}2853\\2385\\\hline 5238\\3285\\\hline 8523\end{array}$$

There are two solutions in which the sum is divisible by 11. These solutions can be found quickly, without first determining all solutions. In order to find them we observe that a number is divisible by 11 when the sum of the digits for the odd places (here the sum of the first and the third digits) differs by a multiple of 11 from the sum of the digits in the even places; this follows immediately from the fact that $10 + 1$, $10^3 + 1$, $10^5 + 1$, and so on, are divisible by 11, as are $10^2 - 1$, $10^4 - 1$, $10^6 - 1$, and so on. If the sum of the four digits is 9 or 27, then the sum of two of the digits cannot differ by a multiple of 11 from the sum of the other two digits (since the two sums would have to be 8 and 19 when the sum of the four digits is 27). If the sum of the four digits is 18, then divisibility by 11 is possible only when the sum of two of the digits is 9. Since only those sets of four deserve consideration in which a bold digit occurs, the cases **0**1**8**9, **04**59, **1**3**6**8, **2**367, and **24**57 alone remain, among which only the first two provide solutions:

$$\begin{array}{r}1089\\8019\\\hline 9108\end{array}\qquad\begin{array}{r}4095\\5409\\\hline 9504\end{array}$$

If we allow initial zeros (but not exclusively zeros, and without the restriction to a sum divisible by 11), then the sets 0045, 0135, 0189, 0225, and 0459 give 6, 4, 10, 2, and 9 solutions, respectively, so that we obtain another 31 solutions. We leave it to the reader to find these.

313. Number written with equal digits. We may pose the problem of writing the number 100 with digits 4 only, using as few such digits as possible. Furthermore, the well-known arithmetical symbols (plus sign, times sign, and so on, and also parentheses) can be used, in unlimited number. With seven fours we can do it in four ways, as follows:

$$\begin{aligned}100 &= 44 + 44 + 4 + 4 + 4 = 44 + 44 + (4 \times 4) - 4\\ &= \frac{4^4}{4} + 44 - 4 - 4 = 4 \times \left(4 + \frac{4}{4}\right) \times \left(4 + \frac{4}{4}\right).\end{aligned}$$

However, it can also be done with six fours, in two ways:

$$100 = 4 \times \{(4 \times 4) + 4 + 4\} + 4 = \frac{444 - 44}{4}.$$

The second expression could equally well use six digits 2, or six digits 3, and so on, and gives an expression in terms of five digits 1: $111 - 11$. With four digits 5, we can write 100 in two ways:

$$100 = 5 \times \{(5 \times 5) - 5\} = (5 + 5) \times (5 + 5).$$

With four digits 9 it can be done as $99\frac{9}{9}$ and as a repeating fraction with three digits 9 thus: $99.\dot{9}$. With four digits 3 it can be done as: $3 \times 33.\dot{3}$.

To have a well-defined problem, we have to know exactly what symbols we are allowed to use. The more symbols we admit, the smaller the number of digits that will suffice. By using the exclamation point ($4! = 1 \times 2 \times 3 \times 4$) we can write 100 with three digits 4 as $(4 \times 4!) + 4$.

With numbers that are not integers, we can admit square brackets as the symbol for "the largest integer that is not larger." Thus we have:

$[3.5] = 3$, $[-3.5] = -4$, $[\sqrt{8}] = 2$, $[-\sqrt{8}] = -3$, and so on.

Making use of this, we can write 100 with two digits 7 as

$$[\sqrt{7! + 7!}],$$

since this is $[\sqrt{10080}]$, and thus 100, since 100^2 is less than 10080, while 101^2 is greater than 10080.

With the symbols so far introduced, 100 can be written with two digits 5 as

$$[\sqrt{5!}] \times [\sqrt{5!}]$$

Because of the later notation, 100 can also be written with two digits 6, since $[\sqrt{\sqrt{6!}}] = 5$, because $6! = 720$, while 5^4 is less than 720, and 6^4 is greater than 720. Hence we have

$$100 = [\sqrt{[\sqrt{\sqrt{6!}}]!}] \times [\sqrt{[\sqrt{\sqrt{6!}}]!}].$$

As it appears from

$$-[-\sqrt{4!}] = 5,\ 3! = 6\ (\text{thus } [\sqrt{\sqrt{3!!}}] = 5),\ \sqrt{9} = 3,$$
$$-[-\sqrt{8}] = 3,$$

the number 100 can also be written with two digits 3, 4, 8, or 9.

However, the resulting notations for 100 are rather complicated, especially the one with two digits 8; this makes them much less interesting than those with two digits 7 or two digits 5.

By admitting further mathematical symbols, the number 100 can be written with one digit, and any digit. However, this is not sufficiently interesting to pursue further.

Chapter XV: PUZZLES IN MECHANICS

I. GENERAL REMARKS ON KINEMATICS

314. Kinematics and dynamics. Mechanics is the study of the motion of objects, which in this context are also called bodies. Since various aspects of this study are concerned with mechanical devices (such as levers, pulleys, and the like), the subject has been called the *theory of mechanisms*, in earlier times, but modern usage replaces this by the term *mechanics*.

When the motion of bodies is studied without reference to causes of the origin or change of the motion, the relevant branch of mechanics is called *kinematics*, a word which means the theory of motion. Kinematics involves the study of geometrical questions, with additional account taken of the concept of time. When we interest ourselves also in the causes of the motion or of the change of motion—and hence in the forces that govern the motion—then we speak of *dynamics*. This subject involves the concept of "force" just mentioned, and introduces another new concept, namely "mass."

If at some moment there is no motion and the situation is such that the forces will not produce any motion, this situation is called *equilibrium*. The study of cases in which motion is absent is called the theory of equilibrium, or *statics*. However, this should not be regarded as a separate part of mechanics, having a place with kinematics alongside dynamics, but rather as a part of dynamics, since absence of motion can be regarded as a special form of motion.

315. Motion of a point along a straight line. If a point moves along a straight line in such a way that equal distances are always traversed in equal intervals of time (so that in an interval twice as long, the distance traversed is twice as large, and so on) then we speak of a uniform motion. The distance, expressed in centimeters, which is traversed in the unit of time, the second, is called the velocity of the motion. In this way the velocity is expressed in centimeters per second. In the case of a uniform motion, the number of centimeters in the distance traversed (s, from the Latin *spatium*) is found by multiplying the velocity (v, *velocitas*) by the number of seconds of elapsed time (t, *tempus*); in short:

$$\text{distance traversed} = \text{velocity} \times \text{time } (s = v \times t).$$

If the motion along the straight line is not uniform, then the velocity at a given moment is found by considering the motion as a uniform motion over a very short time interval, which involves dividing the short distance traversed in the short time interval by the length of that interval, that is, dividing the number of centimeters of the distance by the number of seconds of the time. If the motion is such that equal increments of velocity always occur in equal intervals of time, then we speak of a uniformly accelerated motion. This term is also applied to the motion when the velocity decreases in a similar way; the increment of velocity is negative in this case. The increment of velocity during each second specifies the acceleration of the uniformly accelerated motion, which therefore is expressed as centimeters per second, *per second*. An acceleration can be negative, in which case the velocity reduces in every second by the same amount; this situation is given the—quite superfluous—name of "uniformly retarded motion."

Just as the case of uniform motion requires the distance to be found by multiplying the velocity by the time, the case of uniformly accelerated motion requires the increase of velocity to be obtained as the product of the acceleration and the time elapsed.

Obviously, a uniform motion can be interpreted as a uniformly accelerated motion with acceleration zero, just as "rest" can be interpreted as a uniform motion with velocity zero. In the case of a motion that is not uniformly accelerated, we can also speak of an acceleration, but must add the phrase "at a certain moment." For this purpose, we need only treat the motion as a uniformly accelerated motion during a very short time, which leads us to divide a small increment of velocity by the short time interval.

316. Motion of a point along a curved line. In the case of motion along a curved line, the velocity has to be described not only by its magnitude, but also by its direction. In cases where we have to take into account a magnitude and a direction, we speak of a vector; this is represented by an arrow, whose length is made to indicate the magnitude of the vector. Just as we introduce an acceleration in the case of motion along a straight line when the velocity is subject to change, we do the same thing in the case of motion along a curved line. However, in this case the velocity must also be considered to be changing if its direction changes, as well as when it attains a different magnitude. This means that for the case of a point moving in a circle, there is an acceleration even when the magnitude of the velocity remains the same. This acceleration can also be considered as a vector; that is, it has not only a magnitude but also a direction. Because the point does not describe a straight line, but is, as it were, pulled towards the center of the circle, the acceleration is directed towards this center (if the magnitude of the velocity is constant), for which reason we speak of a "center-seeking," or centripetal, acceleration. The nature of this acceleration was first discovered by the Dutch scientist Huygens (see §132). He found the equation $a = v^2/r$ for the acceleration expressed in centimeters per second per second (a, acceleration), when a circle of radius r cm is traversed with a constant-magnitude velocity of v centimeters per second.

317. Relative character of motion. The motion of a body can never be observed in an absolute way, but only relatively, that is, with regard to another body; to be aware of motion, we have to see some point of one body changing its distance from some point of another body. If, for some bodies, no changes in distance occur, then these bodies are at rest relatively to one another. If distances from a point A to points of those bodies are subject to change, then A is in motion relatively to those bodies.

The bodies which are at rest relatively to one another are called a system of reference for A. When we speak of the velocity and the acceleration of a point A, it must be clear (if these notions are to have any proper meaning) relatively to what system the motion is being considered. In many cases the system is the ground on which we are standing, or equivalently the Earth. If no special system is mentioned, we always imply that motion is relative to the Earth.

As an illustration of a difference in motion relative to one system and to another, we take the case where someone on a moving ship throws a ball straight up and catches it again. From the ship he sees the ball describe a straight line upwards and downwards, but someone from the shore would see the ball describe a curved line (its trajectory), of the type which we call a parabola.

318. Composition of motions. Suppose that some system, say a ship, is in motion relatively to a second system, say the Earth. If we know the motion of the ship relatively to the Earth and the motion of a point A relatively to the ship (for a man walking on the deck of the ship, say) then we can use this to derive the motion of the point relatively to the Earth. This is called composition of motions.

If we keep to the example just mentioned and assume that the ship has such a motion relatively to the earth that the various points of the ship move uniformly along parallel straight lines, all with the same velocity, and that the point A moves uniformly along a straight line drawn on the ship, then it can be easily demonstrated that A, too, moves along a straight line relatively to the Earth, and also uniformly. The velocity with which A moves relatively to the Earth is a diagonal of a parallelogram which can be formed from the velocity of A relative to the ship (the relative velocity) and the velocity of the ship relative to the Earth. This result is called the *parallelogram of velocities* (*Fig. 169*).

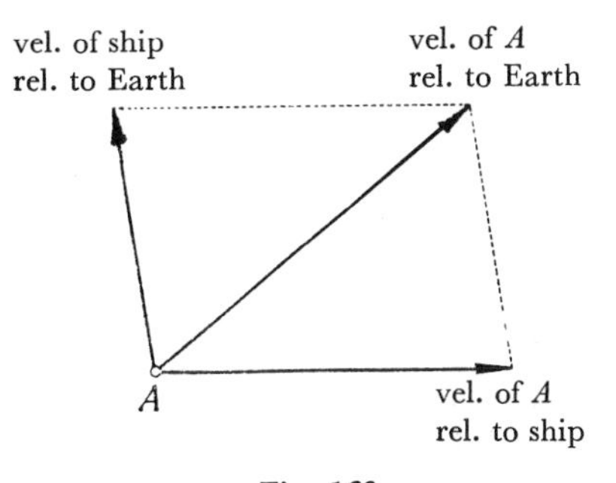

Fig. 169

319. Examples of the composition of motions. It is possible to give quite a number of striking examples of the composition of motions. We shall discuss some of these. On a moving ship the pennant does not show the direction of the velocity of the air relative to the Earth but of the velocity of the air relative to the ship. If someone on board is aware of the magnitude and the direction of the velocity of the ship, and can measure the apparent velocity of the wind (in magnitude and direction once again, of course), then he can derive from this the magnitude and the direction of the velocity of the wind relative to the Earth. This is shown by the parallelogram in Figure 169.

A cyclist travels from south to north while the wind is blowing due east. He returns along the same road and the direction of the wind has not changed. Arriving home he say that he has been riding against the wind—"Not directly against it, admittedly," he adds. The cyclist thought he was riding against the wind because he did not realize that what he felt was the motion of the air relative to himself and not relative to the Earth. We can use the parallelogram construction, to compound the velocity of the cyclist and the apparent velocity of the wind, and so obtain the real velocity of the wind (*Fig. 170*).

A rower wants to cross a river. He leaves from landing stage A and has to arrive at landing stage B. He points his boat in the direction A–B and starts rowing. Because of the current he drifts off and arrives on the opposite bank at B′, instead of at the landing stage B. The velocity that the boat acquires relative to the bank is found by composition of the velocity from rowing and the velocity of the current. A second rower, who has to cross from the landing stage C to the

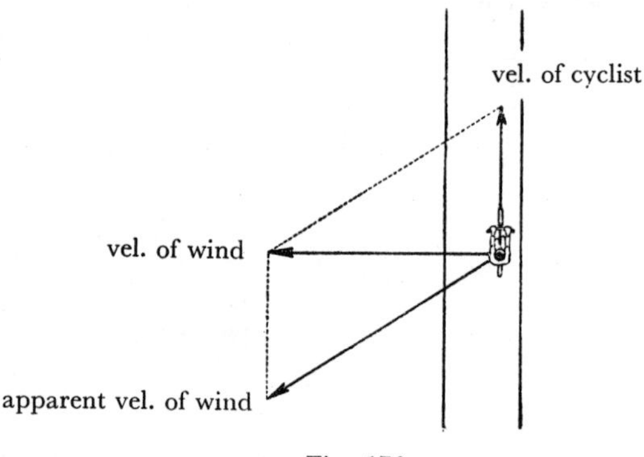

Fig. 170

landing stage D, is more clever. He points his boat obliquely into the current in such a way that by drifting off course he arrives exactly at D. This means that he directs the velocity with which he rows (that is, with which he moves relative to the flowing water) in such a way that composition with the velocity of the current produces a velocity relative to the bank, directed exactly from C to D (*Fig. 171*). In the foregoing it has been assumed that the rowing velocity is greater than the velocity of the current; if this is not the case, it is not possible to find a way of rowing from C to D. Even if you adopt the second rower's method, and row straight against the current, you will remain at rest or drift back more than you row, according as the velocity of rowing is equal to or smaller than the velocity of the current. As an example of the case where the rowing velocity dominates, we take a rowing velocity of 1 meter per second and a current velocity of 80

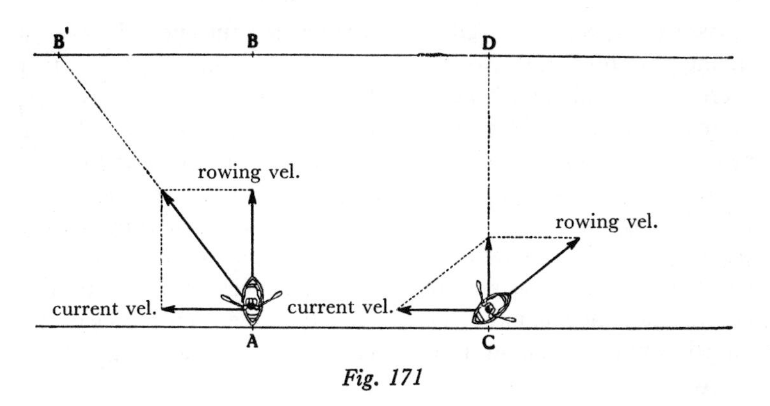

Fig. 171

centimeters per second. From Pythagoras' theorem, the velocity of the second rower relative to the bank is then $\sqrt{100^2 - 80^2} = 60$ centimeters per second. Hence, if the width of the river is 180 meters, the second rower needs 5 minutes to reach point D. The first rower reaches the opposite side in only 3 minutes; however, he arrives at the wrong point, namely at B′, because he has drifted 144 meters downstream in those 3 minutes. If he now wants to row back from B′ to the landing stage B (having a velocity of $100 - 80 = 20$ cm per sec relative to the bank), he will require $4 \times 3 = 12$ minutes, so that he actually takes 15 minutes to row from A to B.

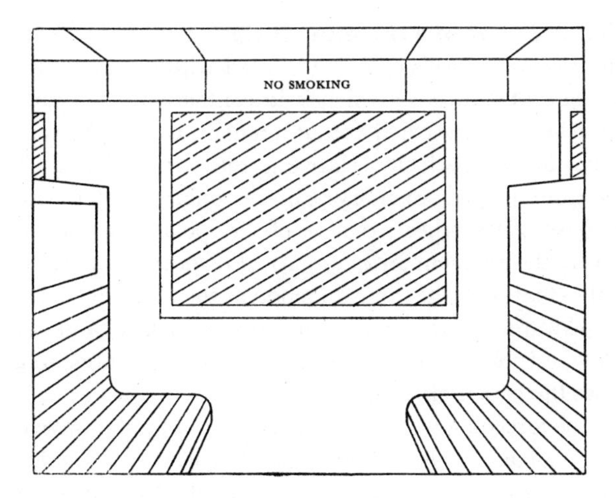

Fig. 172

As a last example we take the case where someone is looking at falling rain through the window of a traveling train. We assume that there is no wind; and hence that the raindrops fall vertically. The passenger in the train sees the raindrops describe parallel sloping straight lines, departing from the vertical in a direction opposite to the direction of the train (*Fig. 172*). The fact that these streaks of rain take the form of straight lines, as viewed from the uniformly moving train, is proof that the raindrops, too, are moving uniformly (this can be ascribed to the resistance of the air); if the velocity of the train is known, then the magnitude of the velocity of the train can be determined from the apparent direction of the streaks of rain (*Fig. 173*).

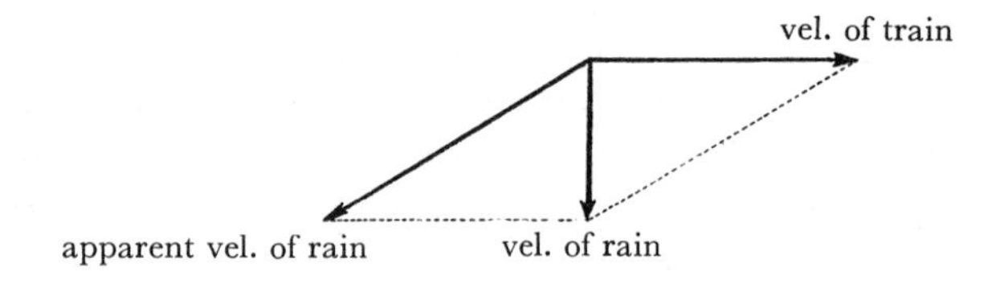

Fig. 173

II. SOME PUZZLES IN KINEMATICS

320. Which points of a traveling train move backwards? When the wheel of a cart rolls over the ground without skidding, the particular point of the wheel which is in contact with the earth has no velocity relative to the earth at that moment. Every other point of the wheel has the same velocity relative to the earth as if the wheel were turning around the momentary point of contact. So the points of the lowest spoke of the wheel (that is, of the spoke that is in the lowest position at the moment in question) also move forward relative to the earth—that is, in the same direction as the cart—even if the points move backwards relative to the cart; the latter fact causes the points of the lowest spoke to move forward more slowly relative to the earth than does the cart itself.

From the foregoing we see that there are points of a travelling train which move backwards. For example, the wheels of the locomotive have an extension beyond their rolling circle, to form the so-called flange. The flanges serve to keep the wheels on the rails, as is shown in

Figure 174. The flange has a somewhat greater distance from the center of the wheel than the periphery of the rolling circle has, so that for every position of the wheel there is a part of the flange which is below the point of contact of wheel and rail. Since the lowest point of the flange has the same velocity as if the wheel were turning around this point of contact, the lowest point in question goes to the left when the train goes to the right. This is shown in *Figure 175*, where the

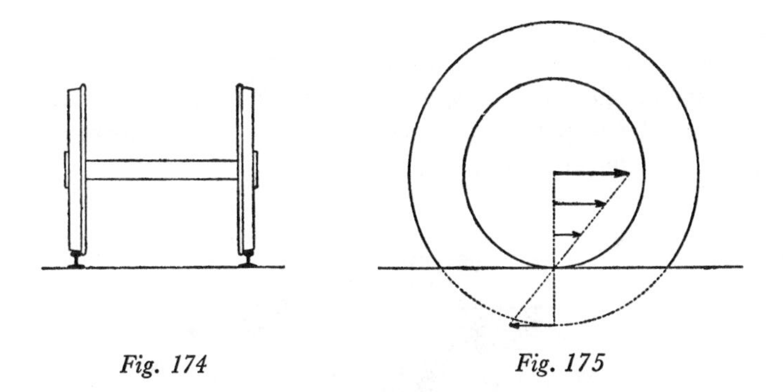

Fig. 174 *Fig. 175*

flange has been drawn exaggeratedly large for the sake of clarity. The heavy arrow indicates the velocity of the center of the wheel; this is the velocity with which the locomotive is moving toward the right. The arrow pointing to the left indicates the velocity of the lowest point of the flange.

Figure 176, in which two trajectories have been drawn, gives a clear idea of the motions of various points of the wheel. A point on the rolling circle describes a so-called cycloid, the dot and dash line in the figure, whereas the outermost point of the flange describes a trochoid, in which the cusp of the cycloid has changed into a loop. The arrows

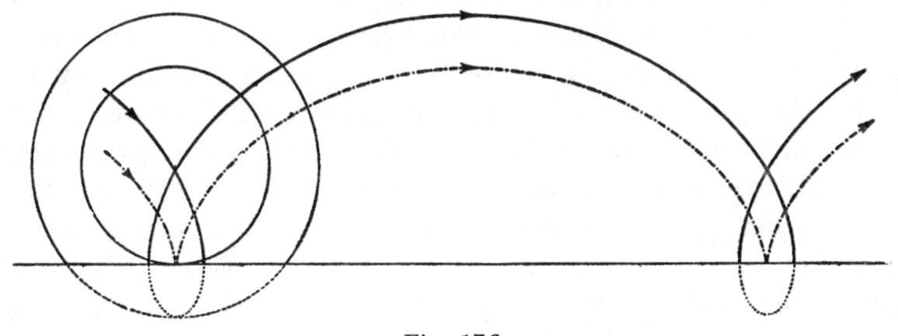

Fig. 176

on the trajectories indicate the direction in which they are traversed; the overall displacement is to the right, but the lowest part of the loop is traversed with a movement to the left.

321. Which way does the bicycle go? We place a bicycle in such a way that one pedal is in its lowest position and the other in its highest position. Standing next to the bicycle, we use our hand to give a backward pull on the pedal that is in the lowest position; in *Figure 177* the force which this applies has been indicated by an arrow. Does the bicycle go forward or backwards?

You might think: forward. If the lower pedal is pulled backwards you might argue (relying on your experience in cycling) that the rear wheel will be made to impart forward motion. turning clockwise in the direction of the curved arrow, in our figure.

Fig. 177

To answer the question without trial and error, you should not in fact pay attention to the interaction of the forces; this question should be considered not dynamically but kinematically. Since you are not cycling now, but standing on the ground, you should not reason as you did above, when you confused the motion of the pedal relative to the Earth with its motion relative to the bicycle.

Of course, when pulled backwards, the pedal must go backwards—that is, relative to the Earth, not necessarily relative to the bicycle. Hence the question is the same as the following: If the lower pedal moves backward relative to the Earth, does the bicycle go forward or backwards? Evidently this question amounts to the same thing as the following: When you are cycling, does the lower pedal move forward or backwards relative to the Earth? Relative to the bicycle, this pedal moves backwards, of course, but that is not the question here. The distance from the pedal to the pivot point of the pedal bar is less than the length of a spoke of the rear wheel; moreover, the pedal bar turns more slowly than the rear wheel because of the gear with which the bicycle is fitted. Both circumstances cause the lower pedal to move backwards more slowly relative to the bicycle than the bicycle moves forward relative to the Earth, so that when you are cycling, this pedal is moving forward relative to the Earth. If now, on the contrary, you move the lower pedal backwards relative to the Earth, then the bicycle moves backwards, too. Just try it out! However, you will see the lower pedal go forward relatively to the bicycle; so the pedal here moves against the direction of progress of the bicycle, just as in ordinary cycling.

You can simplify the question by exerting a force directed backwards (the dotted line in Figure 177) on the lowest spoke of the rear wheel. When you cycle, the points of this spoke have a velocity directed forward relative to the Earth, as appears from the argument in §320; however, these points move backwards relative to the bicycle. Because of the gear of the bicycle, this "going forward relative to the Earth" holds in a still higher degree for the lower pedal. But even if the bicycle had no gear, the answer to the question would be the same. If someone were to construct a bicycle with a gear of such a type that the pedal crank rotated faster than the rear wheel, and if the ratio were made large enough, then when you cycled the lower pedal would move backwards relative to the Earth as well. The answer to the question posed at the beginning of this section would then be: "Forward."[1]

322. Rower and bottle. A rower has to row regularly along a river from a place A to a place B and back. He has got into the habit of exerting himself more when rowing upstream, with the result that he goes twice as fast relative to the water as when rowing downstream. One day he rows upstream and passes a floating bottle. First he does not pay much attention to it, but gradually he becomes curious about the content of the bottle. After 20 minutes he stops rowing and drifts for 15 minutes. Then he rows back to the bottle. After some rowing downstream he thinks his curiosity childish, turns back again and continues his initial route. However, soon his curiosity takes over again and 10 minutes after he thought he had mastered it, he goes after the bottle. This time, too, he is ashamed of his childishness and returns after a while. But after rowing upstream for 5 minutes he can

[1] Here we omit from the translation a paragraph that is concerned solely with Dutch terminology.

no longer master his curiosity and with the firm intention of overtaking the bottle he rows downstream till he picks up the bottle at a distance of exactly 1 kilometer from the place where he passed it. How do we find from this the velocity of the current?

From the moment the rower passes the bottle till the moment when he picks it up, he has rowed $20 + 10 + 5 = 35$ minutes upstream. His total motion relative to the bottle, thus relative to the water, is 0, so that his displacement relative to the water when rowing upstream is equal to that when rowing downstream, but in the reverse direction. Since in the second case he moves half as fast relative to the water as in the first case, the rower has rowed twice as long downstream as upstream, hence for 70 minutes. Because he has been drifting for 15 minutes, he picks up the bottle $35 + 70 + 15 = 120$ minutes, hence exactly 2 hours, after he first saw it. During this time the bottle has covered a distance of 1 kilometer, so that the velocity of the current is $\frac{1}{2}$ kilometer per hour.

Maybe the reader, too, has become curious about the content of the bottle. It turned out to contain a piece of paper with the inscription: "Whoever finds this bottle . . . finds the beer all gone!" So it proved not to be as important as the rower expected.

We are not yet finished with this problem. Before leaving it, we want proof that on his first return the rower rowed downstream for less than 40 minutes, on his second return for less than an hour, and on his third return for more than 10 minutes.

It is obvious that the reason that the rower rows twice as fast upstream as downstream (relatively to the water) is that he wishes to take the same time going there (from A to B) as returning (from B to A). If it is given that this is indeed the case, how much later than usual does the rower arrive at his destination on account of the adventure with the bottle? If v is his rowing velocity relative to the water (in kilometers per hour) when rowing downstream, so that this is $2 \times v$ when rowing upstream, then since the velocity of the current is $\frac{1}{2}$ and the velocities relative to the bank are equal in both cases, this implies that $v + \frac{1}{2} = (2 \times v) - \frac{1}{2}$, and hence that $v = 1$. This makes the velocity of the rower relative to the bank 3/2 kilometers per hour, so that he needs 2/3 hour (hence 40 minutes) to row from the place where he picked up the bottle to the place where he passed the bottle. Hence, because of the adventure he is $120 + 40 = 160$ minutes late, assuming that he suppresses his curiosity until he arrives at his destination and does not drift for a while to examine the content of the bottle.

323. Four points moving along straight lines. Four arbitrary straight lines *a*, *b*, *c*, and *d* lie in a plane. On the line *a* there is a point *A* moving uniformly, hence with a constant velocity. Similarly there is a point *B* which moves uniformly along line *b* with a velocity which is not necessarily the same as the velocity with which *A* moves along *a*. Along *c* too, a point *C* moves uniformly, and likewise along *d* a point *D*. The four straight lines intersect in pairs, in such a way that six points of intersection arise. It is given that five times over (at least) it happens that two of the moving points *A*, *B*, *C*, *D* meet at the common point of intersection of the two lines along which they move. We require to show that this happens the sixth time, too.

Suppose that *A* and *B* meet at the intersection of *a* and *b*. Since *A* and *B* move uniformly, the line *AB* which connects them must move parallel to itself. If *A* and *C* also meet (at the intersection of *a* and *c* this time), then the line joining *A* and *C* must also move parallel to itself. If it is given that *B* and *C* also meet, this implies that there is a position of coincidence for the points *B* and *C*, and thus also for the lines *AB* and *AC*. Because of the parallel shift of *AB* and *AC* these lines (which

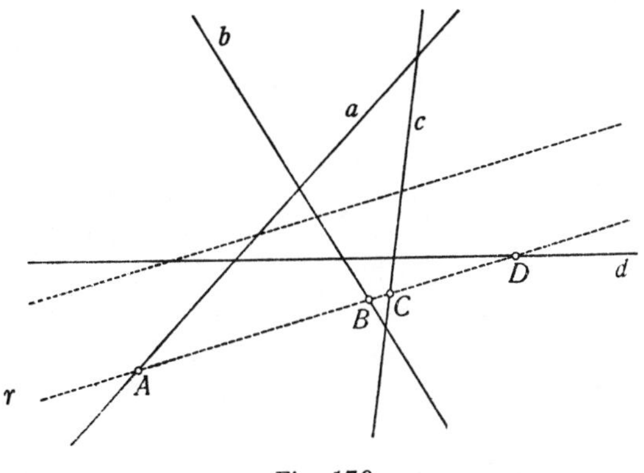

Fig. 178

have the point *A* in common) must coincide at every moment. So the three points *A*, *B*, and *C* always remain on a straight line *r*, and this line moves parallel to itself. If it is further given that points *A* and *D* meet, as well as *B* and *D*, then points *A*, *B*, and *D* must also remain on a straight line; this line is the same as *r*, because these lines have the two points *A* and *B* in common. Hence the points *A*, *B*, *C*, and *D* always remain on the straight line *r*, which moves parallel to itself. So the points C and D must also meet, namely at the moment when the line r passes through the intersection of the lines c and d.

III. PHENOMENA OF INERTIA

324. Principle of inertia. Even after the efforts of the ancient Greeks to explain nature, it still took ages and ages before man had a proper insight into the phenomena of motion. It was the incomparable English mathematician, physicist, and astronomer Newton (1642–1727) who succeeded in making important discoveries of his own which consolidated the painstakingly accumulated concepts into a useful explanation of the motion of bodies, especially the celestial bodies. Newton's work here was based principally on the investigations of Galileo (see §132) and Huygens (see §§132 and 316).

An important contribution to more exact insight was Galileo's principle of inertia, foreshadowings of which are to be found in the work of such of his predecessors as the great painter, sculptor, architect and naturalist Leonardo (1452–1519), often called Leonardo da Vinci after his birthplace. This principle states that every change which occurs in the velocity of a body, whether that change involves the magnitude or the direction of the velocity, must have an external cause; such a cause is called a force. Hence, the law of inertia says that a body that is free from all external influences must move along a straight line, always with the same velocity; a particular case of this is the absence of motion. With the concept of acceleration developed in §315, the principle can also be expressed by saying that where no force is present there can be no acceleration, either. We can also say that an acceleration has a cause in the present; a velocity, on the other hand, has a cause in the past, namely the presence of that velocity some time earlier.

The curious thing about the principle of inertia is that it is susceptible of neither a theoretical nor a direct experimental proof; the latter method fails, if for no other reason, because it is not possible to withdraw a body from the influence of all other bodies. To test the principle, one would have to neutralize gravity in some way or other, which would require invoking a combination with other principles (which themselves could not be proved theoretically or verified experimentally, in isolation). Hence, the confirmation of the principle of inertia can be found only in the fact that, together with other simple principles, it succeeds in giving a valid explanation of all sorts of phenomena.

It is understandable that the principle of inertia was not always recognized and that initially the opinion was held that a moving object has of its own accord the tendency to pass into a state of rest. The friction or resistance which occurs everywhere (friction in the case of one body moving along another, resistance of the air, and so on) was wholly or partly overlooked, and this was not recognized as the external influence that brings the motion to a halt.

325. To what systems does the principle of inertia apply? In §§317–319 we have seen that motion can be considered only relatively, that is, relative to a given system, and that it can be entirely different relative to one system from what it is relative to another system. Hence, the question arises: To what systems does the principle of inertia apply? Answering this question is one of the main difficulties in mechanics. Galileo, who did not discuss it, perhaps thought of motion relative to the ground, hence to the Earth. However, for that system the consequences of the principle of inertia are incorrect, mainly because of the rotation of the Earth about its axis.

As we can regularly observe, the starry heavens seem to rotate about the straight line which runs from the south pole S of the earth to the north pole N, a line which points approximately in the direction of the Pole Star. Looking in the direction SN, we see the rotation in a counterclockwise direction. The time of revolution is called a sidereal day. Because the sun seemingly moves relative to the stars, the sidereal day is shorter than the solar day (that is, the day we are concerned with, in everyday life), by about 4 seconds short of 4 minutes. The fact that this rotation is only apparent, and that, in reality, it is the Earth which rotates around the SN axis in the opposite direction, was clearly stated for the first time by Copernicus (1473–1543), in his major work *De revolutionibus orbium caelestium*, the first copy of which is said to have been handed to him on his deathbed. One of the most fervent adherents of the new theory was Galileo, who, as an old broken man, was forced to abjure that theory in 1633, after which he muttered to himself (though some dispute this) "Eppur si muove" (and yet it moves).

If we imagine a sphere which coincides with the Earth, but which does not take part in the rotation, then we can assume, with a very high degree of accuracy, that this sphere is a system for which the principle of inertia holds. A system which moves relative to that sphere in such a way that its various points all move along parallel straight lines with the same velocity is then also a system for which the principle applies. This follows from the fact that a point which performs a uniform rectilinear motion relative to one system does the same thing relative to the other system.

326. Some consequences of the rotation of the Earth. Although in everyday life the rotation of the Earth cannot be perceived, and it can thus usually be assumed that the principle of inertia applies to the rotating Earth, several phenomena can be mentioned which find their explanation in the rotation of the Earth. These phenomena are related to rapid or long-lasting motions.

If a train moves from south to north in the northern hemisphere, then this train, apart from its velocity relative to the Earth, has an easterly velocity arising from the rotation of the Earth. This easterly velocity is maximum at the Equator (where the radius of the circle described by the rotation is maximum) and decreases to the north. Hence, going north, the train arrives at points where the easterly velocity is smaller, but the train itself would retain the larger easterly velocity if the rails did not prevent an easterly motion relative to the Earth. Because of this, the train is pressed against the right-hand rail. In the same way, this also proves to be the case when the train goes from north to south in the northern hemisphere. Further consideration shows that the phenomenon occurs to the same extent with east-west motion or motion in any direction, but it is less easy to offer a non-technical explanation for this.*

The foregoing shows that a body moving horizontally in the northern hemisphere has a tendency to deviate to the right from a straight line; in the southern hemisphere this deviation is to the left. In the case of the train, this tendency to deviate could be observed, if at all, from the greater wear of the right-hand rail. Now and again, people have thought they had ascertained this, but the correctness of this observation becomes doubtful when it is realized that at the Pole (where it is maximum) the phenomenon can be eliminated merely by giving the rails a curvature with a radius of 190 kilometers, or by laying the right-hand rail 0.59 mm higher than the left-hand rail; in the computation which leads to this, it has been assumed that the velocity of the train is 100 km per hour, while the gauge has been fixed at 1435 mm, the most common one.

---

* [Such an explanation has to take account of changes of direction of velocities, instead of changes of magnitude only.—T.H.O'B.]

The effect of the rotation of the Earth manifests itself very clearly in the trade winds and the ocean currents. On the Atlantic Ocean, where the continental masses have no disturbing influence on the direction of the wind, heated air rises at the Equator and flows to the Poles (antitrades), while a cold undercurrent from the Poles to the Equator arises (trade wind). At sea, only the undercurrent, hence the trade wind, is perceived. In both hemispheres, it has a deviation to the west, so that in the northern hemisphere the northeast trade results, and in the southern hemisphere the southeast trade. A similar thing occurs to the sea water. Since warm water is lighter than cold water, a warm upper current arises from the Equator to the Poles, and a cold undercurrent from the poles to the Equator. In the northern hemisphere, these ocean currents deviate to the right, so that the cold undercurrent passes along North America, and the warm upper current along Europe. As a consequence, it is much colder in New York than in Madrid or Naples, although New York has about the same latitude as these last-mentioned cities.

Air flows from a high-pressure area (high barometer reading) to a low-pressure area, called a depression. In the northern hemisphere, the wind then always deviates to the right. This yields the following law of the Dutch meteorologist Buys Ballot (1817–1890): A person who has his back turned to the wind has the low-pressure area in front of him and to his left. Of course, in the southern hemisphere it is in front of him and to his right. If the air flows from all directions to the area where the pressure is lowest, then in the northern hemisphere a counterclockwise cyclone arises because of the deviation to the right. In the southern hemisphere the cyclone moves clockwise (*Fig. 179*).

The deviation to the right can also be observed (but much less

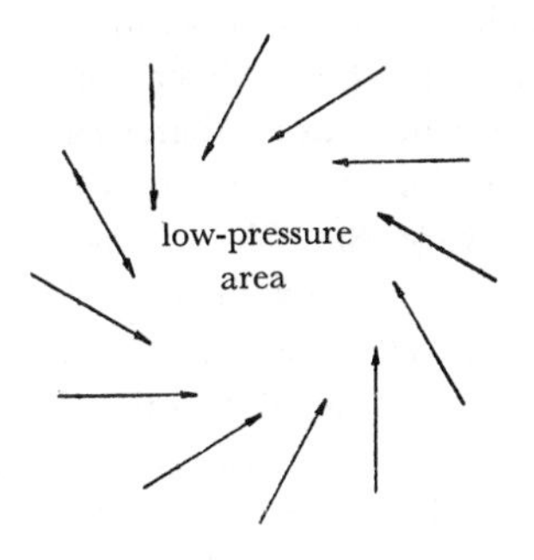

Fig. 179

clearly) in the course of the river at its delta, if the soil is soft and the river is therefore more or less free to choose its bed. With the Volga, it is possible to observe this deviation.

The deviation is demonstrated in a very convincing manner by Foucault's (1819–1868) famous pendulum experiment. A leaden sphere hangs from a long, thin iron wire. The sphere is moved from its state of equilibrium and released without velocity. In the course of the oscillations in a vertical plane (the oscillation plane) which then result, the oscillation plane rotates clockwise (in the northern hemisphere) because of the continuous deviation to the right. At 52° N latitude, the oscillation plane requires more than $7\frac{1}{2}$ hours to reach a position perpendicular to the original one, after rotating through 90°. The phenomenon does not occur at the Equator, while the oscillation plane would rotate most quickly if the experiment were performed at one of the poles of the Earth. With an experiment at the pole it is easy to get an idea of the situation; the Earth rotates under the pendulum, while the oscillation plane is not affected by the rotation at all; hence, the oscillation plane rotates relative to the Earth as quickly as the Earth actually rotates, but in the opposite direction, so that the oscillation plane rotates by 360° in a sidereal day, and hence rotates 90° in nearly 6 hours. It may further be noted that the experimenter has to make sure that the oscillations are performed in a vertical plane. This can be easily achieved by installing a fixed, not too smooth, horizontal ring round the iron wire, so that the wire is bent at its extreme positions (*Fig. 180*). Through the friction of the iron wire against the ring, the possibility of a side-to-side velocity soon disappears.

With a vertical motion, too, the influence of the rotation of the Earth is noticeable. If an object is dropped from a high tower, it deviates to the east. This occurs because, in taking part in the rotation of the Earth, the top of the tower describes a larger circle than the foot of the tower, and hence has a greater easterly velocity. In order not to be hindered by the wind and to achieve a higher drop, the experimenter does not drop the object from a tower, but into a deep mine shaft. For a depth of 100 m, the easterly deviation is 2.2 cm at the Equator; if the drop is 2^2 times as large, the deviation is 2^3 times as large. The last-mentioned phenomenon is the same in the northern as in the southern hemisphere, does not occur at the Poles, and is strongest at the Equator. For the phenomena of horizontal motion, just the reverse was true.

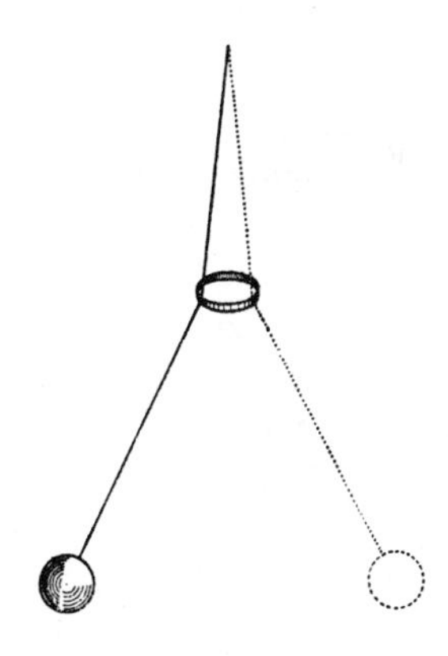

Fig. 180

327. Some phenomena of inertia. Although, as observed in §324, the principle of inertia cannot be proved directly by an experiment, still all sorts of phenomena can be observed that find their explanation in this principle (even though it may be combined with other principles). Place a smooth playing card on a wine glass, and a coin on the card. If you flick away the card with your thumb and finger (*Fig. 181*), then the coin first stays put, then falls into the glass when the card is gone; the card that flies away is not sufficiently rough to exert a horizontal force of any significance, so that the coin remains in its state of rest. The experiment can be arranged differently. Spread out a smooth napkin on a table and on the napkin place a filled wine glass. Take the napkin by a corner and pull it quickly away. The glass will remain standing and not a drop of wine will be spilled. However, you must pull away the napkin so quickly that the small

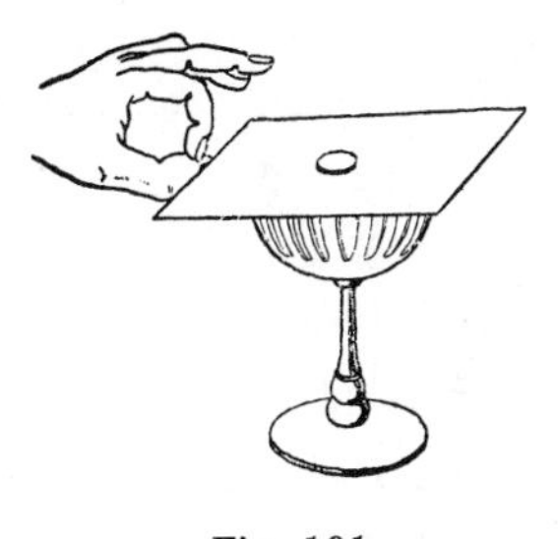

Fig. 181

horizontal force exerted on the glass has no time to cause an appreciable motion of the glass.

Another inertia experiment is the following. The ends of a thin, easily breakable stick are placed on the noses of two clowns. A third clown strikes the stick in the middle with a heavy cudgel, breaking it. The two clowns who were supporting the stick on their noses suffer no harm from the heavy blow. The stick is already broken before its ends have obtained a velocity of any significance. The third clown, however, has to hit so hard that the stick breaks immediately.

An inertia phenomenon that can be observed regularly is the following. If you stand in the corridor of a (European) train that is arriving at a station, you will be pushed forward by the braking, but at the last moment, when the train has already come to a halt, you will get another push backwards. What is the cause of this? Upon braking, the velocity of the train decreases. The person in the corridor, who had the same velocity as the train, has not suddenly lost that velocity, so that he shoots forward if he does not hold tight to something. The push backwards at the last moment is a result of spring action. When the train has come to a halt, the cars spring backwards because the buffer springs were compressed powerfully by the braking and now are released; this springing backwards gives you an impulse forward, but when the springing back comes to a halt, you get a final push backwards. However, it often happens that the phenomenon of the two pushes is obscured by all sorts of additional phenomena, such as the bumping of the cars against each other.

When it rains on a windless day, the drops describe diagonal rather than vertical lines on a train window. This is because the drop, when hitting the pane, does not at once assume the horizontal velocity of the window, but persists in its vertical motion relative to the Earth. Hence, we have here a joint illustration of inertia and of the composition of motions (see §§318 and 319). We often observe that the drop is at rest relative to the window pane for a while, and then starts to move again. The drop has by then assumed the velocity of the train (as a consequence of the friction), and we would expect the drop to move along a vertical line on the window pane (parallel to the vertical edges of the pane). However, again a diagonal line results. This can be explained by the resistance of the air, which quickly reduces the large horizontal velocity which the drop had acquired.

The following experiment shows an interesting phenomenon of inertia. In a closed glass tank which is filled with water to the brim, there are two spheres, one of cork and one of lead. The cork sphere is fixed to the bottom by a thread to prevent it from rising further; the lead sphere hangs from a thread fixed to the top of the tank. The lengths of the threads are such that the lead sphere is hanging just above the cork sphere (*Fig. 182*). If the tank is now suddenly moved to the right, inertia will cause the lead sphere to move to the left relative to the tank. The lead sphere tends to persist in its state of rest. It is true that the water tends to do the same thing, but the sphere, which is so much heavier than the displaced water, dominates the situation. With the cork sphere, which is lighter than the water displaced, the reverse is true; hence, this sphere moves to the right, even relative to the tank. If the tank is moved to and fro, the two spheres will also be observed to move to and fro relative to the tank, the cork sphere in the same direction as the tank every time, the lead

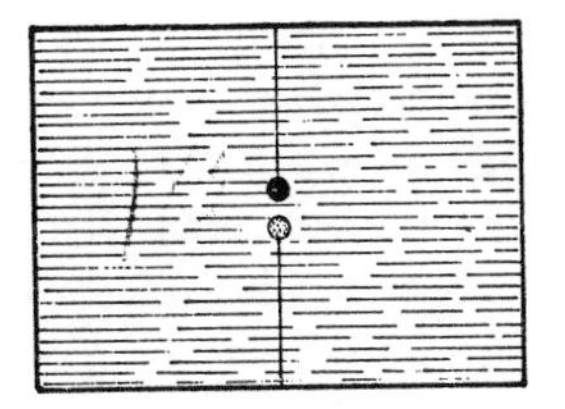

Fig. 182

sphere in the opposite direction. Relative to the Earth, the tank and the two spheres always move to the same side, but for the lead sphere the oscillations are smaller, and for the cork sphere larger, than the oscillations of the tank.

IV. MASS AND WEIGHT

328. Mass and force. A small body for which the dimensions and in particular the differences in velocity may be neglected, is called a material point. If a force acts on such a point, the latter acquires an acceleration in the direction of the force. The magnitude of the acceleration is proportional to that of the force; that is, if the force is made twice (or three times) as large, then the acceleration also becomes twice (three times) as large—provided you take the same material point, because the acceleration also depends on the nature of the material point. If the volume of the body is taken twice (three times) as large, using the same material, then the acceleration becomes twice (three times) as small, provided the same force is applied. We then say that the mass of the material has become two (three) times as large. If we have two bodies of different materials, and if the same force gives the same acceleration to both bodies, then we say that the bodies have the same mass. In this manner, we can also compare the masses of bodies of different materials. By assuming a unit of mass, we can represent the mass of each body by a number. For this unit one takes the gram-mass, the thousandth part of the mass of a piece of metal, the standard kilogram, kept at Sèvres near Paris.

If we take as the unit of force the force that gives an acceleration of a centimeter per second per second to a gram-mass (this unit of force is called a dyne), then the force F, the mass m, and the acceleration a are connected by the relation $F = m \times a$, the so-called fundamental equation of dynamics. It indeed expresses the fact that the acceleration becomes twice as large when the force is taken twice as large, and becomes twice as small with a mass that is twice as large, while the equation also shows that $F = 1$ when $m = 1$ and $a = 1$, in accordance with the definition of the unit of force. If the magnitude and the direction of the force are constant, then (according to $F = m \times a$) the same is true for a; moreover, if the material point then has no velocity initially (or a velocity in the direction of the force), it will have a uniformly accelerated rectilinear motion.

The principle of inertia is included as a special case in $F = m \times a$, since if $F = 0$ all the time, we must have also $a = 0$ all the time; we then have a uniform rectilinear motion, which may possibly be a state of rest. The fundamental equation of dynamics does not hold relatively to any arbitrary system either. The same considerations apply here as were discussed in §325 with respect to the principle of inertia.

329. Gravitation; weight. The force which makes bodies fall downward is called gravitation. In the main, this is a manifestation of the attraction of the Earth. It was one of Newton's important discoveries that the force which makes an apple fall from a tree is a manifestation of the same force of attraction that keeps the moon in its orbit around the Earth. It is said that Newton made this discovery when, on a clear moonlit night, he was resting under an apple tree and a falling apple prompted him to think about the cause; however, this seems to be a legend put into circulation by the French writer Voltaire (1694–1778).

Yet, what we call gravitation is not the same thing as the force of attraction of the Earth. A difference is caused by the rotation of the Earth about its axis, which prevents the Earth from being a system relative to which the fundamental equation of dynamics applies. One might say that gravitation is the apparent attraction of the Earth, the resultant of the real attraction and an imaginary force (a force which is only seemingly present), which is called the centrifugal force (about which we shall say more in §§331 and 332). By substituting the gravitation for the force of attraction, we for the most part nullify the circumstance that the Earth is not a system relative to which the fundamental equation of dynamics applies. By doing so, we can pretend that the Earth is such a system. This is still not completely accurate, but the deviations manifest themselves only in the case of motions like those discussed in §326.

Gravitational force, not as a general phenomenon, but as the force that acts on a certain body, is called the *weight* of that body. A body having a mass that is twice as large has a weight that is twice as large, or, to put it more generally, the weight is proportional to the mass. For bodies of the same material this is practically self-evident, but it is also true for bodies of different materials (see §300). Hence, gravitation has no preference for a certain material, as magnetic force has for soft iron.

The weight of a body is often confused with its mass. Thus, the unit of mass mentioned in §328 is often called a gram for short, while the weight of that mass is also indicated as a gram. However, to preclude a misunderstanding, it is better to speak of a gram-mass in the first case, and of a gram-force in the second case.

330. Acceleration of gravity; free fall. A material point that is dropped without velocity and that is subject only to gravitation has a uniformly accelerated rectilinear motion (if we neglect the deviation discussed at the end of §326, which can be detected only by precise measurements). Its acceleration is called the acceleration of gravity. This proves to be the same for all bodies, or to put it more simply: all bodies fall equally fast. However, this is only the case if the bodies are exclusively subject to gravitation. The resistance of the air has a disturbing influence on the phenomenon, so that, for example, raindrops, which are formed in the higher atmospheric strata, begin to fall with an accelerated motion, but close to the earth acquire a practically uniform one. In fact, the acceleration of gravity is related only to the fall in a vacuum, the so-called free fall, or motion under gravity alone.

As we can observe every day, objects do not fall through the air equally fast. This is because the resistance of the air becomes greater when the area of the body's contact with the air becomes greater. If a small piece of metal and a feather have the same mass, then they have the same weight (in other words, the same force of gravity acts on the two objects); however, because of the much greater area of the feather, it experiences a much greater counteracting resistance from the air. If the air is pumped from a tube in which there are a lead pellet, a scrap of paper, and a feather, then these three objects fall equally fast; this can be ascertained by first placing the tube in a vertical position, and then turning it quickly, so that the tube is again in a vertical position, but with the three objects at the top. If air is reintroduced, then the pellet is observed to fall more quickly, and the feather more slowly, than the piece of paper.

An experiment which can be made more easily, but which is less convincing, is the following. On a coin, placed horizontally, put a piece of paper that does not project beyond the rim of the coin. Drop the coin, and the paper will remain on it. The coin nullifies the resistance of the air to the paper. If you drop the coin and the paper side by side, you will observe a distinct difference in velocity.

In connection with the formula $F = m \times a$, the fact that gravitation gives the same acceleration to all bodies shows that the weight of a body is indeed proportional to its mass.

The acceleration of gravity is not the same for all places on Earth. In Holland this acceleration is approximately 981 centimeters per second per second. This means that a body subject to free fall acquires a velocity of 981 cm per sec after 1 second, hence of almost 10 m per second. At the Equator, the acceleration of gravity is somewhat smaller (as measurements have shown), with a value of 978 cm per sec per sec; at the poles it is somewhat greater, with a value of 983.2 cm per sec per sec. This is partly because the centrifugal force (see §§331 and 332), which counteracts the attraction of the Earth, is maximum at the Equator (because of the greater distance from the axis of the Earth), partly because the Earth is flattened at the poles, so that the distance to the center of the Earth is greatest at the Equator, which makes the attraction a minimum there. The attraction decreases rapidly when the relative distance is reduced.

If a gram-mass is subjected to free fall (dropped in a vacuum), then the force acting on it is a gram-force. In Holland, it gives the body an acceleration of 981 cm per sec per sec. The dyne discussed in §328 gives this gram-mass an acceleration of 1 cm per sec per sec. From this, we see that 1 gram-force = 981 dynes. This gives an idea of the magnitude of a dyne; it is slightly larger than the weight of a milligram. The weight of 1 kg, of which everyone has a fairly clear idea, is approximately one million dynes (actually, a little less).

The gram-force is often used as a unit of force. However, it has the disadvantage of not having an equal magnitude for all places on the Earth, in view of what has been said about the acceleration of gravity. However, in many cases (including everyday life) this objection is insignificant. It can be discounted completely when the acceleration plays no role, and consequently in all cases of equilibrium, where only a comparison of forces is at issue. That is why the dyne is called the dynamic unit of force, and the gram-force the static unit of force.

V. FURTHER CONSIDERATIONS ON FORCES

331. Centrifugal force. Let us imagine the case where a stone tied to a cord is quickly swung about in a vertical plane. Let the velocity of the stone at the highest point by v cm per sec; at the lowest point it is somewhat larger as a consequence of gravitation, but we shall imagine v to be so large that this difference in velocity can be neglected compared with v. According to §316, the stone has an acceleration of v^2/l cm per sec per sec directed toward the center, if the cord has a length of l cm. At the highest point this acceleration is produced by the tension S (in dynes) in the cord and the weight of the stone; if this weight is m grams-force (where m is also the mass of the stone in grams-mass), then in Holland the weight of the stone is $981 \times m$ dynes. Hence, according to the fundamental equation of dynamics we have: $S + (981 \times m) = m \times v^2/l$, and so $S = m\left(\frac{v^2}{l} - 981\right)$ dynes, which is $m\left(0.00102 \times \frac{v^2}{l} - 1\right)$ grams-force. At the lowest point the tension in the cord is greater, first of all because the velocity of the stone is greater there, but also because there the weight of the stone increases the tension instead of reducing it. A simple calculation, which we shall not perform here, shows that this maximum tension amounts to $m\left(0.00102 \times \frac{v^2}{l} + 5\right)$ in grams-force. We take $m = 1000$, hence a stone of one kilogram, $l = 100$, hence a cord of one meter, and $v = 2000$. The smallest tension in the cord is then 39.8 kg and the greatest tension is 45.8 kg, so that the difference is not very large. Hence, in the case in question, the magnitude of the velocity can be fixed approximately at the constant value of 20 m per second. Since the circumference of the circle described by the stone is $2 \times \frac{22}{7} = 6.3$ m, the stone performs over 3 revolutions per second.

Hence, it seems as if the stone experiences a repulsion from the center of the circle; in the example this was about 40 kg. The apparent repulsion, the so-called centrifugal force, becomes 2^2 or (3^2) times as large if the velocity of the stone is taken 2 (or 3) times as large. Hence, this apparent force increases quickly if the velocity is increased.

Elegant applications of these facts are possible. A cup of tea is served on a tray suspended by three cords (*Fig. 183*). If this tray is

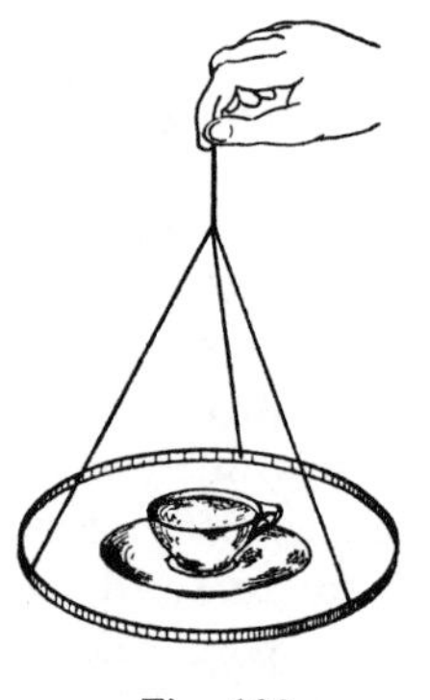

Fig. 183

swung around quickly and adroitly in a vertical plane, so that in a certain position the brim of the cup is directed downward, the cup will stay quietly on the tray and tea in the cup. Something comparable is the following: On a switchback a car can be made to loop the loop (*Fig. 184*). To achieve this, as the calculation shows, the car should come from a height that is above the highest point of the circle by at least a half-radius of the circle to be traversed, if there were no friction. Since a significant loss of velocity results from friction, if you want to avoid accidents, you should choose a considerably higher point for the car to begin its course, for instance, at a height equal to the diameter of the circle to be traversed above the highest point of that circle.

An attraction of some night clubs is a slowly turning circular dance floor. Towards closing time, the dancing couples are removed from the floor in a very simple way by speeding up the motor, to make the floor turn faster and faster. By this means, the customers are seemingly repelled from the center of the dance floor and soon move off the floor. They can still hang on for a while by forming a circle and holding each other by the hand; but finally this does not help, either, because the center of the circle does not precisely coincide with that of the turning disk and because the heavier persons experience a greater centrifugal force than the lighter ones. In the end, the whole party does move off the dance floor.

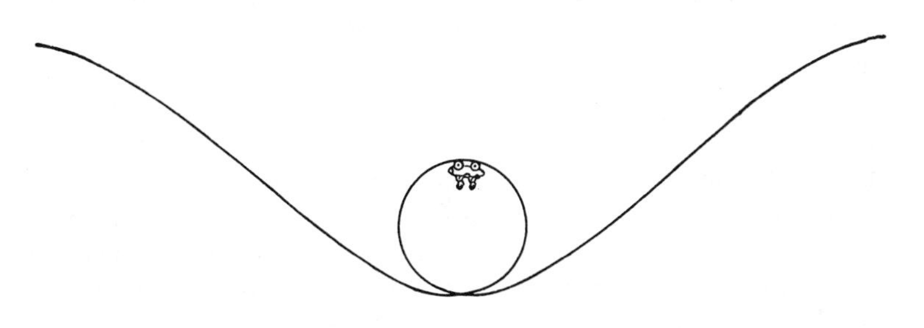

Fig. 184

332. Influence of centrifugal force on gravitation. This topic has been mentioned already in §329. Through the rotation of the Earth, an object on the Earth seemingly experiences a repulsion from the axis of the Earth of $m \times v^2/r$ dynes. Here m is the mass of the object (in grams-mass), v the velocity (in cm per sec) of the Earth at the point in question, as a result of the rotation of the Earth, and r the distance (in cm) of this point to the axis of the Earth. If we take a point on the Equator, then $r = R$, where R is the radius of the Equator. If T is the time of revolution of the Earth (a sidereal day measured in seconds), then $v = 2\pi R/T$, where π is the ratio of the circumference of a circle to its diameter. At the Equator, the centrifugal force is aligned directly against the attraction of the Earth, and hence acts fully as a reduction of the weight. This causes the acceleration of gravity there to be $\dfrac{v^2}{R} = \dfrac{4\pi^2 R}{T^2}$ cm per sec per sec less than that of the attraction. Now $T = 24 \times 60 \times 60$, if we neglect the difference between a sidereal day and a solar day. Furthermore, $R = 638 \times 10^6$ and (since $\pi = 3.1416$) $\pi^2 = 9.87$, from which we find that at the Equator the acceleration of gravity is diminished, as a result of the centrifugal force, by 3.4 cm per sec per sec. According to the measurements referred to in §330, the acceleration of gravity at the poles (where there is no centrifugal force) is 5.1 cm per sec per sec greater than at the Equator. According to what we have just found, two thirds of this can be ascribed to the centrifugal force, leaving one third to be ascribed to the flattening of the Earth.

If the velocity of the rotation of the Earth were 2 or 3 times as great, then its influence on the acceleration of gravity should be 4 or 9 times as great, respectively. If the Earth rotated 17 times as fast as it does in reality, the acceleration of gravity would be zero at the Equator. An object released there would then remain floating in the air.

The flattening of the Earth, too, can be ascribed to the centrifugal force. This flattening occurred in days long past, when the Earth was a hot, viscous liquid. This can be demonstrated with a so-called centrifugal machine, as shown in *Figure 185*, as seen from above. A driving-belt runs round a large and a small revolving disk. When the large disk is rotated by means of a handle, the small disk will turn much more quickly. A vertical drum with a clamping screw has been

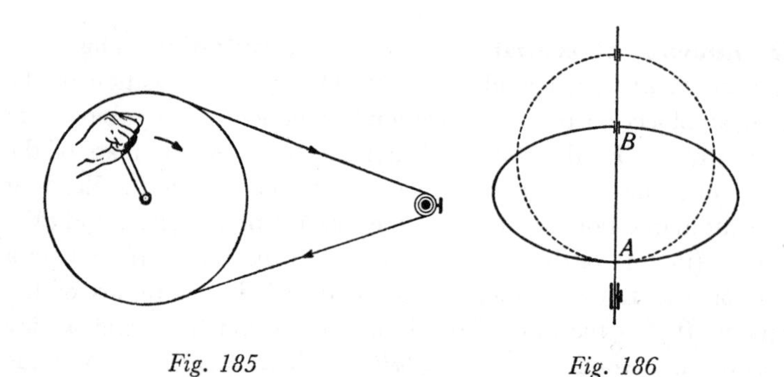

Fig. 185 *Fig. 186*

attached to the small disk, so that various apparatus can be fastened to the small disk and rotated rapidly. We place a metal spring bent to form a circle on the centrifugal machine. Its lowest point A is fixed to a vertical bar that is clamped in the drum. At the highest point B there is a small hole through which the vertical bar is placed (*Fig. 186*). If the bar with the spring is rotated quickly, the circle will be stretched out in the horizontal direction, and will consequently be flattened in the vertical direction. The point B of the spring moves down along the bar. As a consequence of the swift rotation, the images of the spring in the various positions blend, because the impression of light on the eye has an after-effect of about $\frac{1}{10}$ second. Therefore, you observe not a rotating flattened circle, but a flattened body. The experiment gives a highly exaggerated image of the flattening of the Earth, since the major and minor axes of the Earth have lengths in the proportion of 297 and 296.

333. Principle of action and reaction. The general force of attraction discovered by Newton, which acts between all objects and governs the motion of the celestial bodies, made him realize that the force which one object exerts on the other is as great as the force that the second object exerts on the first, but directed the other way. Newton developed this into a general principle, that of action and reaction (force and counterforce). According to this principle, forces occur in pairs. As is the case with the general force of attraction, the forces of the same pair (action and reaction), for different material points, act along the same straight line and are equally great, but have opposite directions. It is immaterial which force is called the action and which is called the reaction. You should not look upon the action as the cause and the reaction as the effect. Both forces arise and disappear simultaneously.

I put an object weighing 5 kg on my hand. What is the reaction to the weight of this object? You might respond: the pressure that the hand exerts on the object. Apparently that pressure meets all requirements. The pressure is as great as the weight, acts along the same line, and has the opposite direction. However, it cannot be the reaction to the weight of the object, if only because it also acts on the body, whereas the reaction always acts upon another body. But there is more. The equality of weight and pressure is accidental, a consequence of the circumstance that the body is in equilibrium, and hence of the fact that I keep my hand still. If I move my hand up in an accelerated motion, so that the body too acquires an acceleration directed upwards, then a total force, directed upwards, acts upon this body; this force is equal to the pressure that my hand exerts on the body, diminished by the weight of the body. Hence, this pressure is now greater than the weight; this can be clearly perceived from the effort it takes to move the hand up quickly. The object then seems to have become heavier suddenly. However, it is not the upward velocity of my hand that causes this phenomenon, but the acceleration. If, after suddenly giving my hand a velocity (for which a large acceleration is required),

I then move my hand upwards uniformly, then the apparent increase in weight disappears again. If I then move my hand upwards more slowly, then I even experience an apparent decrease in the weight.

The foregoing shows that "the pressure of my hand" is not the correct answer. To answer the question as to the reaction to the weight of the object, we must first neglect the rotation of the earth, thus leaving out of consideration the difference between the weight of the object and the attraction that the Earth exerts on it. For the sake of convenience, we further assume that the entire mass of the Earth is concentrated in its center, and we imagine the attraction to originate from this center. Then the answer to the question is: the force that the object exerts on the Earth. Because we cannot perceive this force, it is readily overlooked. It is only from theoretical considerations—we might also say, through analogy—that we assume the existence of a force exerted on the Earth by a body. When we realize that the force exerted by the Earth on a terrestrial object is of the same nature as the attraction exerted on the Moon and that conversely the Moon attracts the Earth (which can be observed and manifests itself chiefly in the phenomenon of tidal ebb and flow), then it is an obvious assumption that a much smaller terrestrial object also attracts the Earth. That we do not perceive this force is no objection against assuming its existence; compared to the mass of the Earth, the force assumed to be present is so small, that we are unable to perceive its effect.

It will be clear, furthermore, that the reaction to the pressure exerted by my hand on the object is the pressure that the object conversely exerts on my hand; the latter pressure is caused by the weight of the body, it is true, but it is something quite different. When I move my hand, both pressures become greater or smaller simultaneously. If I move my hand downwards with an acceleration larger than that of gravity, so that the object leaves my hand (unless it is tied on), then the pressure that my hand exerts on the body disappears but so does the pressure that the object exerts on my hand.

This question still remains: What is the reaction to the centrifugal force that is incorporated in the gravitation? The answer to this is that this reaction does not exist. The principle of action and reaction does not hold for a fictitious force.

334. Motion of the center of gravity. If A and B are two material points of which the masses are 5 and 3, say, then by the center of gravity of A and B is meant the point Z placed on the straight line AB

between A and B in such a way that the distances of Z from A and B are inversely proportional to the masses, hence here as 3 to 5 (*Fig. 187*). Hence, the center of gravity is closer to the larger mass. For various purposes, the center of gravity can conveniently be thought of as a material point of which the mass is equal to the sum of the masses (hence 8 in the example). Sometimes one comes across the misconception that the center of gravity is of importance only in connection with gravitation (as the point through which the force of gravitation exerted on the system passes). In order to counter this

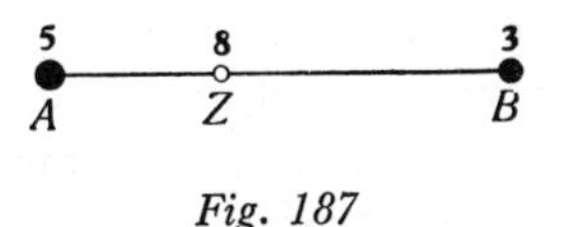

Fig. 187

misconception, which is encouraged by the name "center of gravity," we sometimes speak of "center of mass."

If two equal parallel forces having opposite directions act on the material points A and B, and if A and B have no velocity initially, then after a certain time these points arrive at A' and B', respectively, where AA' and BB' are parallel (and have opposite directions) and are in the proportion of 3 and 5. Hence, the center of gravity has stayed where it was, as *Figure 188* clearly shows. Hence, the center of gravity acts as if it were the point of impact of the forces exerted on A and B, since in that case Z would be affected by two equal forces of opposite direction which cancel each other.

Next we take the case where a force acts only on B, while again A

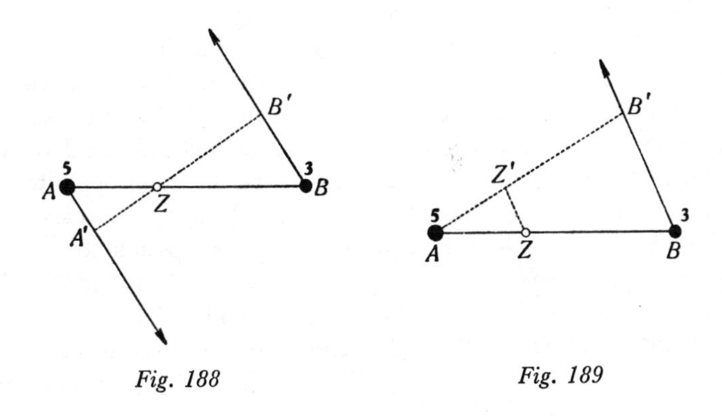

Fig. 188

Fig. 189

and B have no initial velocity. A now stays where it is, while B arrives at B' after a certain time. This causes the center of gravity to move from Z to Z', where ZZ' is parallel to BB' and equal to $3/8\ BB'$ (*Fig. 189*). Z would have been displaced by the same distance if the force had acted on Z, if we imagine Z (as noted before) to be a material point with a mass $5 + 3$.

The same results also hold when there are initial velocities and when arbitrary forces act on A and B. In that case, too, the center of gravity of the material points A and B moves as if it were a material point (with a mass equal to the sum of the masses of A and B) to which the forces acting on A and B have been transmitted (by parallel shifts). This proposition (of great importance for the entire field of dynamics), which, as you see, has nothing to do with gravity, can be extended to several material points and also to objects that cannot be considered as material points, and to systems of objects. To this end, the definition of center of gravity has to be extended to several material points; but we shall not go any further into this subject.

VI. SOME PUZZLES IN DYNAMICS

335. Enchanted cage. In one of his fascinating little books, the French mathematician Poincaré (1854–1912) describes the following remarkable situation. Some people are in a cage which a sorçerer is pulling through space in a uniformly accelerated motion with the acceleration of gravity. The cage is at so great a distance from all heavenly bodies that they do not exert any attraction on it. We assume the problem of the supply of air and food to be solved somehow. The question now is: What happens in the cage?

Relative to space (but not to the cage), a released object persists in the same velocity that it had at the moment of release. The cage, however, moves more and more quickly, so that the object lags behind the cage. Hence, relative to the cage, it performs a uniformly accelerated motion with the acceleration of gravity in the direction opposite to that in which the cage is pulled by the sorcerer. Hence, in the cage the situation is the same as if there were gravity (as a pseudo-force), so that the people in the cage feel completely at ease. To them "upwards" means "the direction in which the sorcerer is pulling."

336. Problem of the falling elevator. The cable supporting an elevator breaks, and the elevator falls. It is assumed that this occurs without friction, so that the elevator falls with the acceleration of gravity. On the bottom of the elevator lies a drop of mercury. On the wall of the elevator a burning candle has been fixed. What happens to the drop of mercury and the burning candle after the cable breaks?

If an object which initially has no velocity relative to the elevator is released in the elevator, then this object moves downward with uniform acceleration relative to the elevator shaft, and hence has exactly the same motion as the falling elevator; it thus remains floating in the elevator. Therefore, as a consequence of the breaking of the cable, a situation has arisen in the elevator just as if gravity had disappeared, and as if the bodies no longer had any weight; however, their mass has not been diminished. Therefore, the situation is completely the opposite to that of the enchanted cage of §335; there, through the motion of the cage, gravity had appeared; now, through the motion of the elevator, gravity has disappeared.

The drop of mercury, which was flattened by gravity, resumes its spherical shape since its surface tension still exists. Thus, the center of gravity of the drop of mercury moves upward relative to the elevator and retains the upward velocity thus acquired, since there are no forces that reduce this velocity. Hence the drop of mercury moves upwards in its entirety, always relative to the elevator. The form of the drop of mercury fluctuates between an oblate and a prolate spheroid. When the drop of mercury has reached the ceiling of the elevator, it is impelled back down again, and it keeps oscillating in this way between the floor and the ceiling.

The candle goes out for want of oxygen. The gases formed by the combustion (carbon dioxide and water vapor) are strongly heated, and as a result, lighter than air (although carbon dioxide is heavier than air of the same temperature). Therefore the combustion products rise in normal circumstances, which gives fresh air an opportunity of access. However, in the falling elevator, differences in specific gravity play no role, because gravity has disappeared, and so the gases in question remain hanging around the wick of the candle for a long time, and spread only slowly through the elevator (by diffusion).

337. Will the body topple over onto the smooth inclined plane? On a perfectly smooth inclined plane is placed a high cylindrical wooden object, weighted on the top by a lead hemisphere, so that it has a high center of gravity Z (*Fig. 190*). This body is released from a state of rest. Will it then topple over?

On the face of it, it seems that, although the body will slide down because of the absence of friction, it will also topple over because the perpendicular through the center of gravity Z intersects the inclined plane outside the supporting surface. However, the body will do nothing but slide; in doing so, it will remain erect. This can be proved with the aid of certain propositions of dynamics which we cannot discuss here. Yet, in the following way you can obtain a good idea of the matter (without the propositions in question, although they may be more convincing). Just as in the falling elevator of §336 gravity was completely eliminated, here the component of gravity parallel to the inclined plane is, as it were, rendered inoperative as a result of the body's sliding down unhampered.

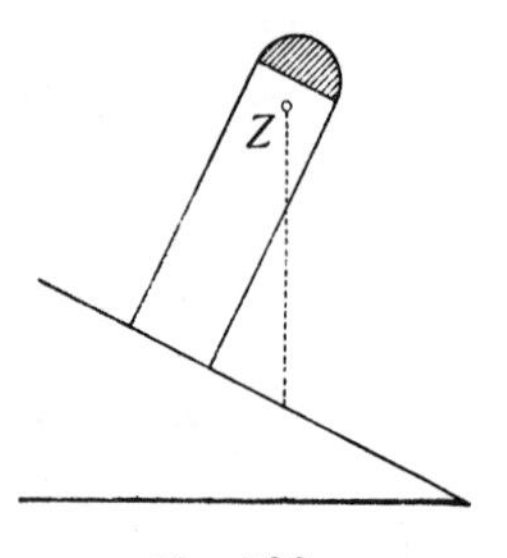

Fig. 190

This experiment gives a confirmation of the theory only when the center of gravity of the object is not raised too high, that is, when the perpendicular through the center of gravity does not meet the plane too far outside the supporting surface. This is because it is impossible to make the inclined plane sufficiently smooth.

338. How do I get off a smooth table? Let us imagine I am in a state of rest on a large, perfectly smooth table. By perfectly smooth I mean that the table can exert only a vertical force on me. The forces that act on me then are the downward-directed vertical weights of the material points of which I consist, and the upward-directed vertical pressures of the table. Apart from that, only so-called internal forces act on me, that is, forces that are acting upon my various components, or, to put it differently, forces whose reaction also acts on me. If we now apply the proposition of §334 about the motion of the center of gravity, then we have to transfer my weight and the pressure of the table to my center of gravity. This is not necessary for the internal forces just mentioned, since these occur in pairs and hence annul each other by pairs after being transferred to my center of gravity. Hence, only vertical forces are acting on my center of gravity, and since we have assumed that I was at rest originally, I cannot give a horizontal velocity to my center of gravity, however much I struggle. I can only move my center of gravity up and down, and hence cannot crawl on the table, let alone walk. Supposing now that the table is so large that I cannot seize its edge, how do I get off the table without help?

The solution is simply this: I throw away an object in a horizontal direction. The common center of gravity of myself and this object will still remain on the same vertical line. Hence, if I have thrown away the object to the right, say, then my center of gravity (exclusive of the object) will move to the left, although more slowly than the object, which no doubt will have a much smaller mass; however, slowly but surely I slide off the table. Hence, I can indeed get off the smooth table, since I will always have something to throw away, even if it were only the air that I exhale.

Of course, in reality it is not possible to make a table so smooth that I could not even crawl on it. It is possible, though, to make the table so smooth that, when throwing away a not too light object, I will acquire a horizontal velocity.

VII. SOME OTHER DYNAMIC PROBLEMS

339. Problem of the climbing monkey. A rope is slung round a fixed pulley (that is, a pulley with a fixed center). A monkey is hanging on one end of the rope, and on the other end, at the same height, there is an equally heavy counterweight. Hence, the monkey and the counterweight can remain hanging without the introduction of any motion (*Fig. 191*). Now the monkey starts to climb the rope. By pulling itself up, it pulls the rope down. This causes the tension in the rope to increase, and the counterweight to rise, which makes the monkey go down again. The question is: As the monkey climbs the rope, does it get any higher relative to the earth, and if so, how fast does it rise?

In order to facilitate the answer to this question, we neglect the weight of the rope. For the sake of simplicity, the pulley is thought of as a perfectly smooth sheave that cannot turn; however, with a well-oiled and light pulley the experiment would be much more accurate, since a sheave cannot be made sufficiently smooth. The monkey is considered as a material point. This point then moves up along the rope in a given manner.

As noted above, the tension in the rope becomes greater than the weight of the climbing monkey. The sheave being perfectly smooth, and the mass of the rope being negligible, the tension is the same in both parts of the rope. The forces that act on the monkey from outside (the external forces, the forces that have no reaction that affects the monkey) are the upward-directed tension S of the rope and the downward-directed weight G of the monkey. The counterweight, which has the same mass as the monkey, is subject to the same forces. As a consequence, both the monkey and the counterweight rise with the same acceleration, and hence with the same velocity, and therefore they constantly remain at the same relative height. The point of

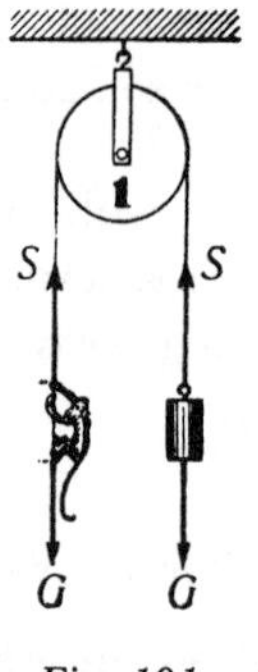

Fig. 191

the rope where the monkey originally was goes down as fast as the counterweight goes up, so that this point goes down as fast as the monkey goes up. Therefore, relative to the rope, the monkey has covered a distance that is equal to twice the rise of the monkey (and that of the counterweight) relative to the Earth.

The foregoing shows that the monkey rises, but the distance covered is half the distance climbed, that is, half the distance that the monkey would have risen if the top end of the rope had been fixed.

340. Extension of the problem of the climbing monkey. Now let us modify the problem of §339 in the sense that the pulley marked 1 is fastened to a rope which is slung round a second pulley (marked 2 in *Figure 192*), which is fixed. The other end of the rope that is slung round pulley 2, carries a counterweight that has the same weight as the monkey and the first counterweight (that of §339) combined.

The weights of the pulleys and those of the ropes are neglected. As long as the monkey does not move, the system is in equilibrium; we assume, moreover, that the monkey and both counterweights are at the same height. Now the monkey starts climbing. How quickly does it rise in this case?

For convenience we again think of the pulleys as smooth sheaves that cannot turn. Since the weight (and hence the mass) of sheave 1 and of the ropes is neglected, the tension in the rope that runs from sheave 1 to sheave 2 is equal to the sum $2 \times S$ of the two downward-directed tensions S that act on sheave 1. Hence an upward-directed tension $2 \times S$, and the downward-directed weight $2 \times G$, are acting on the second counterweight. These forces are twice as large as those acting on the first counterweight (or on the monkey). Since the second counterweight also has a mass that is twice as large, it acquires the same acceleration as the first one, and as the monkey. Hence the monkey and both counterweights rise equally fast and remain at the same relative height. When the three bodies have risen a distance p, sheave 1 has gone down a distance p. So the rope between the first counterweight and sheave 1 has been shortened by $2 \times p$, so that the point of the rope where the monkey originally was has sunk a distance of $(2 \times p) + p = 3 \times p$. Since the monkey has risen by an amount p, it has climbed a distance $(3 \times p) + p = 4 \times p$. This shows that the monkey rises, but over a distance that is only a quarter of the distance climbed.

This result can be obtained more quickly by considering the monkey and the first counterweight (bodies which rise at the same rate) as one body that climbs the rope running over sheave 2, with a velocity equal to half the velocity with which the monkey climbs. Again, the velocity with which this new body rises relative to the Earth is half the velocity with which it climbs the rope (running to sheave 2), hence a quarter of the velocity with which the monkey climbs.

Next we can fasten a rope to sheave 2, pass it round a third (fixed) perfectly smooth sheave (marked 3 in *Figure 193*), and hang a counterweight $4 \times G$ on the other end of this rope. Then there is a state of equilibrium as long as the monkey does not move. We assume that the four bodies are at the same height, and that the monkey then starts to climb. Neglecting, as before, the masses of ropes and sheaves, the forces that act on the bodies have been indicated in Figure 193 in a self-explanatory manner. The third counterweight has a mass that is four times as great as that of the first one, but it is subject to forces

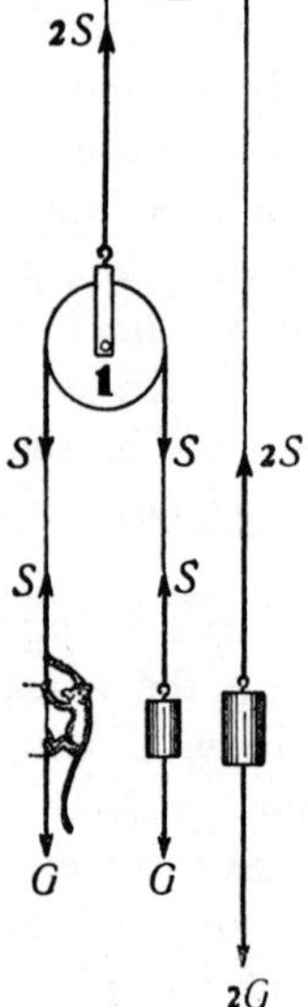

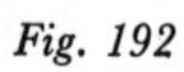
Fig. 192

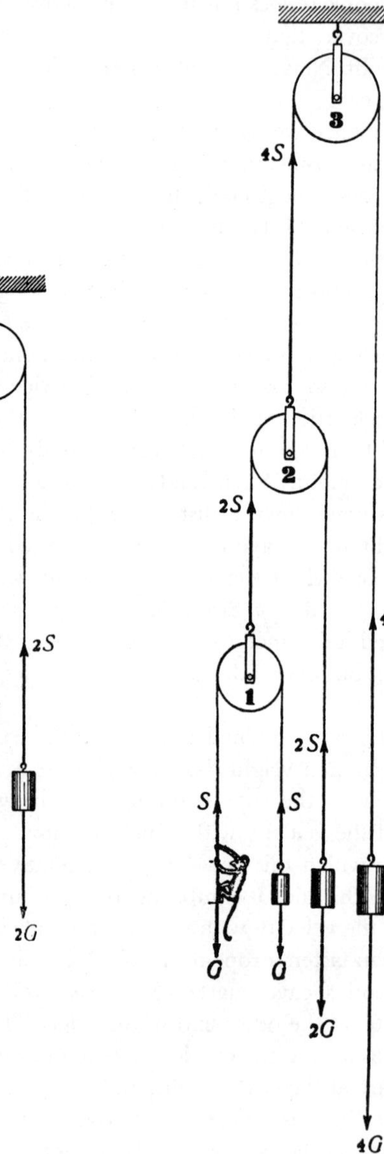

Fig. 193

that are four times as great. Hence, as the monkey climbs, all bodies rise with the same velocity. By considering the monkey and the first two counterweights (G and $2 \times G$) together, as one body that climbs the rope running to sheave 3 with a velocity that is a quarter of the velocity with which the monkey climbs, we find that the velocity with which the bodies rise relative to the Earth is again obtained by a halving procedure. Hence, in this case the monkey rises by an amount that is the eighth part of the distance that it climbs.

It is perhaps more convincing to express the various rises and drops in terms of the amount p that the four bodies rise. Sheave 2 sinks an amount p, and the rope between the second counterweight and sheave 2 becomes shorter by $2 \times p$, so that sheave 1 sinks an amount $(2 \times p) + p = 3 \times p$. The rope between the first counterweight and sheave 1 gets shorter by $(3 \times p) + p = 4 \times p$, so that the point of the rope where the monkey originally was has sunk an amount $(4 \times p) + (3 \times p) = 7 \times p$. Hence, the monkey has climbed an amount $(7 \times p) + p = 8 \times p$.

It is clear that if a fourth sheave is introduced, the monkey rises still more slowly relative to the Earth, namely a sixteenth part of the amount it climbs. And so on; so that in the case of many sheaves the monkey's climbing does not bring it appreciably higher.

341. Related problem. Round the perfectly smooth sheave 1 (a pulley in reality) a rope is slung. On one of its ends a weight of P grams is hanging, and on the other end a lighter weight of Q grams. Sheave 1 hangs on a rope which is slung round a perfectly smooth immobile sheave 2 (a fixed pulley in reality); on the other end of this rope, a weight of $P + Q$ grams is hanging (*see Figure 194*). The weight of sheave 1 and that of the ropes is neglected. If the first-mentioned rope were fixed to sheave 1, then everything would be in equilibrium if the bodies were released without velocity. However, now that this rope can slide on sheave 1, the weight P moves down and the weight Q moves up. We ask the question: Will the weight $P + Q$ and the sheave 1 still be at rest?

To answer this question, we first take the simpler case where sheave 1 is connected by a rope to a fixed point (see *Figure 195*). The tension S (expressed in grams) in the rope which is slung round the sheave has a magnitude which lies between P and Q. A net downward-directed force $P - S$ is acting on the weight P, and a net upward-directed force $S - Q$ is acting on the weight Q. Since both bodies acquire the same acceleration, P downwards and Q upwards, the forces $P - S$

A CATALOGUE OF SELECTED DOVER BOOKS IN ALL FIELDS OF INTEREST

WHAT IS SCIENCE?, *N. Campbell*
The role of experiment and measurement, the function of mathematics, the nature of scientific laws, the difference between laws and theories, the limitations of science, and many similarly provocative topics are treated clearly and without technicalities by an eminent scientist. "Still an excellent introduction to scientific philosophy," H. Margenau in *Physics Today*. "A first-rate primer . . . deserves a wide audience," *Scientific American*. 192pp. 5⅜ x 8.
60043-2 Paperbound $1.25

THE NATURE OF LIGHT AND COLOUR IN THE OPEN AIR, *M. Minnaert*
Why are shadows sometimes blue, sometimes green, or other colors depending on the light and surroundings? What causes mirages? Why do multiple suns and moons appear in the sky? Professor Minnaert explains these unusual phenomena and hundreds of others in simple, easy-to-understand terms based on optical laws and the properties of light and color. No mathematics is required but artists, scientists, students, and everyone fascinated by these "tricks" of nature will find thousands of useful and amazing pieces of information. Hundreds of observational experiments are suggested which require no special equipment. 200 illustrations; 42 photos. xvi + 362pp. 5⅜ x 8.
20196-1 Paperbound $2.00

THE STRANGE STORY OF THE QUANTUM, AN ACCOUNT FOR THE GENERAL READER OF THE GROWTH OF IDEAS UNDERLYING OUR PRESENT ATOMIC KNOWLEDGE, *B. Hoffmann*
Presents lucidly and expertly, with barest amount of mathematics, the problems and theories which led to modern quantum physics. Dr. Hoffmann begins with the closing years of the 19th century, when certain trifling discrepancies were noticed, and with illuminating analogies and examples takes you through the brilliant concepts of Planck, Einstein, Pauli, Broglie, Bohr, Schroedinger, Heisenberg, Dirac, Sommerfeld, Feynman, etc. This edition includes a new, long postscript carrying the story through 1958. "Of the books attempting an account of the history and contents of our modern atomic physics which have come to my attention, this is the best," H. Margenau, Yale University, in *American Journal of Physics*. 32 tables and line illustrations. Index. 275pp. 5⅜ x 8.
20518-5 Paperbound $2.00

GREAT IDEAS OF MODERN MATHEMATICS: THEIR NATURE AND USE, *Jagjit Singh*
Reader with only high school math will understand main mathematical ideas of modern physics, astronomy, genetics, psychology, evolution, etc. better than many who use them as tools, but comprehend little of their basic structure. Author uses his wide knowledge of non-mathematical fields in brilliant exposition of differential equations, matrices, group theory, logic, statistics, problems of mathematical foundations, imaginary numbers, vectors, etc. Original publication. 2 appendixes. 2 indexes. 65 ills. 322pp. 5⅜ x 8.
20587-8 Paperbound $2.25

In the case of the two sheaves 1 and 2, the weights P and Q move in such a way that the center of gravity of the combination of the weights P and Q goes down as quickly as the weight $P + Q$. This follows from the proposition of §334 on the motion of the center of gravity; the same downward-directed force $(P + Q - S - S \times P + Q - (2 = S))$ is acting on the combination of the weights P and Q as on the weight $P + Q$. Since the combination in question also has the same mass as the weight $P + Q$, it follows that the center of gravity of this combination will get the same acceleration as the weight $P + Q$. Hence, if originally the three weights are at the same relative height, then the center of gravity of the weights P and Q always remains at the same height as the weight $P + Q$. In Figure 194, where we have taken $P = 2Q$, this has been taken into account.

In the problems of this section, the forces remain constant, so that the motions are uniformly accelerated. In the problems of §§339 and 340 (the ones with the monkey), nothing can be said about the motion. The tension S introduced there is variable and hence refers to a certain moment; the nature of the motion depends entirely on what the monkey does.

***342. Some calculations concerning the problems of §341.** In Figure 195 $(P - S)/(S - Q) = P/Q$. From this it follows that

$$S = \frac{2PQ}{P + Q},$$

so that the tension $2S$ in the upper rope is equal to $\frac{4PQ}{P + Q}$. Hence, in Figure 194 we must increase the weight of sheave 1 by

$$P + Q - \frac{4PQ}{P + Q} = \frac{(P - Q)^2}{P + Q}$$

grams to keep it in equilibrium.

Applying the fundamental equation of dynamics (see §328) to both weights in Figure 195, we find

$$(P - S)g = Pa, \qquad (S - Q)g = Qa,$$

where g is the acceleration of gravity, and a the acceleration with which P moves down and Q up (in cm per sec per sec in each case).

By addition, we find from this that

$$a = \frac{P - Q}{P + Q}g$$

In the case of Figure 194, we find (again from the fundamental equation of dynamics) that when b is the acceleration with which sheave 1 moves upwards, and a the acceleration with which the weight

P goes down relative to sheave 1, both in cm per sec per sec, we then have the results

$$(P - S)g = P(a - b),$$
$$(S - Q)g = Q(a + b),$$
$$(P + Q - 2S)g = (P + Q)b.$$

From this we find the following formulae for the unknowns a, b, and S:

$$a = \frac{2(P^2 - Q^2)}{P^2 + Q^2 + 6PQ} g,$$
$$b = \frac{(P - Q)^2}{P^2 + Q^2 + 6PQ} g,$$
$$S = \frac{4PQ(P + Q)}{P^2 + Q^2 + 6PQ}.$$

The accelerations of P (downwards) and of Q (upwards) are given by

$$a - b = \frac{(P - Q)(P + 3Q)}{P^2 + Q^2 + 6PQ} g,$$
$$a + b = \frac{(P - Q)(3P + Q)}{P^2 + Q^2 + 6PQ} g,$$

respectively.

The accelerations of P (downwards) and of Q (upwards) relative to the weight $P + Q$ are

$$a - 2b = \frac{4Q(P - Q)}{P^2 + Q^2 + 6PQ} g,$$
$$a + 2b = \frac{4P(P - Q)}{P^2 + Q^2 + 6PQ} g,$$

respectively.

Hence, these accelerations are in the proportion of Q/P, which confirms the fact that the center of gravity of the weights P and Q has no acceleration relative to the weight $P + Q$, hence that this center of gravity goes down as fast as the weight $P + Q$.

From the results obtained, we derive the fact that the tension in the rope slung round sheave 1 and the accelerations of the weights P and Q relative to sheave 1 are greater in Figure 194 than in Figure 195. Hence, the same is true for the acceleration of the weight Q relative to the Earth. However, the acceleration of the weight P relative to the Earth is smaller in Figure 194 than it is in Figure 195. We leave it to the reader to verify this statement.

(The end)

THE MUSIC OF THE SPHERES: THE MATERIAL UNIVERSE—FROM ATOM TO QUASAR, SIMPLY EXPLAINED, *Guy Murchie*
Vast compendium of fact, modern concept and theory, observed and calculated data, historical background guides intelligent layman through the material universe. Brilliant exposition of earth's construction, explanations for moon's craters, atmospheric components of Venus and Mars (with data from recent fly-by's), sun spots, sequences of star birth and death, neighboring galaxies, contributions of Galileo, Tycho Brahe, Kepler, etc.; and (Vol. 2) construction of the atom (describing newly discovered sigma and xi subatomic particles), theories of sound, color and light, space and time, including relativity theory, quantum theory, wave theory, probability theory, work of Newton, Maxwell, Faraday, Einstein, de Broglie, etc. "Best presentation yet offered to the intelligent general reader," *Saturday Review*. Revised (1967). Index. 319 illustrations by the author. Total of xx + 644pp. 5⅜ x 8½.
21809-0, 21810-4 Two volume set, paperbound $5.00

FOUR LECTURES ON RELATIVITY AND SPACE, *Charles Proteus Steinmetz*
Lecture series, given by great mathematician and electrical engineer, generally considered one of the best popular-level expositions of special and general relativity theories and related questions. Steinmetz translates complex mathematical reasoning into language accessible to laymen through analogy, example and comparison. Among topics covered are relativity of motion, location, time; of mass; acceleration; 4-dimensional time-space; geometry of the gravitational field; curvature and bending of space; non-Euclidean geometry. Index. 40 illustrations. x + 142pp. 5⅜ x 8½. 61771-8 Paperbound $1.35

HOW TO KNOW THE WILD FLOWERS, *Mrs. William Starr Dana*
Classic nature book that has introduced thousands to wonders of American wild flowers. Color-season principle of organization is easy to use, even by those with no botanical training, and the genial, refreshing discussions of history, folklore, uses of over 1,000 native and escape flowers, foliage plants are informative as well as fun to read. Over 170 full-page plates, collected from several editions, may be colored in to make permanent records of finds. Revised to conform with 1950 edition of Gray's Manual of Botany. xlii + 438pp. 5⅜ x 8½. 20332-8 Paperbound $2.50

MANUAL OF THE TREES OF NORTH AMERICA, *Charles Sprague Sargent*
Still unsurpassed as most comprehensive, reliable study of North American tree characteristics, precise locations and distribution. By dean of American dendrologists. Every tree native to U.S., Canada, Alaska; 185 genera, 717 species, described in detail—leaves, flowers, fruit, winterbuds, bark, wood, growth habits, etc. plus discussion of varieties and local variants, immaturity variations. Over 100 keys, including unusual 11-page analytical key to genera, aid in identification. 783 clear illustrations of flowers, fruit, leaves. An unmatched permanent reference work for all nature lovers. Second enlarged (1926) edition. Synopsis of families. Analytical key to genera. Glossary of technical terms. Index. 783 illustrations, 1 map. Total of 982pp. 5⅜ x 8.
20277-1, 20278-X Two volume set, paperbound $6.00

It's Fun to Make Things From Scrap Materials, *Evelyn Glantz Hershoff*

What use are empty spools, tin cans, bottle tops? What can be made from rubber bands, clothes pins, paper clips, and buttons? This book provides simply worded instructions and large diagrams showing you how to make cookie cutters, toy trucks, paper turkeys, Halloween masks, telephone sets, aprons, linoleum block- and spatter prints — in all 399 projects! Many are easy enough for young children to figure out for themselves; some challenging enough to entertain adults; all are remarkably ingenious ways to make things from materials that cost pennies or less! Formerly "Scrap Fun for Everyone." Index. 214 illustrations. 373pp. 5⅜ x 8½. 21251-3 Paperbound $1.75

Symbolic Logic and The Game of Logic, *Lewis Carroll*

"Symbolic Logic" is not concerned with modern symbolic logic, but is instead a collection of over 380 problems posed with charm and imagination, using the syllogism and a fascinating diagrammatic method of drawing conclusions. In "The Game of Logic" Carroll's whimsical imagination devises a logical game played with 2 diagrams and counters (included) to manipulate hundreds of tricky syllogisms. The final section, "Hit or Miss" is a lagniappe of 101 additional puzzles in the delightful Carroll manner. Until this reprint edition, both of these books were rarities costing up to $15 each. Symbolic Logic: Index. xxxi + 199pp. The Game of Logic: 96pp. 2 vols. bound as one. 5⅜ x 8. 20492-8 Paperbound $2.50

Mathematical Puzzles of Sam Loyd, Part I
selected and edited by M. Gardner

Choice puzzles by the greatest American puzzle creator and innovator. Selected from his famous collection, "Cyclopedia of Puzzles," they retain the unique style and historical flavor of the originals. There are posers based on arithmetic, algebra, probability, game theory, route tracing, topology, counter and sliding block, operations research, geometrical dissection. Includes the famous "14-15" puzzle which was a national craze, and his "Horse of a Different Color" which sold millions of copies. 117 of his most ingenious puzzles in all. 120 line drawings and diagrams. Solutions. Selected references. xx + 167pp. 5⅜ x 8. 20498-7 Paperbound $1.35

String Figures and How to Make Them, *Caroline Furness Jayne*

107 string figures plus variations selected from the best primitive and modern examples developed by Navajo, Apache, pygmies of Africa, Eskimo, in Europe, Australia, China, etc. The most readily understandable, easy-to-follow book in English on perennially popular recreation. Crystal-clear exposition; step-by-step diagrams. Everyone from kindergarten children to adults looking for unusual diversion will be endlessly amused. Index. Bibliography. Introduction by A. C. Haddon. 17 full-page plates, 960 illustrations. xxiii + 401pp. 5⅜ x 8½. 20152-X Paperbound $2.25

Paper Folding for Beginners, *W. D. Murray and F. J. Rigney*

A delightful introduction to the varied and entertaining Japanese art of origami (paper folding), with a full, crystal-clear text that anticipates every difficulty; over 275 clearly labeled diagrams of all important stages in creation. You get results at each stage, since complex figures are logically developed from simpler ones. 43 different pieces are explained: sailboats, frogs, roosters, etc. 6 photographic plates. 279 diagrams. 95pp. 5⅝ x 8⅜. 20713-7 Paperbound $1.00

PRINCIPLES OF ART HISTORY,
H. Wölfflin
Analyzing such terms as "baroque," "classic," "neoclassic," "primitive," "picturesque," and 164 different works by artists like Botticelli, van Cleve, Dürer, Hobbema, Holbein, Hals, Rembrandt, Titian, Brueghel, Vermeer, and many others, the author establishes the classifications of art history and style on a firm, concrete basis. This classic of art criticism shows what really occurred between the 14th-century primitives and the sophistication of the 18th century in terms of basic attitudes and philosophies. "A remarkable lesson in the art of seeing," *Sat. Rev. of Literature.* Translated from the 7th German edition. 150 illustrations. 254pp. 6⅛ x 9¼. 20276-3 Paperbound $2.25

PRIMITIVE ART,
Franz Boas
This authoritative and exhaustive work by a great American anthropologist covers the entire gamut of primitive art. Pottery, leatherwork, metal work, stone work, wood, basketry, are treated in detail. Theories of primitive art, historical depth in art history, technical virtuosity, unconscious levels of patterning, symbolism, styles, literature, music, dance, etc. A must book for the interested layman, the anthropologist, artist, handicrafter (hundreds of unusual motifs), and the historian. Over 900 illustrations (50 ceramic vessels, 12 totem poles, etc.). 376pp. 5⅜ x 8. 20025-6 Paperbound $2.50

THE GENTLEMAN AND CABINET MAKER'S DIRECTOR,
Thomas Chippendale
A reprint of the 1762 catalogue of furniture designs that went on to influence generations of English and Colonial and Early Republic American furniture makers. The 200 plates, most of them full-page sized, show Chippendale's designs for French (Louis XV), Gothic, and Chinese-manner chairs, sofas, canopy and dome beds, cornices, chamber organs, cabinets, shaving tables, commodes, picture frames, frets, candle stands, chimney pieces, decorations, etc. The drawings are all elegant and highly detailed; many include construction diagrams and elevations. A supplement of 24 photographs shows surviving pieces of original and Chippendale-style pieces of furniture. Brief biography of Chippendale by N. I. Bienenstock, editor of *Furniture World.* Reproduced from the 1762 edition. 200 plates, plus 19 photographic plates. vi + 249pp. 9⅛ x 12¼. 21601-2 Paperbound $3.50

AMERICAN ANTIQUE FURNITURE: A BOOK FOR AMATEURS,
Edgar G. Miller, Jr.
Standard introduction and practical guide to identification of valuable American antique furniture. 2115 illustrations, mostly photographs taken by the author in 148 private homes, are arranged in chronological order in extensive chapters on chairs, sofas, chests, desks, bedsteads, mirrors, tables, clocks, and other articles. Focus is on furniture accessible to the collector, including simpler pieces and a larger than usual coverage of Empire style. Introductory chapters identify structural elements, characteristics of various styles, how to avoid fakes, etc. "We are frequently asked to name some book on American furniture that will meet the requirements of the novice collector, the beginning dealer, and . . . the general public. . . . We believe Mr. Miller's two volumes more completely satisfy this specification than any other work," *Antiques.* Appendix. Index. Total of vi + 1106pp. 7⅞ x 10¾.
21599-7, 21600-4 Two volume set, paperbound $7.50

THE BAD CHILD'S BOOK OF BEASTS, MORE BEASTS FOR WORSE CHILDREN, and A MORAL ALPHABET, *H. Belloc*
Hardly and anthology of humorous verse has appeared in the last 50 years without at least a couple of these famous nonsense verses. But one must see the entire volumes — with all the delightful original illustrations by Sir Basil Blackwood — to appreciate fully Belloc's charming and witty verses that play so subacidly on the platitudes of life and morals that beset his day — and ours. A great humor classic. Three books in one. Total of 157pp. 5⅜ x 8.
20749-8 Paperbound $1.00

THE DEVIL'S DICTIONARY, *Ambrose Bierce*
Sardonic and irreverent barbs puncturing the pomposities and absurdities of American politics, business, religion, literature, and arts, by the country's greatest satirist in the classic tradition. Epigrammatic as Shaw, piercing as Swift, American as Mark Twain, Will Rogers, and Fred Allen, Bierce will always remain the favorite of a small coterie of enthusiasts, and of writers and speakers whom he supplies with "some of the most gorgeous witticisms of the English language" (H. L. Mencken). Over 1000 entries in alphabetical order. 144pp. 5⅜ x 8. 20487-1 Paperbound $1.00

THE COMPLETE NONSENSE OF EDWARD LEAR.
This is the only complete edition of this master of gentle madness available at a popular price. *A Book of Nonsense, Nonsense Songs, More Nonsense Songs and Stories* in their entirety with all the old favorites that have delighted children and adults for years. The Dong With A Luminous Nose, The Jumblies, The Owl and the Pussycat, and hundreds of other bits of wonderful nonsense. 214 limericks, 3 sets of Nonsense Botany, 5 Nonsense Alphabets, 546 drawings by Lear himself, and much more. 320pp. 5⅜ x 8. 20167-8 Paperbound $1.75

THE WIT AND HUMOR OF OSCAR WILDE, *ed. by Alvin Redman*
Wilde at his most brilliant, in 1000 epigrams exposing weaknesses and hypocrisies of "civilized" society. Divided into 49 categories—sin, wealth, women, America, etc.—to aid writers, speakers. Includes excerpts from his trials, books, plays, criticism. Formerly "The Epigrams of Oscar Wilde." Introduction by Vyvyan Holland, Wilde's only living son. Introductory essay by editor. 260pp. 5⅜ x 8. 20602-5 Paperbound $1.50

A CHILD'S PRIMER OF NATURAL HISTORY, *Oliver Herford*
Scarcely an anthology of whimsy and humor has appeared in the last 50 years without a contribution from Oliver Herford. Yet the works from which these examples are drawn have been almost impossible to obtain! Here at last are Herford's improbable definitions of a menagerie of familiar and weird animals, each verse illustrated by the author's own drawings. 24 drawings in 2 colors; 24 additional drawings. vii + 95pp. 6½ x 6. 21647-0 Paperbound $1.00

THE BROWNIES: THEIR BOOK, *Palmer Cox*
The book that made the Brownies a household word. Generations of readers have enjoyed the antics, predicaments and adventures of these jovial sprites, who emerge from the forest at night to play or to come to the aid of a deserving human. Delightful illustrations by the author decorate nearly every page. 24 short verse tales with 266 illustrations. 155pp. 6⅝ x 9¼.
21265-3 Paperbound $1.50

THE PRINCIPLES OF PSYCHOLOGY,
William James
The full long-course, unabridged, of one of the great classics of Western literature and science. Wonderfully lucid descriptions of human mental activity, the stream of thought, consciousness, time perception, memory, imagination, emotions, reason, abnormal phenomena, and similar topics. Original contributions are integrated with the work of such men as Berkeley, Binet, Mills, Darwin, Hume, Kant, Royce, Schopenhauer, Spinoza, Locke, Descartes, Galton, Wundt, Lotze, Herbart, Fechner, and scores of others. All contrasting interpretations of mental phenomena are examined in detail—introspective analysis, philosophical interpretation, and experimental research. "A classic," *Journal of Consulting Psychology*. "The main lines are as valid as ever," *Psychoanalytical Quarterly*. "Standard reading . . . a classic of interpretation," *Psychiatric Quarterly*. 94 illustrations. 1408pp. 5⅜ x 8.
20381-6, 20382-4 Two volume set, paperbound $6.00

VISUAL ILLUSIONS: THEIR CAUSES, CHARACTERISTICS AND APPLICATIONS,
M. Luckiesh
"Seeing is deceiving," asserts the author of this introduction to virtually every type of optical illusion known. The text both describes and explains the principles involved in color illusions, figure-ground, distance illusions, etc. 100 photographs, drawings and diagrams prove how easy it is to fool the sense: circles that aren't round, parallel lines that seem to bend, stationary figures that seem to move as you stare at them – illustration after illustration strains our credulity at what we see. Fascinating book from many points of view, from applications for artists, in camouflage, etc. to the psychology of vision. New introduction by William Ittleson, Dept. of Psychology, Queens College. Index. Bibliography. xxi + 252pp. 5⅜ x 8½. 21530-X Paperbound $1.50

FADS AND FALLACIES IN THE NAME OF SCIENCE,
Martin Gardner
This is the standard account of various cults, quack systems, and delusions which have masqueraded as science: hollow earth fanatics. Reich and orgone sex energy, dianetics, Atlantis, multiple moons, Forteanism, flying saucers, medical fallacies like iridiagnosis, zone therapy, etc. A new chapter has been added on Bridey Murphy, psionics, and other recent manifestations in this field. This is a fair, reasoned appraisal of eccentric theory which provides excellent inoculation against cleverly masked nonsense. "Should be read by everyone, scientist and non-scientist alike," R. T. Birge, Prof. Emeritus of Physics, Univ. of California; Former President, American Physical Society. Index. x + 365pp. 5⅜ x 8. 20394-8 Paperbound $2.00

ILLUSIONS AND DELUSIONS OF THE SUPERNATURAL AND THE OCCULT,
D. H. Rawcliffe
Holds up to rational examination hundreds of persistent delusions including crystal gazing, automatic writing, table turning, mediumistic trances, mental healing, stigmata, lycanthropy, live burial, the Indian Rope Trick, spiritualism, dowsing, telepathy, clairvoyance, ghosts, ESP, etc. The author explains and exposes the mental and physical deceptions involved, making this not only an exposé of supernatural phenomena, but a valuable exposition of characteristic types of abnormal psychology. Originally titled "The Psychology of the Occult." 14 illustrations. Index. 551pp. 5⅜ x 8. 20503-7 Paperbound $3.50

FAIRY TALE COLLECTIONS, *edited by Andrew Lang*
Andrew Lang's fairy tale collections make up the richest shelf-full of traditional children's stories anywhere available. Lang supervised the translation of stories from all over the world—familiar European tales collected by Grimm, animal stories from Negro Africa, myths of primitive Australia, stories from Russia, Hungary, Iceland, Japan, and many other countries. Lang's selection of translations are unusually high; many authorities consider that the most familiar tales find their best versions in these volumes. All collections are richly decorated and illustrated by H. J. Ford and other artists.

THE BLUE FAIRY BOOK. 37 stories. 138 illustrations. ix + 390pp. 5⅜ x 8½.
21437-0 Paperbound $1.95

THE GREEN FAIRY BOOK. 42 stories. 100 illustrations. xiii + 366pp. 5⅜ x 8½.
21439-7 Paperbound $1.75

THE BROWN FAIRY BOOK. 32 stories. 50 illustrations, 8 in color. xii + 350pp. 5⅜ x 8½.
21438-9 Paperbound $1.95

THE BEST TALES OF HOFFMANN, *edited by E. F. Bleiler*
10 stories by E. T. A. Hoffmann, one of the greatest of all writers of fantasy. The tales include "The Golden Flower Pot," "Automata," "A New Year's Eve Adventure," "Nutcracker and the King of Mice," "Sand-Man," and others. Vigorous characterizations of highly eccentric personalities, remarkably imaginative situations, and intensely fast pacing has made these tales popular all over the world for 150 years. Editor's introduction. 7 drawings by Hoffmann. xxxiii + 419pp. 5⅜ x 8½.
21793-0 Paperbound $2.25

GHOST AND HORROR STORIES OF AMBROSE BIERCE,
edited by E. F. Bleiler
Morbid, eerie, horrifying tales of possessed poets, shabby aristocrats, revived corpses, and haunted malefactors. Widely acknowledged as the best of their kind between Poe and the moderns, reflecting their author's inner torment and bitter view of life. Includes "Damned Thing," "The Middle Toe of the Right Foot," "The Eyes of the Panther," "Visions of the Night," "Moxon's Master," and over a dozen others. Editor's introduction. xxii + 199pp. 5⅜ x 8½.
20767-6 Paperbound $1.50

THREE GOTHIC NOVELS, *edited by E. F. Bleiler*
Originators of the still popular Gothic novel form, influential in ushering in early 19th-century Romanticism. Horace Walpole's *Castle of Otranto*, William Beckford's *Vathek*, John Polidori's *The Vampyre*, and a *Fragment* by Lord Byron are enjoyable as exciting reading or as documents in the history of English literature. Editor's introduction. xi + 291pp. 5⅜ x 8½.
21232-7 Paperbound $2.00

BEST GHOST STORIES OF LEFANU, *edited by E. F. Bleiler*
Though admired by such critics as V. S. Pritchett, Charles Dickens and Henry James, ghost stories by the Irish novelist Joseph Sheridan LeFanu have never become as widely known as his detective fiction. About half of the 16 stories in this collection have never before been available in America. Collection includes "Carmilla" (perhaps the best vampire story ever written), "The Haunted Baronet," "The Fortunes of Sir Robert Ardagh," and the classic "Green Tea." Editor's introduction. 7 contemporary illustrations. Portrait of LeFanu. xii + 467pp. 5⅜ x 8.
20415-4 Paperbound $2.50

EASY-TO-DO ENTERTAINMENTS AND DIVERSIONS WITH COINS, CARDS, STRING, PAPER AND MATCHES, *R. M. Abraham*
Over 300 tricks, games and puzzles will provide young readers with absorbing fun. Sections on card games; paper-folding; tricks with coins, matches and pieces of string; games for the agile; toy-making from common household objects; mathematical recreations; and 50 miscellaneous pastimes. Anyone in charge of groups of youngsters, including hard-pressed parents, and in need of suggestions on how to keep children sensibly amused and quietly content will find this book indispensable. Clear, simple text, copious number of delightful line drawings and illustrative diagrams. Originally titled "Winter Nights' Entertainments." Introduction by Lord Baden Powell. 329 illustrations. v + 186pp. 5⅜ x 8½. 20921-0 Paperbound $1.00

AN INTRODUCTION TO CHESS MOVES AND TACTICS SIMPLY EXPLAINED, *Leonard Barden*
Beginner's introduction to the royal game. Names, possible moves of the pieces, definitions of essential terms, how games are won, etc. explained in 30-odd pages. With this background you'll be able to sit right down and play. Balance of book teaches strategy — openings, middle game, typical endgame play, and suggestions for improving your game. A sample game is fully analyzed. True middle-level introduction, teaching you all the essentials without oversimplifying or losing you in a maze of detail. 58 figures. 102pp. 5⅜ x 8½. 21210-6 Paperbound $1.25

LASKER'S MANUAL OF CHESS, *Dr. Emanuel Lasker*
Probably the greatest chess player of modern times, Dr. Emanuel Lasker held the world championship 28 years, independent of passing schools or fashions. This unmatched study of the game, chiefly for intermediate to skilled players, analyzes basic methods, combinations, position play, the aesthetics of chess, dozens of different openings, etc., with constant reference to great modern games. Contains a brilliant exposition of Steinitz's important theories. Introduction by Fred Reinfeld. Tables of Lasker's tournament record. 3 indices. 308 diagrams. 1 photograph. xxx + 349pp. 5⅜ x 8.20640-8Paperbound $2.50

COMBINATIONS: THE HEART OF CHESS, *Irving Chernev*
Step-by-step from simple combinations to complex, this book, by a well-known chess writer, shows you the intricacies of pins, counter-pins, knight forks, and smothered mates. Other chapters show alternate lines of play to those taken in actual championship games; boomerang combinations; classic examples of brilliant combination play by Nimzovich, Rubinstein, Tarrasch, Botvinnik, Alekhine and Capablanca. Index. 356 diagrams. ix + 245pp. 5⅜ x 8½. 21744-2 Paperbound $2.00

HOW TO SOLVE CHESS PROBLEMS, *K. S. Howard*
Full of practical suggestions for the fan or the beginner — who knows only the moves of the chessmen. Contains preliminary section and 58 two-move, 46 three-move, and 8 four-move problems composed by 27 outstanding American problem creators in the last 30 years. Explanation of all terms and exhaustive index. "Just what is wanted for the student," Brian Harley. 112 problems, solutions. vi + 171pp. 5⅜ x 8. 20748-X Paperbound $1.50

SOCIAL THOUGHT FROM LORE TO SCIENCE,
H. E. Barnes and H. Becker
An immense survey of sociological thought and ways of viewing, studying, planning, and reforming society from earliest times to the present. Includes thought on society of preliterate peoples, ancient non-Western cultures, and every great movement in Europe, America, and modern Japan. Analyzes hundreds of great thinkers: Plato, Augustine, Bodin, Vico, Montesquieu, Herder, Comte, Marx, etc. Weighs the contributions of utopians, sophists, fascists and communists; economists, jurists, philosophers, ecclesiastics, and every 19th and 20th century school of scientific sociology, anthropology, and social psychology throughout the world. Combines topical, chronological, and regional approaches, treating the evolution of social thought as a process rather than as a series of mere topics. "Impressive accuracy, competence, and discrimination . . . easily the best single survey," *Nation*. Thoroughly revised, with new material up to 1960. 2 indexes. Over 2200 bibliographical notes. Three volume set. Total of 1586pp. 5⅜ x 8.
20901-6, 20902-4, 20903-2 Three volume set, paperbound $9.00

A HISTORY OF HISTORICAL WRITING, *Harry Elmer Barnes*
Virtually the only adequate survey of the whole course of historical writing in a single volume. Surveys developments from the beginnings of historiography in the ancient Near East and the Classical World, up through the Cold War. Covers major historians in detail, shows interrelationship with cultural background, makes clear individual contributions, evaluates and estimates importance; also enormously rich upon minor authors and thinkers who are usually passed over. Packed with scholarship and learning, clear, easily written. Indispensable to every student of history. Revised and enlarged up to 1961. Index and bibliography. xv + 442pp. 5⅜ x 8½.
20104-X Paperbound $2.75

JOHANN SEBASTIAN BACH, *Philipp Spitta*
The complete and unabridged text of the definitive study of Bach. Written some 70 years ago, it is still unsurpassed for its coverage of nearly all aspects of Bach's life and work. There could hardly be a finer non-technical introduction to Bach's music than the detailed, lucid analyses which Spitta provides for hundreds of individual pieces. 26 solid pages are devoted to the B minor mass, for example, and 30 pages to the glorious St. Matthew Passion. This monumental set also includes a major analysis of the music of the 18th century: Buxtehude, Pachelbel, etc. "Unchallenged as the last word on one of the supreme geniuses of music," John Barkham, *Saturday Review Syndicate*. Total of 1819pp. Heavy cloth binding. 5⅜ x 8.
22278-0, 22279-9 Two volume set, clothbound $15.00

BEETHOVEN AND HIS NINE SYMPHONIES, *George Grove*
In this modern middle-level classic of musicology Grove not only analyzes all nine of Beethoven's symphonies very thoroughly in terms of their musical structure, but also discusses the circumstances under which they were written, Beethoven's stylistic development, and much other background material. This is an extremely rich book, yet very easily followed; it is highly recommended to anyone seriously interested in music. Over 250 musical passages. Index. viii + 407pp. 5⅜ x 8.
20334-4 Paperbound $2.25

THREE SCIENCE FICTION NOVELS,
John Taine
Acknowledged by many as the best SF writer of the 1920's, Taine (under the name Eric Temple Bell) was also a Professor of Mathematics of considerable renown. Reprinted here are *The Time Stream,* generally considered Taine's best, *The Greatest Game,* a biological-fiction novel, and *The Purple Sapphire,* involving a supercivilization of the past. Taine's stories tie fantastic narratives to frameworks of original and logical scientific concepts. Speculation is often profound on such questions as the nature of time, concept of entropy, cyclical universes, etc. 4 contemporary illustrations. v + 532pp. 5⅜ x 8⅜.
21180-0 Paperbound $2.50

SEVEN SCIENCE FICTION NOVELS,
H. G. Wells
Full unabridged texts of 7 science-fiction novels of the master. Ranging from biology, physics, chemistry, astronomy, to sociology and other studies, Mr. Wells extrapolates whole worlds of strange and intriguing character. "One will have to go far to match this for entertainment, excitement, and sheer pleasure . . ."*New York Times.* Contents: The Time Machine, The Island of Dr. Moreau, The First Men in the Moon, The Invisible Man, The War of the Worlds, The Food of the Gods, In The Days of the Comet. 1015pp. 5⅜ x 8.
20264-X Clothbound $5.00

28 SCIENCE FICTION STORIES OF H. G. WELLS.
Two full, unabridged novels, *Men Like Gods* and *Star Begotten,* plus 26 short stories by the master science-fiction writer of all time! Stories of space, time, invention, exploration, futuristic adventure. Partial contents: *The Country of the Blind, In the Abyss, The Crystal Egg, The Man Who Could Work Miracles, A Story of Days to Come, The Empire of the Ants, The Magic Shop, The Valley of the Spiders, A Story of the Stone Age, Under the Knife, Sea Raiders,* etc. An indispensable collection for the library of anyone interested in science fiction adventure. 928pp. 5⅜ x 8. 20265-8 Clothbound $5.00

THREE MARTIAN NOVELS,
Edgar Rice Burroughs
Complete, unabridged reprinting, in one volume, of Thuvia, Maid of Mars; Chessmen of Mars; The Master Mind of Mars. Hours of science-fiction adventure by a modern master storyteller. Reset in large clear type for easy reading. 16 illustrations by J. Allen St. John. vi + 490pp. 5⅜ x 8½.
20039-6 Paperbound $2.50

AN INTELLECTUAL AND CULTURAL HISTORY OF THE WESTERN WORLD,
Harry Elmer Barnes
Monumental 3-volume survey of intellectual development of Europe from primitive cultures to the present day. Every significant product of human intellect traced through history: art, literature, mathematics, physical sciences, medicine, music, technology, social sciences, religions, jurisprudence, education, etc. Presentation is lucid and specific, analyzing in detail specific discoveries, theories, literary works, and so on. Revised (1965) by recognized scholars in specialized fields under the direction of Prof. Barnes. Revised bibliography. Indexes. 24 illustrations. Total of xxix + 1318pp.
21275-0, 21276-9, 21277-7 Three volume set, paperbound $8.25

HEAR ME TALKIN' TO YA, *edited by Nat Shapiro and Nat Hentoff*
In their own words, Louis Armstrong, King Oliver, Fletcher Henderson, Bunk Johnson, Bix Beiderbecke, Billy Holiday, Fats Waller, Jelly Roll Morton, Duke Ellington, and many others comment on the origins of jazz in New Orleans and its growth in Chicago's South Side, Kansas City's jam sessions, Depression Harlem, and the modernism of the West Coast schools. Taken from taped conversations, letters, magazine articles, other first-hand sources. Editors' introduction. xvi + 429pp. 5⅜ x 8½. 21726-4 Paperbound $2.00

THE JOURNAL OF HENRY D. THOREAU
A 25-year record by the great American observer and critic, as complete a record of a great man's inner life as is anywhere available. Thoreau's Journals served him as raw material for his formal pieces, as a place where he could develop his ideas, as an outlet for his interests in wild life and plants, in writing as an art, in classics of literature, Walt Whitman and other contemporaries, in politics, slavery, individual's relation to the State, etc. The Journals present a portrait of a remarkable man, and are an observant social history. Unabridged republication of 1906 edition, Bradford Torrey and Francis H. Allen, editors. Illustrations. Total of 1888pp. 8⅜ x 12¼.
20312-3, 20313-1 Two volume set, clothbound $30.00

A SHAKESPEARIAN GRAMMAR, *E. A. Abbott*
Basic reference to Shakespeare and his contemporaries, explaining through thousands of quotations from Shakespeare, Jonson, Beaumont and Fletcher, North's *Plutarch* and other sources the grammatical usage differing from the modern. First published in 1870 and written by a scholar who spent much of his life isolating principles of Elizabethan language, the book is unlikely ever to be superseded. Indexes. xxiv + 511pp. 5⅜ x 8½. 21582-2 Paperbound $3.00

FOLK-LORE OF SHAKESPEARE, *T. F. Thistelton Dyer*
Classic study, drawing from Shakespeare a large body of references to supernatural beliefs, terminology of falconry and hunting, games and sports, good luck charms, marriage customs, folk medicines, superstitions about plants, animals, birds, argot of the underworld, sexual slang of London, proverbs, drinking customs, weather lore, and much else. From full compilation comes a mirror of the 17th-century popular mind. Index. ix + 526pp. 5⅜ x 8½.
21614-4 Paperbound $2.75

THE NEW VARIORUM SHAKESPEARE, *edited by H. H. Furness*
By far the richest editions of the plays ever produced in any country or language. Each volume contains complete text (usually First Folio) of the play, all variants in Quarto and other Folio texts, editorial changes by every major editor to Furness's own time (1900), footnotes to obscure references or language, extensive quotes from literature of Shakespearian criticism, essays on plot sources (often reprinting sources in full), and much more.

HAMLET, *edited by H. H. Furness*
Total of xxvi + 905pp. 5⅜ x 8½.
21004-9, 21005-7 Two volume set, paperbound $5.25

TWELFTH NIGHT, *edited by H. H. Furness*
Index. xxii + 434pp. 5⅜ x 8½. 21189-4 Paperbound $2.75

LA BOHEME BY GIACOMO PUCCINI,
translated and introduced by Ellen H. Bleiler
Complete handbook for the operagoer, with everything needed for full enjoyment except the musical score itself. Complete Italian libretto, with new, modern English line-by-line translation—the only libretto printing all repeats; biography of Puccini; the librettists; background to the opera, Murger's La Boheme, etc.; circumstances of composition and performances; plot summary; and pictorial section of 73 illustrations showing Puccini, famous singers and performances, etc. Large clear type for easy reading. 124pp. 5⅜ x 8½.
20404-9 Paperbound $1.25

ANTONIO STRADIVARI: HIS LIFE AND WORK (1644-1737),
W. Henry Hill, Arthur F. Hill, and Alfred E. Hill
Still the only book that really delves into life and art of the incomparable Italian craftsman, maker of the finest musical instruments in the world today. The authors, expert violin-makers themselves, discuss Stradivari's ancestry, his construction and finishing techniques, distinguished characteristics of many of his instruments and their locations. Included, too, is story of introduction of his instruments into France, England, first revelation of their supreme merit, and information on his labels, number of instruments made, prices, mystery of ingredients of his varnish, tone of pre-1684 Stradivari violin and changes between 1684 and 1690. An extremely interesting, informative account for all music lovers, from craftsman to concert-goer. Republication of original (1902) edition. New introduction by Sydney Beck, Head of Rare Book and Manuscript Collections, Music Division, New York Public Library. Analytical index by Rembert Wurlitzer. Appendixes. 68 illustrations. 30 full-page plates. 4 in color. xxvi + 315pp. 5⅜ x 8½. 20425-1 Paperbound $2.25

MUSICAL AUTOGRAPHS FROM MONTEVERDI TO HINDEMITH,
Emanuel Winternitz
For beauty, for intrinsic interest, for perspective on the composer's personality, for subtleties of phrasing, shading, emphasis indicated in the autograph but suppressed in the printed score, the mss. of musical composition are fascinating documents which repay close study in many different ways. This 2-volume work reprints facsimiles of mss. by virtually every major composer, and many minor figures—196 examples in all. A full text points out what can be learned from mss., analyzes each sample. Index. Bibliography. 18 figures. 196 plates. Total of 170pp. of text. 7⅞ x 10¾.
21312-9, 21313-7 Two volume set, paperbound $5.00

J. S. BACH,
Albert Schweitzer
One of the few great full-length studies of Bach's life and work, and the study upon which Schweitzer's renown as a musicologist rests. On first appearance (1911), revolutionized Bach performance. The only writer on Bach to be musicologist, performing musician, and student of history, theology and philosophy, Schweitzer contributes particularly full sections on history of German Protestant church music, theories on motivic pictorial representations in vocal music, and practical suggestions for performance. Translated by Ernest Newman. Indexes. 5 illustrations. 650 musical examples. Total of xix + 928pp. 5⅜ x 8½. 21631-4, 21632-2 Two volume set, paperbound $4.50

THE METHODS OF ETHICS, *Henry Sidgwick*
Propounding no organized system of its own, study subjects every major methodological approach to ethics to rigorous, objective analysis. Study discusses and relates ethical thought of Plato, Aristotle, Bentham, Clarke, Butler, Hobbes, Hume, Mill, Spencer, Kant, and dozens of others. Sidgwick retains conclusions from each system which follow from ethical premises, rejecting the faulty. Considered by many in the field to be among the most important treatises on ethical philosophy. Appendix. Index. xlvii + 528pp. 5⅜ x 8½.
21608-X Paperbound $2.50

TEUTONIC MYTHOLOGY, *Jakob Grimm*
A milestone in Western culture; the work which established on a modern basis the study of history of religions and comparative religions. 4-volume work assembles and interprets everything available on religious and folkloristic beliefs of Germanic people (including Scandinavians, Anglo-Saxons, etc.). Assembling material from such sources as Tacitus, surviving Old Norse and Icelandic texts, archeological remains, folktales, surviving superstitions, comparative traditions, linguistic analysis, etc. Grimm explores pagan deities, heroes, folklore of nature, religious practices, and every other area of pagan German belief. To this day, the unrivaled, definitive, exhaustive study. Translated by J. S. Stallybrass from 4th (1883) German edition. Indexes. Total of lxxvii + 1887pp. 5⅜ x 8½.
21602-0, 21603-9, 21604-7, 21605-5 Four volume set, paperbound $11.00

THE I CHING, *translated by James Legge*
Called "The Book of Changes" in English, this is one of the Five Classics edited by Confucius, basic and central to Chinese thought. Explains perhaps the most complex system of divination known, founded on the theory that all things happening at any one time have characteristic features which can be isolated and related. Significant in Oriental studies, in history of religions and philosophy, and also to Jungian psychoanalysis and other areas of modern European thought. Index. Appendixes. 6 plates. xxi + 448pp. 5⅜ x 8½.
21062-6 Paperbound $2.75

HISTORY OF ANCIENT PHILOSOPHY, *W. Windelband*
One of the clearest, most accurate comprehensive surveys of Greek and Roman philosophy. Discusses ancient philosophy in general, intellectual life in Greece in the 7th and 6th centuries B.C., Thales, Anaximander, Anaximenes, Heraclitus, the Eleatics, Empedocles, Anaxagoras, Leucippus, the Pythagoreans, the Sophists, Socrates, Democritus (20 pages), Plato (50 pages), Aristotle (70 pages), the Peripatetics, Stoics, Epicureans, Sceptics, Neo-platonists, Christian Apologists, etc. 2nd German edition translated by H. E. Cushman. xv + 393pp. 5⅜ x 8.
20357-3 Paperbound $2.25

THE PALACE OF PLEASURE, *William Painter*
Elizabethan versions of Italian and French novels from *The Decameron*, Cinthio, Straparola, Queen Margaret of Navarre, and other continental sources — the very work that provided Shakespeare and dozens of his contemporaries with many of their plots and sub-plots and, therefore, justly considered one of the most influential books in all English literature. It is also a book that any reader will still enjoy. Total of cviii + 1,224pp.
21691-8, 21692-6, 21693-4 Three volume set, paperbound $6.75

THE WONDERFUL WIZARD OF OZ, *L. F. Baum*
All the original W. W. Denslow illustrations in full color—as much a part of "The Wizard" as Tenniel's drawings are of "Alice in Wonderland." "The Wizard" is still America's best-loved fairy tale, in which, as the author expresses it, "The wonderment and joy are retained and the heartaches and nightmares left out." Now today's young readers can enjoy every word and wonderful picture of the original book. New introduction by Martin Gardner. A Baum bibliography. 23 full-page color plates. viii + 268pp. 5⅜ x 8.
20691-2 Paperbound $1.95

THE MARVELOUS LAND OF OZ, *L. F. Baum*
This is the equally enchanting sequel to the "Wizard," continuing the adventures of the Scarecrow and the Tin Woodman. The hero this time is a little boy named Tip, and all the delightful Oz magic is still present. This is the Oz book with the Animated Saw-Horse, the Woggle-Bug, and Jack Pumpkinhead. All the original John R. Neill illustrations, 10 in full color. 287pp. 5⅜ x 8.
20692-0 Paperbound $1.75

ALICE'S ADVENTURES UNDER GROUND, *Lewis Carroll*
The original *Alice in Wonderland*, hand-lettered and illustrated by Carroll himself, and originally presented as a Christmas gift to a child-friend. Adults as well as children will enjoy this charming volume, reproduced faithfully in this Dover edition. While the story is essentially the same, there are slight changes, and Carroll's spritely drawings present an intriguing alternative to the famous Tenniel illustrations. One of the most popular books in Dover's catalogue. Introduction by Martin Gardner. 38 illustrations. 128pp. 5⅜ x 8½.
21482-6 Paperbound $1.00

THE NURSERY "ALICE," *Lewis Carroll*
While most of us consider *Alice in Wonderland* a story for children of all ages, Carroll himself felt it was beyond younger children. He therefore provided this simplified version, illustrated with the famous Tenniel drawings enlarged and colored in delicate tints, for children aged "from Nought to Five." Dover's edition of this now rare classic is a faithful copy of the 1889 printing, including 20 illustrations by Tenniel, and front and back covers reproduced in full color. Introduction by Martin Gardner. xxiii + 67pp. 6⅛ x 9¼.
21610-1 Paperbound $1.75

THE STORY OF KING ARTHUR AND HIS KNIGHTS, *Howard Pyle*
A fast-paced, exciting retelling of the best known Arthurian legends for young readers by one of America's best story tellers and illustrators. The sword Excalibur, wooing of Guinevere, Merlin and his downfall, adventures of Sir Pellias and Gawaine, and others. The pen and ink illustrations are vividly imagined and wonderfully drawn. 41 illustrations. xviii + 313pp. 6⅛ x 9¼.
21445-1 Paperbound $2.00

Prices subject to change without notice.

Available at your book dealer or write for free catalogue to Dept. Adsci, Dover Publications, Inc., 180 Varick St., N.Y., N.Y. 10014. Dover publishes more than 150 books each year on science, elementary and advanced mathematics, biology, music, art, literary history, social sciences and other areas.